Further MECHANICS

BRIAN JEFFERSON
TONY BEADSWORTH

UNIVERSITY PRESS

Great Clarendon Street, Oxford OX2 6DP

Oxford University Press is a department of the University of Oxford.
It furthers the University's objective of excellence in research, scholarship,
and education by publishing worldwide in

Oxford New York

Auckland Cape Town Dar es Salaam Hong Kong Karachi
Kuala Lumpur Madrid Melbourne Mexico City Nairobi
New Delhi Shanghai Taipei Toronto

With offices in

Argentina Austria Brazil Chile Czech Republic France Greece
Guatemala Hungary Italy Japan Poland Portugal Singapore
South Korea Switzerland Thailand Turkey Ukraine Vietnam

© Brian Jefferson and Tony Beadsworth 2001

The moral rights of the author have been asserted

Database right Oxford University Press (maker)

First published 2001

All rights reserved. No part of this publication may be reproduced, stored
in a retrieval system, or transmitted, in any form or by any means,
without prior permission in writing from Oxford University Press, or as
expressly permitted by law, or under terms agreed with the appropriate
reprographics rights organisation. Enquiries concerning reproduction
outside the scope of the above should be sent to the Rights Department,
Oxford University Press, at the address above.

You must not circulate this book in any order binding or cover and you
must impose this same condition on any acquirer

British Library Cataloguing in Publication Data

Data available

ISBN 978 0 19 914738 0

20 19 18 17 16

Typeset and illustrated by Tech-Set Ltd, Gateshead, Tyne and Wear
Printed in Great Britain by CPI Group (UK) Ltd., Croydon CR0 4YY

Contents

Preface v

1 Vectors I — 1
Specifying a vector — 1
Properties of vectors — 3
Describing space using vectors — 5

2 Vectors II — 14
Vector product — 14
Scalar triple product — 19
Other geometrical uses of the vector product and the scalar triple product — 22
Vector equation of a curve — 25
Moments — 27
Equivalent force systems and equilibrium — 29

3 Differential Equations — 35
Definitions and classification — 35
Forming differential equations — 36
First-order differential equations — 38
Separation of variables — 43
Homogeneous equations — 45
First-order linear equations — 48
Linear equations with constant coefficients — 50
The complementary function and the particular integral — 52
Second-order linear equations with constant coefficients — 56
Vector differential equations — 67
Simultaneous linear differential equations — 70
Numerical methods — 76

4 General Motion in One Dimension — 81
Motion with variable mass — 81
Variable forces — 88

Examination Questions: Chapters 1 to 4 — 97

5 Impulses and Impulsive Tensions — 103
Oblique impact — 103
Impulsive tensions — 109

6 Oscillating Systems — 115
Damped oscillations — 115
Forced oscillations — 126
Investigation — 132
Resonance — 132

7 Projectiles — 137
Projectiles on an inclined plane — 137
Air resistance — 145

8 Motion of a Particle in a Plane — 155
Cartesian coordinates — 155
Polar resolutes of acceleration — 159
Motion under a central force — 165
Intrinsic coordinates — 175

Examination Questions: Chapters 5 to 8 — 182

9 Stability of Equilibrium — 188
Centre of mass — 188
Stability of equilibrium — 196

10 Moment of Inertia — 201
Kinetic energy of rotation — 201
Moment of inertia — 201
Equation of rotational motion — 203
Angular momentum — 203
Uniform angular acceleration — 204
Calculating the moment of inertia — 209
Parallel and perpendicular axes theorems — 219
Radius of gyration — 221
Summary of standard results — 222

11 Rotation about a Fixed Axis — 224
Total energy of a rotating body — 224
Work done by a couple — 225
Relationship between work and energy — 225
Compound pendulum — 231
Reaction at the axis — 235
Impulse and momentum — 238
Conservation of angular momentum — 242

12 General Motion of a Rigid Body in Two Dimensions — 246
Particle model — 246
Translational and rotational motion of the lamina — 247
Angular momentum about a fixed point — 249
Kinetic energy — 249
Small oscillations of a system — 255

Examination Questions: Chapters 9 to 12 — 261

Answers — 267

Index — 277

Preface

This book is the companion volume to *Introducing Mechanics* and completes the coverage of the mechanics content of the A-level mathematics specifications which came into force in September 2000. The text does not follow the syllabus of any one examination board, but seeks to develop the subject in such a way as to be accessible to all students of further mathematics.

As with *Introducing Mechanics*, emphasis is given to the modelling aspects of the subject and, where possible, reference is made to real-world situations and to practical investigations which may complement the theoretical development of topics. Because of the nature of the topics covered, this has proved much more difficult than with the previous volume. It is not possible, for example, to devise a simple experiment to illustrate the general motion of a body in two dimensions, and so a good deal of the work is necessarily rather theoretical. No new use has been made of spreadsheets, although the sheet which explores damped harmonic motion is again relevant, and can be downloaded from the Oxford University Press website (http://www.oup.co.uk/mechanics).

There is a good deal of variation in the emphasis given to different topics by the various syllabuses across the single-subject/further-mathematics divide. For the sake of coherence, we have therefore found it necessary to repeat in this volume some of the text and exercises from *Introducing Mechanics*. The topics are introduced in what we believe is a helpful order, but a degree of cross-referencing has been included to help those who wish to take a different approach in developing the subject.

The first three chapters are devoted to revising and extending the basic mathematical tools of the trade – vectors and differential equations. Chapters 4 to 7 develop further some of the topics covered in *Introducing Mechanics*, and deal with motion under a variable force or with variable mass, oblique impulses, damped and forced oscillations and the motion of a projectile on an inclined plane.

Chapter 8 explores the motion of a particle in a plane, described in cartesian, polar or intrinsic coordinates. Chapter 9 introduces the notion of stability of equilibrium and its relationship with potential energy.

A good deal of further mechanics consists of a fuller exploration of ideas already encountered in the earlier book. The major new area of study is rotational motion, and the final three chapters are devoted to this, covering moment of inertia, rotation about a fixed axis and the general motion of a body in two dimensions.

Most students will use this book with the guidance of a teacher, but every effort has been made to make it readable and accessible to those using it for self-study or for revision. A large number of worked examples are included, with exercises designed both to provide rote practice and to stretch the more able student. There are also three sections containing a selection of recent examination questions.

We are grateful to AEB, EDEXCEL, MEI, NEAB, NICCEA, OCR, SQA and WJEC for permission to use their questions. The answers provided for these questions are the sole responsibility of the authors.

Thanks are due to James Nicholson for his helpful suggestions when reviewing the text and for his diligence in checking the answers to the exercises and examination questions. Heartfelt thanks also go to John Day for his untiring attention to detail in the editing of this book. Finally, thanks are due to our wives for their patience over the last several years during which these two books have consumed so much of our time.

Brian Jefferson
Tony Beadsworth
September 2000

1 Vectors I

A line is length without breadth.
EUCLID

We will start with a resumé and some slight formalisation of the vector properties dealt with in *Introducing Mechanics*.

- To define a **vector** quantity requires both magnitude and direction (a **scalar** quantity has only magnitude).
- Vectors **a** and **b** are equal if and only if they have the same magnitude **and** the same direction.
- Vectors **a** and **b** are parallel if and only if $\mathbf{a} = \lambda\mathbf{b}$ for some scalar λ. If λ is positive, the vectors have the same direction (they are **like** parallel vectors). If λ is negative, the vectors are in opposite directions (they are **unlike** parallel vectors).
- If $\mathbf{a} = -\mathbf{b}$, vectors **a** and **b** have equal magnitudes but opposite directions.
- Vectors can be represented by directed line segments. We can then represent vector addition as the single displacement equivalent to the successive displacements representing the individual vectors.

In the diagram, the line segment AD represents the **resultant** or **vector sum** of the vectors **a**, **b** and **c**.

- The subtraction of vectors can be regarded as adding a negative vector, so that $\mathbf{a} - \mathbf{b}$ is effectively $\mathbf{a} + (-\mathbf{b})$.

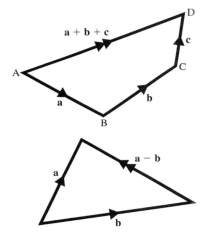

Specifying a vector

To specify a vector, we need a frame of reference. This would normally be the standard cartesian axes in two or three dimensions.

In two dimensions, we can state the magnitude of **r** ($|\mathbf{r}|$ or r) and its angle, θ, with the positive x-direction.

Alternatively, we state the x- and y-**components** of **r**.

Here, $\mathbf{r} = \begin{pmatrix} a \\ b \end{pmatrix}$ or $\mathbf{r} = a\mathbf{i} + b\mathbf{j}$, where **i** and **j** are **unit vectors** in the x- and y-directions.

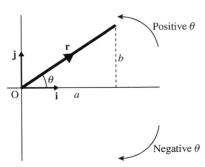

Clearly, we have
$$a = r\cos\theta \quad b = r\sin\theta$$
and $\quad r = \sqrt{a^2 + b^2} \quad \theta = \arctan\left(\dfrac{b}{a}\right)$

In three dimensions, we would normally use the *x*-, *y*- and *z*-components of the vector, as shown below.

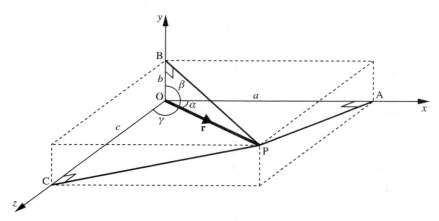

Here, $\mathbf{r} = \begin{pmatrix} a \\ b \\ c \end{pmatrix}$ or $r = a\mathbf{i} + b\mathbf{j} + c\mathbf{k}$, where **i**, **j** and **k** are unit vectors in the *x*-, *y*- and *z*-directions.

If **r** makes angles α, β and γ with the *x*-, *y*- and *z*-directions, then we have
$$a = r\cos\alpha \quad b = r\cos\beta \quad c = r\cos\gamma$$
$$r = \sqrt{a^2 + b^2 + c^2}$$

The direction of vector **r** is specified by the ratios $a:b:c$. We can see this because if vector $\mathbf{s} = d\mathbf{i} + e\mathbf{j} + f\mathbf{k}$ is parallel to **r** then
$$\mathbf{s} = \lambda\mathbf{r}$$
$\Rightarrow \quad d = \lambda a \quad e = \lambda b \quad f = \lambda c$
$\Rightarrow \quad d:e:f = \lambda a : \lambda b : \lambda c = a:b:c$

The ratios $a:b:c$ are called the **direction ratios** of the vector. Two vectors having the same directions ratios are parallel (either like or unlike).

The angles α, β and γ specify the direction of **r**. The cosines of these angles are called the **direction cosines** of the vector. These are frequently given the symbols *l*, *m* and *n*. Hence, we have
$$\cos\alpha = \frac{a}{r} = l \quad \cos\beta = \frac{b}{r} = m \quad \cos\gamma = \frac{c}{r} = n$$

If **s** is a like parallel vector to **r**, it will have direction cosines *l*, *m*, *n*.

If **s** is an unlike parallel vector to **r**, it will have direction cosines $-l$, $-m$, $-n$.

The directions cosines are related because

$$l^2 + m^2 + n^2 = \frac{a^2 + b^2 + c^2}{r^2}$$

But $r^2 = a^2 + b^2 + c^2$, which gives

$$l^2 + m^2 + n^2 = 1$$

Properties of vectors

Combining vectors in component form

Addition, subtraction and multiplication by a scalar can be done component-wise. Hence, we have

$$\begin{pmatrix} a \\ b \\ c \end{pmatrix} \pm \begin{pmatrix} d \\ e \\ f \end{pmatrix} = \begin{pmatrix} a \pm d \\ b \pm e \\ c \pm f \end{pmatrix}$$

and

$$\lambda \begin{pmatrix} a \\ b \\ c \end{pmatrix} = \begin{pmatrix} \lambda a \\ \lambda b \\ \lambda c \end{pmatrix}$$

Unit vectors

A unit vector has a magnitude of 1. It is often indicated by a 'hat' over the symbol.

The unit vector in the direction of $\mathbf{r} = a\mathbf{i} + b\mathbf{j} + c\mathbf{k}$ is given by

$$\hat{\mathbf{r}} = \frac{\mathbf{r}}{|\mathbf{r}|} = \frac{a\mathbf{i} + b\mathbf{j} + c\mathbf{k}}{\sqrt{a^2 + b^2 + c^2}}$$

Scalar product

Also known as the **dot product**, this is the first of the two ways in which vectors can be combined in a form reminiscent of multiplication. It is called the scalar product because the result is a scalar quantity. The symbol used to denote a scalar product is a dot. (See *Introducing Mechanics*, pages 25–6.)

Consider the two vectors

$$\mathbf{a} = \begin{pmatrix} a_1 \\ a_2 \\ a_3 \end{pmatrix} \text{ and } \mathbf{b} = \begin{pmatrix} b_1 \\ b_2 \\ b_3 \end{pmatrix}$$

where $a = \sqrt{a_1^2 + a_2^2 + a_3^2}$ and $b = \sqrt{b_1^2 + b_2^2 + b_3^2}$.

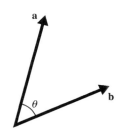

We can calculate their scalar product in two ways:

$\mathbf{a} \cdot \mathbf{b} = ab \cos \theta$, where θ is the angle between the vectors

and $\mathbf{a} \cdot \mathbf{b} = a_1 b_1 + a_2 b_2 + a_3 b_3$

Notice in particular:
- $\mathbf{a}.\mathbf{a} = a^2 \cos 0° = a^2$
  (In fact, some authors write $\mathbf{a}.\mathbf{a}$ as $\mathbf{a}^2$)
- If $\mathbf{a}.\mathbf{b} = 0$, where $\mathbf{a}$ and $\mathbf{b}$ are non-zero vectors, then $\cos \theta = 0$ and $\theta = 90°$.
  So, vectors are **perpendicular if and only if their scalar product is zero**.

Properties of scalar product

We can show that
- the scalar product is commutative:
$$\mathbf{a}.\mathbf{b} = \mathbf{b}.\mathbf{a}$$
- the scalar product is distributive over vector addition:
$$\mathbf{a}.(\mathbf{b}+\mathbf{c}) = \mathbf{a}.\mathbf{b}+\mathbf{a}.\mathbf{c}$$

But note that $\mathbf{a}.\mathbf{b} = \mathbf{a}.\mathbf{c}$ does not mean that $\mathbf{b} = \mathbf{c}$. For example, we have

$$\begin{pmatrix} 1 \\ 2 \\ 3 \end{pmatrix} \cdot \begin{pmatrix} 2 \\ -1 \\ 1 \end{pmatrix} = 3 \quad \text{and} \quad \begin{pmatrix} 1 \\ 2 \\ 3 \end{pmatrix} \cdot \begin{pmatrix} 1 \\ 1 \\ 0 \end{pmatrix} = 3$$

Angle between vectors

Using the fact that $ab \cos \theta = \mathbf{a}.\mathbf{b}$, we have

$$\cos \theta = \frac{\mathbf{a}.\mathbf{b}}{ab}$$

Hence, we can find the angle between two vectors.

Example 1 If $\mathbf{a} = 2\mathbf{i} + 4\mathbf{j} - \mathbf{k}$ and $\mathbf{b} = \mathbf{i} + 6\mathbf{j} + 3\mathbf{k}$, find the angle θ between them.

SOLUTION

Using $\mathbf{a}.\mathbf{b} = a_1 b_1 + a_2 b_2 + a_3 b_3$, we have

$$\cos \theta = \frac{(2 \times 1) + (4 \times 6) - (1 \times 3)}{\sqrt{2^2 + 4^2 + (-1)^2} \times \sqrt{1^2 + 6^2 + 3^2}} = 0.74$$

$$\Rightarrow \quad \theta = 42.3°$$

The angle between two vectors can also be expressed in terms of direction cosines.

If $\mathbf{a} = a_1\mathbf{i} + a_2\mathbf{j} + a_3\mathbf{k}$ has direction cosines

$$l_a = \frac{a_1}{a} \quad m_a = \frac{a_2}{a} \quad n_a = \frac{a_3}{a}$$

and $\mathbf{b} = b_1\mathbf{i} + b_2\mathbf{j} + b_3\mathbf{k}$ has direction cosines

$$l_b = \frac{b_1}{b} \quad m_b = \frac{b_2}{b} \quad n_b = \frac{b_3}{b}$$

we have

$$\cos\theta = \frac{a_1 b_1 + a_2 b_2 + a_3 b_3}{ab}$$

$$\Rightarrow \cos\theta = \frac{a_1}{a}\frac{b_1}{b} + \frac{a_2}{a}\frac{b_2}{b} + \frac{a_3}{a}\frac{b_3}{b}$$

which gives

$$\cos\theta = l_a l_b + m_a m_b + n_a n_b$$

Exercise 1A

1 Given vector $\mathbf{V} = 2\mathbf{i} + 3\mathbf{j} - 6\mathbf{k}$, find

  a) the magnitude of $\mathbf{V}$
  b) the unit vector in the direction of $\mathbf{V}$
  c) the vector in the opposite direction to $\mathbf{V}$ and with magnitude 10.5 units
  d) the angles which $\mathbf{V}$ makes with the coordinate axes.

2 Octavius the spider spins two successive threads along vectors $\mathbf{a} = \mathbf{i} - 3\mathbf{j} + 2\mathbf{k}$ and $\mathbf{b} = 4\mathbf{i} + \mathbf{j} - 3\mathbf{k}$. Find

  a) the angle through which Octavius turns at the junction of the two threads
  b) the vector he must then travel to form a triangular web
  c) the angle this third side makes with the x-axis.

3 Vector $\mathbf{V}$ has direction ratios $4:-4:7$ and magnitude 36 units. Find the two possible values of $\mathbf{V}$.

4 Vector $\mathbf{V}$ makes angles of $50°$ and $70°$ with the positive x- and y-directions. Find the two possible angles it could make with the z-direction and hence find the possible values of $\mathbf{V}$ if $|\mathbf{V}| = 6$ units.

5 Find the possible vectors which make equal angles with the positive x-, y- and z-directions and which have magnitude 12 units.

6 A force $\mathbf{F}$ is of magnitude 40 N. Its direction makes an angle of $75°$ with the positive x-axis and makes equal, acute angles with the positive y- and z-directions. Find $\mathbf{F}$ in component form.

Describing space using vectors

You are used to describing two-dimensional space using cartesian coordinates, equations of straight lines, etc. We can do the same using vector notation, and in this form the ideas are more easily extended to three dimensions (and beyond).

To specify points, we need an origin O and a frame of reference, which is normally the standard cartesian axes. The point A with coordinates (a_1, a_2, a_3) is then specified by its **position vector** $\overrightarrow{OA} = a_1\mathbf{i} + a_2\mathbf{j} + a_3\mathbf{k}$.

CHAPTER 1 VECTORS I

Conventionally, the position vector of point A is labelled **a**, of point B is labelled **b** and so on. (The only obvious exception to this is that it is common for the general point P to be given the position vector **r**.)

Distance between points

Given two points A and B, the displacement of B from A is given by the vector $\overrightarrow{AB}$, where

$$\overrightarrow{AB} = \mathbf{b} - \mathbf{a}$$

So, the distance $AB = |\overrightarrow{AB}|$ is given by

$$|\overrightarrow{AB}| = |\mathbf{b} - \mathbf{a}|$$

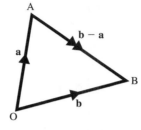

Example 2 Find the distance between A(2, 3, −1) and B(4, 2, 3).

SOLUTION

The position vectors are $\mathbf{a} = 2\mathbf{i} + 3\mathbf{j} - \mathbf{k}$ and $\mathbf{b} = 4\mathbf{i} + 2\mathbf{j} + 3\mathbf{k}$, which give

$$\mathbf{b} - \mathbf{a} = 2\mathbf{i} - \mathbf{j} + 4\mathbf{k}$$
$$\Rightarrow \quad AB = |\mathbf{b} - \mathbf{a}| = \sqrt{2^2 + (-1)^2 + 4^2} = \sqrt{21}$$

The ratio formula

Suppose we have two points, A and B, and we wish to find the point C on AB such that

$$AC:CB = m:n$$

We have

$$\overrightarrow{AC} = \mathbf{c} - \mathbf{a} \quad \text{and} \quad \overrightarrow{CB} = \mathbf{b} - \mathbf{c}$$

We know that

$$\frac{AC}{CB} = \frac{m}{n}$$
$$\Rightarrow \quad n\overrightarrow{AC} = m\overrightarrow{CB} \qquad \text{(since ABC is a straight line)}$$
$$\Rightarrow \quad n(\mathbf{c} - \mathbf{a}) = m(\mathbf{b} - \mathbf{c})$$
$$\Rightarrow \quad \mathbf{c} = \frac{n\mathbf{a} + m\mathbf{b}}{m + n}$$

This formula gives a point C on AB for all ratios $m:n$. If the ratio is positive, then C lies between A and B (C divides AB **internally**). If the ratio is negative, then C lies on the line AB extended (C divides AB **externally**).

In particular, if C is the **mid-point** of AB, then $m = n = 1$ and we have

$$\mathbf{c} = \tfrac{1}{2}(\mathbf{a} + \mathbf{b})$$

Vector equation of a line

In general, the equation of a line or of any curve can be regarded as a formula which will generate all the points on the line or curve. In vector terms, this means it must generate the position vectors of the points.

There are two ways in which we may define a line. We can either specify the direction of the line and the position of one point through which it passes, or we can specify the positions of two points on the line.

Equation of a line through a given point and in a given direction

Suppose the line passes through point A with position vector **a** relative to some origin O and is in the direction of vector **b**.

The position vector **r** of a general point P on the line is given by

$$\mathbf{r} = \overrightarrow{OA} + \overrightarrow{AP}$$
$$= \mathbf{a} + \overrightarrow{AP}$$

But $\overrightarrow{AP}$ is a multiple of **b**, so we get

$$\mathbf{r} = \mathbf{a} + \lambda\mathbf{b}$$

By varying the value of λ, this equation will give all the points on the line, and so it is the vector equation of the line.

Equation of a line through two given points

Suppose the line passes through points A and B with position vectors **a** and **b** relative to some origin O.

The direction of the line is given by the vector $\overrightarrow{AB} = \mathbf{b} - \mathbf{a}$.

From the result in the previous section, the equation of the line is therefore

$$\mathbf{r} = \mathbf{a} + \lambda(\mathbf{b} - \mathbf{a})$$

Example 3 Find the vector equation of the line through the point A(2, 5, −1) in the direction $\mathbf{i} + \mathbf{j} - 2\mathbf{k}$, and show that it passes through the point P(−1, 2, −5).

SOLUTION

The position vector of A is $2\mathbf{i} + 5\mathbf{j} - \mathbf{k}$. Therefore, the equation of the line is

$$\mathbf{r} = (2\mathbf{i} + 5\mathbf{j} - \mathbf{k}) + \lambda(\mathbf{i} + \mathbf{j} - 2\mathbf{k})$$

This is sometimes given in the form

$$\mathbf{r} = (2 + \lambda)\mathbf{i} + (5 + \lambda)\mathbf{j} - (1 + 2\lambda)\mathbf{k}$$

For the line to pass through P, we need

$$2 + \lambda = -1$$
$$5 + \lambda = 2$$
$$1 + 2\lambda = -5$$

These are all satisfied by $\lambda = -3$, and so P is on the line.

Example 4 Find the equation of the line passing through the points A(3, 4, 1) and B(−2, 4, 0).

SOLUTION

The position vector of A is $\mathbf{a} = \begin{pmatrix} 3 \\ 4 \\ 1 \end{pmatrix}$ and of B is $\mathbf{b} = \begin{pmatrix} -2 \\ 4 \\ 0 \end{pmatrix}$.

The direction vector of the line is $\mathbf{b} - \mathbf{a} = \begin{pmatrix} -5 \\ 0 \\ -1 \end{pmatrix}$.

The equation of the line is then

$$\mathbf{r} = \begin{pmatrix} 3 \\ 4 \\ 1 \end{pmatrix} + \lambda \begin{pmatrix} -5 \\ 0 \\ -1 \end{pmatrix} \quad \text{or} \quad \mathbf{r} = \begin{pmatrix} 3 - 5\lambda \\ 4 \\ 1 - \lambda \end{pmatrix}$$

Note Any equation of the form $\mathbf{r} = \mathbf{a} + \lambda \mathbf{b}$ is a straight line, where $\mathbf{b}$ indicates the direction of the line.

Example 5 Show that the lines $\mathbf{r} = (3\mathbf{i} + \mathbf{j} - \mathbf{k}) + \lambda(2\mathbf{i} - 4\mathbf{j} - 3\mathbf{k})$ and $\mathbf{r} = (7 - 4\mu)\mathbf{i} + (1 + 8\mu)\mathbf{j} + (2 + 6\mu)\mathbf{k}$ are parallel.

SOLUTION

The second line can be written as

$$\mathbf{r} = (7\mathbf{i} + \mathbf{j} + 2\mathbf{k}) - 2\mu(2\mathbf{i} - 4\mathbf{j} - 3\mathbf{k})$$

$(2\mathbf{i} - 4\mathbf{j} - 3\mathbf{k})$ is therefore a direction vector for both lines and so they are parallel.

Example 6 Find the equation of the line through the point A(2, 2, 1) which is parallel to the line $\mathbf{r} = (1 - \lambda)\mathbf{i} + 2\lambda\mathbf{j} + (3 + 2\lambda)\mathbf{k}$

SOLUTION

The given line can be written

$$\mathbf{r} = (\mathbf{i} + 3\mathbf{k}) + \lambda(-\mathbf{i} + 2\mathbf{j} + 2\mathbf{k})$$

$(-\mathbf{i} + 2\mathbf{j} + 2\mathbf{k})$ is therefore a direction vector for this line and all lines parallel to it. Hence, the required line is

$$\mathbf{r} = (2\mathbf{i} + 2\mathbf{j} + \mathbf{k}) + \mu(-\mathbf{i} + 2\mathbf{j} + 2\mathbf{k})$$

Cartesian equations for a line in three dimensions

The vector equation of a line can be expressed in cartesian form.

Suppose the line is

$$\mathbf{r} = \mathbf{a} + \lambda \mathbf{b}$$

where $\mathbf{a} = a_1\mathbf{i} + a_2\mathbf{j} + a_3\mathbf{k}$, $\mathbf{b} = b_1\mathbf{i} + b_2\mathbf{j} + b_3\mathbf{k}$ and $\mathbf{r}$ is the general point $x\mathbf{i} + y\mathbf{j} + z\mathbf{k}$. We then have

$$x = a_1 + \lambda b_1$$
$$y = a_2 + \lambda b_2$$
$$z = a_3 + \lambda b_3$$

Solving each of these for λ, we obtain the equations

$$\frac{x - a_1}{b_1} = \frac{y - a_2}{b_2} = \frac{z - a_3}{b_3}$$

which give the cartesian representation of the line.

Notice that it requires two cartesian equations to define a line in three dimensions.

Notice also that for a line given in this form, we can still easily identify the direction vector $\begin{pmatrix} b_1 \\ b_2 \\ b_3 \end{pmatrix}$ of the line.

Example 7 Find the vector and cartesian equations of

a) the line through $(2, 1, 3)$ with direction ratios $4 : 3 : -1$
b) the line through $(5, 2, -3)$ and $(2, 2, 1)$

SOLUTION

a) The vector equation is

$$\mathbf{r} = \begin{pmatrix} 2 \\ 1 \\ 3 \end{pmatrix} + \lambda \begin{pmatrix} 4 \\ 3 \\ -1 \end{pmatrix}$$

The cartesian equations are

$$\frac{x - 2}{4} = \frac{y - 1}{3} = \frac{z - 3}{-1}$$

b) The direction of the line is $\begin{pmatrix} 3 \\ 0 \\ -4 \end{pmatrix}$.

The vector equation can be written

$$\mathbf{r} = \begin{pmatrix} 5 \\ 2 \\ -3 \end{pmatrix} + \lambda \begin{pmatrix} 3 \\ 0 \\ -4 \end{pmatrix}$$

Because the y component of the direction is zero, we cannot express λ in terms of y.

The cartesian equations are therefore

$$\frac{x - 5}{3} = \frac{z + 3}{-4} \quad \text{and} \quad y = 2$$

CHAPTER 1 VECTORS I

Intersection of two straight lines

In **two dimensions**, two straight lines are either parallel or have a single point of intersection.

Parallel lines are easily spotted because the direction vector of one is a multiple of the direction vector of the other, as in Example 5 (page 8).

Example 8 Find the point of intersection of the lines
$$\mathbf{r} = (3 - \lambda)\mathbf{i} + (1 + 2\lambda)\mathbf{j} \quad \text{and} \quad \mathbf{r} = (2 + \mu)\mathbf{i} + (1 + \mu)\mathbf{j}$$

SOLUTION

If the point (x, y) is on both lines, then we have
$$x = 3 - \lambda \quad \text{and} \quad x = 2 + \mu \quad \Rightarrow \quad 3 - \lambda = 2 + \mu$$
$$y = 1 + 2\lambda \quad \text{and} \quad y = 1 + \mu \quad \Rightarrow \quad 1 + 2\lambda = 1 + \mu$$

This leads to the simultaneous equations
$$\lambda + \mu = 1 \quad \text{and} \quad 2\lambda = \mu$$

Solving them, we get $\lambda = \frac{1}{3}$, $\mu = \frac{2}{3}$.

Substituting for λ in the first line equation or for μ in the second, we find the point of intersection of the lines is $(2\frac{2}{3}, 1\frac{2}{3})$.

In **three dimensions**, the situation is slightly more involved. A pair of straight lines may be parallel, may have a point of intersection or may 'miss each other' like telephone wires going in different directions. In the last case, they are called **skew lines**.

Example 9 Investigate whether the following pairs of lines intersect.

a) $\mathbf{r} = (\lambda - 2)\mathbf{i} + 2\lambda\mathbf{j} + (3 - \lambda)\mathbf{k}$ and $\mathbf{r} = (5 - 2\mu)\mathbf{i} - 4\mu\mathbf{j} + (2\mu + 1)\mathbf{k}$
b) $\mathbf{r} = (2\lambda - 1)\mathbf{i} + (\lambda + 1)\mathbf{j} + \lambda\mathbf{k}$ and $\mathbf{r} = (2 - \mu)\mathbf{i} + (1 - 2\mu)\mathbf{j} - 2\mu\mathbf{k}$
c) $\mathbf{r} = (\lambda - 4)\mathbf{i} + \lambda\mathbf{j} + (5 - 2\lambda)\mathbf{k}$ and $\mathbf{r} = (3\mu + 1)\mathbf{i} + (\mu - 1)\mathbf{j} + (2 - \mu)\mathbf{k}$

SOLUTION

a) Rewriting the lines as
$$\mathbf{r} = (-2\mathbf{i} + 3\mathbf{k}) + \lambda(\mathbf{i} + 2\mathbf{j} - \mathbf{k})$$
and
$$\mathbf{r} = (5\mathbf{i} + \mathbf{k}) + \mu(-2\mathbf{i} - 4\mathbf{j} + 2\mathbf{k})$$

we can see that the direction vector of the second line is $-2 \times$ that of the first. The lines are therefore parallel.

b) The lines are not parallel. If they have a point of intersection (x, y, z), then we have
$$x = 2\lambda - 1 \quad \text{and} \quad x = 2 - \mu \quad \Rightarrow \quad 2\lambda - 1 = 2 - \mu \quad [1]$$
$$y = \lambda + 1 \quad \text{and} \quad y = 1 - 2\mu \quad \Rightarrow \quad \lambda + 1 = 1 - 2\mu \quad [2]$$
$$z = \lambda \quad \text{and} \quad z = -2\mu \quad \Rightarrow \quad \lambda = -2\mu \quad [3]$$

Solving equations [1] and [2], we get $\lambda = 2$, $\mu = -1$. These values also satisfy equation [3], so the lines intersect.

Substituting for λ in the first line equation or for μ in the second, we find that the point of intersection of the lines is (3, 3, 2).

c) The lines are not parallel. If they have a point of intersection (x, y, z), then we have

$$\begin{aligned} x = \lambda - 4 \quad \text{and} \quad x = 3\mu + 1 &\Rightarrow \lambda - 4 = 3\mu + 1 & [1] \\ y = \lambda \quad \text{and} \quad y = \mu - 1 &\Rightarrow \lambda = \mu - 1 & [2] \\ z = 5 - 2\lambda \quad \text{and} \quad z = 2 - \mu &\Rightarrow 5 - 2\lambda = 2 - \mu & [3] \end{aligned}$$

Solving equations [1] and [2], we get $\lambda = -4$, $\mu = -3$.

If we try these values in equation [3], we find that the LHS = 13 and the RHS = 5. Hence, the lines do not have a point of intersection – they are skew lines.

Angle between two straight lines

We can define the angle between a pair of straight lines as being the angle between their direction vectors. This makes sense even for skew lines, as in this case we are effectively finding the angle between a pair of lines through an arbitrary point which are parallel to the given lines.

The angle is then found using the scalar product as described on page 4.

Example 10 Find the angle between the lines

$$\mathbf{r} = \begin{pmatrix} 4 - \lambda \\ \lambda + 2 \\ 2\lambda - 1 \end{pmatrix} \quad \text{and} \quad \mathbf{r} = \begin{pmatrix} 3\mu + 2 \\ 5 - 2\mu \\ 1 - 3\mu \end{pmatrix}$$

SOLUTION

We can rewrite the equations as

$$\mathbf{r} = \begin{pmatrix} 4 \\ 2 \\ -1 \end{pmatrix} + \lambda \begin{pmatrix} -1 \\ 1 \\ 2 \end{pmatrix} \quad \text{and} \quad \mathbf{r} = \begin{pmatrix} 2 \\ 5 \\ 1 \end{pmatrix} + \mu \begin{pmatrix} 3 \\ -2 \\ -3 \end{pmatrix}$$

and so the direction vectors are $\mathbf{p} = \begin{pmatrix} -1 \\ 1 \\ 2 \end{pmatrix}$ and $\mathbf{q} = \begin{pmatrix} 3 \\ -2 \\ -3 \end{pmatrix}$, which give

$$|\mathbf{p}| = \sqrt{(-1)^2 + 1^2 + 2^2} = \sqrt{6}$$

$$|\mathbf{q}| = \sqrt{3^2 + (-2)^2 + (-3)^2} = \sqrt{22}$$

and

$$\mathbf{p} \cdot \mathbf{q} = (-1) \times 3 + 1 \times (-2) + 2 \times (-3) = -11$$

Hence, if θ is the angle between the lines, we have

$$\cos \theta = \frac{-11}{\sqrt{6} \times \sqrt{22}} = -0.957 \quad \Rightarrow \quad \theta = 163.2°$$

Exercise 1B

1 Points A, B and C have position vectors $2\mathbf{i}+3\mathbf{j}-3\mathbf{k}$, $\mathbf{i}-4\mathbf{j}+2\mathbf{k}$ and $3\mathbf{i}-5\mathbf{j}+\mathbf{k}$ respectively. Prove that ABC is a right-angled triangle

  a) by using Pythagoras' theorem
  b) by using the scalar product.

  Forces **P** and **Q** act along AB and AC respectively. **P** has magnitude $20\sqrt{3}$ N and **Q** has magnitude 18 N. Find the magnitude of the resultant force and the angle it makes with AB.

2 A has position vector $\mathbf{a}=\mathbf{i}+2\mathbf{j}-2\mathbf{k}$ and B has position vector $\mathbf{b}=2\mathbf{i}-3\mathbf{j}+6\mathbf{k}$. The point C divides AB in the ratio $a:b$.

  a) Find the position vector of C.
  b) Show that OC bisects the angle AOB.

3 Points A, B and C have position vectors $\begin{pmatrix}4\\2\\1\end{pmatrix}$, $\begin{pmatrix}1\\-1\\2\end{pmatrix}$ and $\begin{pmatrix}2\\3\\-3\end{pmatrix}$ respectively. The point P divides AB in the ratio $1:k$, and the point Q divides AC in the ratio $1:k$

  a) Find the position vectors of P and Q.
  b) Show that PQ is parallel to BC.
  c) Find the ratio PQ:BC.

4 Find in both vector and cartesian form the equations of the following lines.

  a) Through (2, 1, 1) in the direction $3\mathbf{i}+\mathbf{j}+2\mathbf{k}$.
  b) Through $(-3, 4, 2)$ and $(1, 0, 3)$.
  c) Through $(4, -2, 2)$ and parallel to the line $\dfrac{x-3}{5}=\dfrac{y+1}{2}=z-2$.
  d) Through (3, 3, 2) in the direction $\mathbf{j}+2\mathbf{k}$.
  e) Through $(-4, -2, 3)$ and parallel to the y-axis.

5 For each of the lines in Question **4**, find the coordinates of the point (if it exists) at which the line intersects the xz-plane.

6 Find the angle between the lines

  a) $\mathbf{r}=(2+\lambda)\mathbf{i}+(3-2\lambda)\mathbf{j}+3\lambda\mathbf{k}$ and $\mathbf{r}=(\mu-2)\mathbf{i}+(\mu+1)\mathbf{j}-(2\mu+3)\mathbf{k}$
  b) $\dfrac{x-4}{3}=\dfrac{y-2}{2}=\dfrac{1-z}{2}$ and $\dfrac{2x-1}{4}=\dfrac{y+5}{3}=\dfrac{z-1}{6}$

7 Determine whether the following pairs of lines are parallel, intersecting or skew. If they intersect, find the position vector of their point of intersection.

  a) $\mathbf{r}=(3\lambda+1)\mathbf{i}+(1-2\lambda)\mathbf{j}+2(2\lambda+1)\mathbf{k}$ and $\mathbf{r}=2(1-3\mu)\mathbf{i}+(4\mu-1)\mathbf{j}+(3-8\mu)\mathbf{k}$
  b) $\mathbf{r}=\begin{pmatrix}1\\0\\2\end{pmatrix}+\lambda\begin{pmatrix}2\\-1\\1\end{pmatrix}$ and $\mathbf{r}=\begin{pmatrix}3\\1\\1\end{pmatrix}+\mu\begin{pmatrix}1\\2\\-2\end{pmatrix}$

c) $\dfrac{x-9}{3} = \dfrac{y-1}{5} = \dfrac{18-z}{7}$ and $\dfrac{x-5}{13} = \dfrac{y-11}{5} = \dfrac{z+6}{3}$

d) $\mathbf{r} = (-3\mathbf{i} + 6\mathbf{j}) + \lambda(-4\mathbf{i} + 3\mathbf{j} + 2\mathbf{k})$ and $\mathbf{r} = (-2\mathbf{i} + 7\mathbf{k}) + \mu(-4\mathbf{i} + \mathbf{j} + \mathbf{k})$

e) $\mathbf{r} = (-\mathbf{i} + \mathbf{j} + 3\mathbf{k}) + \lambda(\mathbf{i} + 2\mathbf{j} + 3\mathbf{k})$ and $\mathbf{r} = (\mathbf{i} - \mathbf{j} - 3\mathbf{k}) + \mu(3\mathbf{i} + 2\mathbf{j} + \mathbf{k})$

8 Prove that the lines

$$\mathbf{r} = \begin{pmatrix} a \\ a+b \\ a+2b \end{pmatrix} + \lambda \begin{pmatrix} c \\ c+d \\ c+2d \end{pmatrix} \quad \text{and} \quad \mathbf{r} = \begin{pmatrix} e \\ e+f \\ e+2f \end{pmatrix} + \mu \begin{pmatrix} g \\ g+h \\ g+2h \end{pmatrix}$$

are always either parallel or intersecting (that is, they lie in the same plane).

2 Vectors II

Beauty is as useful as usefulness. More perhaps.
VICTOR HUGO

Vector product

We will now look at the second method of combining vectors which is reminiscent of multiplication. This is called the **vector product** or sometimes the **cross product**.

If we have vectors **a** and **b**, the vector product is written $\mathbf{a} \times \mathbf{b}$ and is spoken as '**a** cross **b**'.

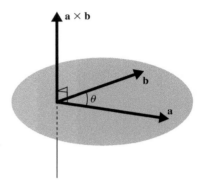

If the angle between **a** and **b** is θ, then $\mathbf{a} \times \mathbf{b}$ is the vector with magnitude $ab \sin \theta$ and with direction perpendicular to both **a** and **b**, such that a right-hand screw turning from **a** to **b** would travel in the direction $\mathbf{a} \times \mathbf{b}$.

This means that $\mathbf{b} \times \mathbf{a}$ also has magnitude $ab \sin \theta$ and is also perpendicular to both **a** and **b**.

Commutativity

A right-hand screw turning from **b** to **a** would travel in the opposite direction from $\mathbf{a} \times \mathbf{b}$. Therefore, it follows that

$$\mathbf{b} \times \mathbf{a} = -\mathbf{a} \times \mathbf{b}$$

This means that, unlike the scalar product, the vector product is **not commutative**.

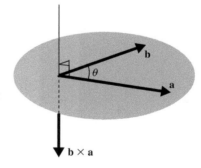

Parallel and perpendicular vectors

If **a** and **b** are **parallel** vectors, the angle between them is $0°$. The magnitude of $\mathbf{a} \times \mathbf{b}$ is, therefore, $ab \sin 0° = 0$ and hence we have

$$\mathbf{a} \times \mathbf{b} = \mathbf{0}$$

where **0** is known as the zero vector.

If **a** and **b** are **perpendicular**, the magnitude of $\mathbf{a} \times \mathbf{b}$ is $ab \sin 90° = ab$.

We can use these facts to find the vector products of the unit vectors **i**, **j** and **k**.

By the nature of parallel vectors, we have

$$\mathbf{i} \times \mathbf{i} = \mathbf{j} \times \mathbf{j} = \mathbf{k} \times \mathbf{k} = \mathbf{0}$$

By the nature of perpendicular vectors, and using the fact that **i**, **j** and **k** form a right-hand set, we also have

$$\mathbf{i} \times \mathbf{j} = \mathbf{k} \quad \mathbf{j} \times \mathbf{k} = \mathbf{i} \quad \mathbf{k} \times \mathbf{i} = \mathbf{j}$$

and, of course,

$$\mathbf{j} \times \mathbf{i} = -\mathbf{k} \quad \mathbf{k} \times \mathbf{j} = -\mathbf{i} \quad \mathbf{i} \times \mathbf{k} = -\mathbf{j}$$

Associativity

We should now examine two other properties of the vector product. First, is it associative – that is, is the following true?

$$\mathbf{a} \times (\mathbf{b} \times \mathbf{c}) = (\mathbf{a} \times \mathbf{b}) \times \mathbf{c}$$

We can readily see that this is **not** true by considering $\mathbf{a} \times (\mathbf{b} \times \mathbf{b})$ and $(\mathbf{a} \times \mathbf{b}) \times \mathbf{b}$.

As $\mathbf{b} \times \mathbf{b} = \mathbf{0}$, we have

$$|\mathbf{a} \times (\mathbf{b} \times \mathbf{b})| = a \times 0 = 0$$

If **a** and **b** are inclined at angle θ, $|\mathbf{a} \times \mathbf{b}| = ab \sin \theta$. Also, $\mathbf{a} \times \mathbf{b}$ is perpendicular to **b**. Hence, it follows that

$$|(\mathbf{a} \times \mathbf{b}) \times \mathbf{b}| = ab \sin \theta \times b \neq 0$$

and so we have

$$\mathbf{a} \times (\mathbf{b} \times \mathbf{b}) \neq (\mathbf{a} \times \mathbf{b}) \times \mathbf{b}$$

This means that, in general,

$$\mathbf{a} \times (\mathbf{b} \times \mathbf{c}) \neq (\mathbf{a} \times \mathbf{b}) \times \mathbf{c}$$

Distributivity

The other property we need to examine is whether the vector product is distributive over vector addition. That is, is the following true?

$$\mathbf{a} \times (\mathbf{b} + \mathbf{c}) = \mathbf{a} \times \mathbf{b} + \mathbf{a} \times \mathbf{c}$$

The statement is, in fact, **true**, but the argument is slightly more involved than those above. It will be enough to prove it for the unit vector $\hat{\mathbf{a}}$.

CHAPTER 2 VECTORS II

Proof

We have to prove that

$$\hat{\mathbf{a}} \times (\mathbf{b} + \mathbf{c}) = \hat{\mathbf{a}} \times \mathbf{b} + \hat{\mathbf{a}} \times \mathbf{c}$$

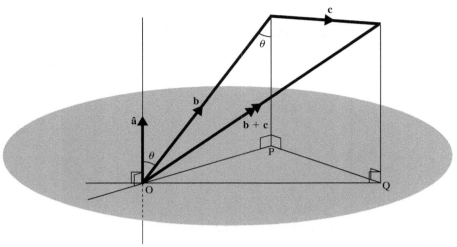

The diagram shows vectors $\hat{\mathbf{a}}$, $\mathbf{b}$, $\mathbf{c}$ and $\mathbf{b} + \mathbf{c}$. The shaded area represents a plane perpendicular to $\hat{\mathbf{a}}$. OP, PQ and OQ are the projections of $\mathbf{b}$, $\mathbf{c}$ and $\mathbf{b} + \mathbf{c}$ onto the plane.

$\hat{\mathbf{a}} \times \mathbf{b}$ is a vector perpendicular to $\hat{\mathbf{a}}$ and $\mathbf{b}$. This means that it must be parallel to the shaded plane and be perpendicular to OP.

We also have

$$|\hat{\mathbf{a}} \times \mathbf{b}| = 1 \times b \times \sin\theta = b \sin\theta$$

But $b \sin\theta = \text{OP}$. So, we have

$$|\hat{\mathbf{a}} \times \mathbf{b}| = \text{OP}$$

By a similar argument, we find that

$\hat{\mathbf{a}} \times \mathbf{c}$ is in the plane and perpendicular to PQ
$|\hat{\mathbf{a}} \times \mathbf{c}| = \text{PQ}$

and

$\hat{\mathbf{a}} \times (\mathbf{b} + \mathbf{c})$ is in the plane and perpendicular to OQ
$|\hat{\mathbf{a}} \times (\mathbf{b} + \mathbf{c})| = \text{OQ}$

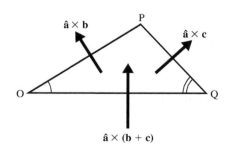 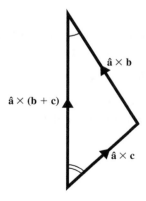

Because the three vectors have magnitudes equal to the sides of triangle OPQ and are perpendicular to their respective sides, we can readily see that drawing the vectors as directed line segments gives a triangle congruent to OPQ. It follows that

$$\hat{\mathbf{a}} \times (\mathbf{b} + \mathbf{c}) = \hat{\mathbf{a}} \times \mathbf{b} + \hat{\mathbf{a}} \times \mathbf{c}$$

Vector product in component form

Because the vector product is distributive over addition, we are justified in 'multiplying out' brackets and can use this to find the vector product in component form.

Suppose

$$\mathbf{a} = a_1\mathbf{i} + a_2\mathbf{j} + a_3\mathbf{k} \quad \text{and} \quad \mathbf{b} = b_1\mathbf{i} + b_2\mathbf{j} + b_3\mathbf{k}$$

then we have

$$\mathbf{a} \times \mathbf{b} = (a_1\mathbf{i} + a_2\mathbf{j} + a_3\mathbf{k}) \times (b_1\mathbf{i} + b_2\mathbf{j} + b_3\mathbf{k})$$

Because $\mathbf{i} \times \mathbf{i} = \mathbf{j} \times \mathbf{j} = \mathbf{k} \times \mathbf{k} = \mathbf{0}$, expanding these brackets gives

$$\mathbf{a} \times \mathbf{b} = a_1b_2(\mathbf{i} \times \mathbf{j}) + a_1b_3(\mathbf{i} \times \mathbf{k}) + a_2b_1(\mathbf{j} \times \mathbf{i}) + a_2b_3(\mathbf{j} \times \mathbf{k}) + \\ + a_3b_1(\mathbf{k} \times \mathbf{i}) + a_3b_2(\mathbf{k} \times \mathbf{j})$$

Using the results from page 15, we obtain

$$\mathbf{a} \times \mathbf{b} = a_1b_2\mathbf{k} - a_1b_3\mathbf{j} - a_2b_1\mathbf{k} + a_2b_3\mathbf{i} + a_3b_1\mathbf{j} - a_3b_2\mathbf{i}$$

$$\Rightarrow \quad \mathbf{a} \times \mathbf{b} = (a_2b_3 - a_3b_2)\mathbf{i} + (a_3b_1 - a_1b_3)\mathbf{j} + (a_1b_2 - a_2b_1)\mathbf{k}$$

This expression is most conveniently remembered in the form of a determinant. (If you are unfamiliar with determinants, see *Further Pure Mathematics*, pages 80–93.)

$$\mathbf{a} \times \mathbf{b} = \begin{vmatrix} \mathbf{i} & \mathbf{j} & \mathbf{k} \\ a_1 & a_2 & a_3 \\ b_1 & b_2 & b_3 \end{vmatrix}$$

For example, if $\mathbf{a} = 2\mathbf{i} + 3\mathbf{j} - 2\mathbf{k}$ and $\mathbf{b} = \mathbf{i} - 4\mathbf{j} + 3\mathbf{k}$, then

$$\mathbf{a} \times \mathbf{b} = \begin{vmatrix} \mathbf{i} & \mathbf{j} & \mathbf{k} \\ 2 & 3 & -2 \\ 1 & -4 & 3 \end{vmatrix}$$

$$= \begin{vmatrix} 3 & -2 \\ -4 & 3 \end{vmatrix}\mathbf{i} - \begin{vmatrix} 2 & -2 \\ 1 & 3 \end{vmatrix}\mathbf{j} + \begin{vmatrix} 2 & 3 \\ 1 & -4 \end{vmatrix}\mathbf{k}$$

$$= \mathbf{i} - 8\mathbf{j} - 11\mathbf{k}$$

Interpreting the vector product

We will see on pages 27–34 that the vector product has applications when we consider moments of forces, but for now there is a direct geometrical interpretation of it.

CHAPTER 2 VECTORS II

The diagram on the right shows a parallelogram PQRS, with $\overrightarrow{PS} = \mathbf{a}$ and $\overrightarrow{PQ} = \mathbf{b}$. The height of the parallelogram is h. We therefore have

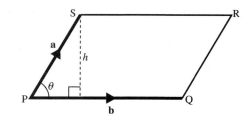

$$\text{Area PQRS} = bh$$

As $h = a \sin \theta$, we get

$$\text{Area PQRS} = ab \sin \theta$$

But $|\mathbf{a} \times \mathbf{b}| = ab \sin \theta$, so we have

$$\text{Area PQRS} = |\mathbf{a} \times \mathbf{b}|$$

We can therefore see that the magnitude of the vector product of two vectors gives the area of the parallelogram generated by the vectors.

The direction of the vector product indicates the orientation of the parallelogram by giving the normal direction to the plane.

In this sense, a plane area can be regarded as a vector quantity.

Example 1 Find the area of the triangle formed by the vectors $\mathbf{a} = 2\mathbf{i} + \mathbf{j} - 3\mathbf{k}$ and $\mathbf{b} = -2\mathbf{i} + 3\mathbf{j} + \mathbf{k}$.

SOLUTION

The area of the triangle is half that of the parallelogram defined by the vectors. That is,

$$\text{Area} = \tfrac{1}{2}|\mathbf{a} \times \mathbf{b}|$$

Now, $\mathbf{a} \times \mathbf{b}$ is given by

$$\mathbf{a} \times \mathbf{b} = \begin{vmatrix} \mathbf{i} & \mathbf{j} & \mathbf{k} \\ 2 & 1 & -3 \\ -2 & 3 & 1 \end{vmatrix} = 10\mathbf{i} + 4\mathbf{j} + 8\mathbf{k}$$

Hence, we have

$$\text{Area} = \tfrac{1}{2} \times \sqrt{10^2 + 4^2 + 8^2} = 3\sqrt{5} \text{ units}^2$$

Exercise 2A

1. Given vectors $\mathbf{a} = 4\mathbf{i} + \mathbf{j} + 2\mathbf{k}$, $\mathbf{b} = \mathbf{i} + 2\mathbf{j} - \mathbf{k}$ and $\mathbf{c} = 3\mathbf{i} - 3\mathbf{j} + \mathbf{k}$.

   a) Verify that $\mathbf{a} \times (\mathbf{b} + \mathbf{c}) = \mathbf{a} \times \mathbf{b} + \mathbf{a} \times \mathbf{c}$
   b) Show that $\mathbf{a} \times (\mathbf{b} \times \mathbf{c}) \neq (\mathbf{a} \times \mathbf{b}) \times \mathbf{c}$

2. Given vectors $\mathbf{a} = 2\mathbf{i} + 2\mathbf{j} + \mathbf{k}$ and $\mathbf{b} = 4\mathbf{i} + 3\mathbf{j} - 7\mathbf{k}$, find a unit vector perpendicular to both $\mathbf{a}$ and $\mathbf{b}$.

3. Given vectors $\mathbf{a} = \mathbf{i} - 3\mathbf{j} + \mathbf{k}$, $\mathbf{b} = 2\mathbf{i} + \mathbf{j} - 2\mathbf{k}$ and $\mathbf{c} = -3\mathbf{i} + 2\mathbf{j} + 3\mathbf{k}$, verify that

   $$\mathbf{a} . \mathbf{b} \times \mathbf{c} = \mathbf{a} \times \mathbf{b} . \mathbf{c}$$

   [**Note** There is no need for brackets in these expressions, as to be meaningful the vector product must be done first.]

4 Prove that $(\mathbf{a} - \mathbf{b}) \times (\mathbf{a} + \mathbf{b}) = 2\mathbf{a} \times \mathbf{b}$

5 Given $\mathbf{a} = 2\mathbf{i} + \mathbf{j} - 3\mathbf{k}$ and $\mathbf{c} = 14\mathbf{i} - \mathbf{j} + 9\mathbf{k}$, find vector $\mathbf{b}$ such that $\mathbf{a} \cdot \mathbf{b} = -4$ and $\mathbf{a} \times \mathbf{b} = \mathbf{c}$.

6 Triangle ABC has vertices $A(2, 1, 1)$, $B(4, 0, -2)$ and $C(-1, -2, 3)$. Using the vector product, find

a) the area of ABC
b) a unit vector perpendicular to the plane of the triangle.

7 Triangle ABC has vertices A, B and C with position vectors $\mathbf{a}$, $\mathbf{b}$ and $\mathbf{c}$ respectively. By considering $\overrightarrow{AB} \times \overrightarrow{AC}$ or otherwise, show that the area of ABC is given by

$$\tfrac{1}{2}|\mathbf{a} \times \mathbf{b} + \mathbf{b} \times \mathbf{c} + \mathbf{c} \times \mathbf{a}|$$

What could you say about the points A, B and C if $\mathbf{a} \times \mathbf{b} + \mathbf{b} \times \mathbf{c} = \mathbf{a} \times \mathbf{c}$?

8 The diagram on the right shows a cuboid in which OA is 3 units, OC is 4 units and OG is 2 units. H is the mid-point of GD.

Use the vector product to find the area of the triangle BFH.

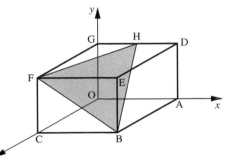

9 The triangle ABC has sides formed by vectors $\mathbf{a}$, $\mathbf{b}$ and $\mathbf{c}$, as shown in the diagram on the right.

Show that $\mathbf{a} \times \mathbf{b} = \mathbf{b} \times \mathbf{c} = \mathbf{c} \times \mathbf{a}$ and hence derive the sine rule for the triangle.

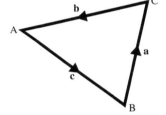

10 Three vectors $\mathbf{a}$, $\mathbf{b}$ and $\mathbf{c}$ are such that $\mathbf{a} \times \mathbf{b} = \mathbf{a} \times \mathbf{c}$.

Explain why this means that the three vectors lie in the same plane. Show further that $\mathbf{b} - \mathbf{c}$ is parallel to $\mathbf{a}$.

Scalar triple product

There are three ways in which three vectors $\mathbf{a}$, $\mathbf{b}$ and $\mathbf{c}$ may be combined as a product.

First, there is the rather trivial $(\mathbf{a} \cdot \mathbf{b})\mathbf{c}$. As $\mathbf{a} \cdot \mathbf{b}$ is a scalar, this is just a scalar multiple of vector $\mathbf{c}$.

Second, there is the **vector triple product**, $\mathbf{a} \times (\mathbf{b} \times \mathbf{c})$, so called because the result is a vector. This is more interesting and important, but is outside the A-level syllabus.

CHAPTER 2 VECTORS II

Finally, we have the **scalar triple product**, $\mathbf{a} \cdot \mathbf{b} \times \mathbf{c}$, which results in a scalar quantity. As was pointed out in Exercise 2A, there is no ambiguity in leaving out brackets, because for the second operation to be meaningful the first operation must be the vector product.

Example 2 Given $\mathbf{a} = 3\mathbf{i} + 2\mathbf{j} + 2\mathbf{k}$, $\mathbf{b} = \mathbf{i} + \mathbf{j} + 3\mathbf{k}$ and $\mathbf{c} = \mathbf{i} - 2\mathbf{j} - \mathbf{k}$, verify that $\mathbf{a} \cdot \mathbf{b} \times \mathbf{c} = \mathbf{a} \times \mathbf{b} \cdot \mathbf{c}$

SOLUTION

We have

$$\mathbf{b} \times \mathbf{c} = \begin{vmatrix} \mathbf{i} & \mathbf{j} & \mathbf{k} \\ 1 & 1 & 3 \\ 1 & -2 & -1 \end{vmatrix} = 5\mathbf{i} + 4\mathbf{j} - 3\mathbf{k}$$

$$\Rightarrow \quad \mathbf{a} \cdot \mathbf{b} \times \mathbf{c} = (3\mathbf{i} + 2\mathbf{j} + 2\mathbf{k}) \cdot (5\mathbf{i} + 4\mathbf{j} - 3\mathbf{k})$$
$$= 3 \times 5 + 2 \times 4 + 2 \times (-3)$$
$$= 17$$

We also have

$$\mathbf{a} \times \mathbf{b} = \begin{vmatrix} \mathbf{i} & \mathbf{j} & \mathbf{k} \\ 3 & 2 & 2 \\ 1 & 1 & 3 \end{vmatrix} = 4\mathbf{i} - 7\mathbf{j} + \mathbf{k}$$

$$\Rightarrow \quad \mathbf{a} \times \mathbf{b} \cdot \mathbf{c} = (4\mathbf{i} - 7\mathbf{j} + \mathbf{k}) \cdot (\mathbf{i} - 2\mathbf{j} - \mathbf{k})$$
$$= 4 \times 1 + (-7) \times (-2) + 1 \times (-1)$$
$$= 17$$

Hence, we have verified that

$$\mathbf{a} \cdot \mathbf{b} \times \mathbf{c} = \mathbf{a} \times \mathbf{b} \cdot \mathbf{c}$$

Component form of the scalar triple product

Suppose we have vectors

$$\mathbf{a} = \begin{pmatrix} a_1 \\ a_2 \\ a_3 \end{pmatrix} \quad \mathbf{b} = \begin{pmatrix} b_1 \\ b_2 \\ b_3 \end{pmatrix} \quad \mathbf{c} = \begin{pmatrix} c_1 \\ c_2 \\ c_3 \end{pmatrix}$$

We know that

$$\mathbf{b} \times \mathbf{c} = (b_2 c_3 - b_3 c_2)\mathbf{i} + (b_3 c_1 - b_1 c_3)\mathbf{j} + (b_1 c_2 - b_2 c_1)\mathbf{k}$$

which gives

$$\mathbf{a} \cdot \mathbf{b} \times \mathbf{c} = a_1(b_2 c_3 - b_3 c_2) + a_2(b_3 c_1 - b_1 c_3) + a_3(b_1 c_2 - b_2 c_1)$$

This can be written in determinant form as

$$\mathbf{a} \cdot \mathbf{b} \times \mathbf{c} = \begin{vmatrix} a_1 & a_2 & a_3 \\ b_1 & b_2 & b_3 \\ c_1 & c_2 & c_3 \end{vmatrix}$$

SCALAR TRIPLE PRODUCT

One of the features of a determinant is that any cyclic repositioning of the rows, as depicted on the right, leaves the value unchanged, while swapping two rows changes the sign of the determinant. From this we deduce that

$$\mathbf{a}.\mathbf{b} \times \mathbf{c} = \mathbf{b}.\mathbf{c} \times \mathbf{a} = \mathbf{c}.\mathbf{a} \times \mathbf{b}$$

and, for example, that

$$\mathbf{a}.\mathbf{b} \times \mathbf{c} = -\mathbf{b}.\mathbf{a} \times \mathbf{c}$$

As scalar product is commutative, we also have, for example,

$$\mathbf{a}.\mathbf{b} \times \mathbf{c} = \mathbf{a} \times \mathbf{b}.\mathbf{c}$$

Geometrical interpretation

The diagram on the right shows a solid called a **parallelepiped** (pronounced *parallele-pie-ped*), whose faces are pairs of congruent parallelograms.

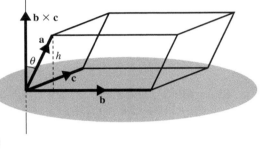

The solid is generated by three vectors, **a**, **b** and **c**, as shown. Vectors **b** and **c** form the base, and $\mathbf{b} \times \mathbf{c}$, which is perpendicular to the base, makes an angle θ with vector **a**. Hence, we have

$$\mathbf{a}.\mathbf{b} \times \mathbf{c} = |\mathbf{a}|\,|\mathbf{b} \times \mathbf{c}|\cos\theta$$

But $|\mathbf{b} \times \mathbf{c}|$ gives the area of the base, and $|\mathbf{a}|\cos\theta = h$, the perpendicular height of the parallelepiped. So, we have

$$\mathbf{a}.\mathbf{b} \times \mathbf{c} = \text{Base area} \times \text{Height} = \text{Volume of parallelepiped}$$

We can now see why reordering **a**, **b** and **c** cyclically, such as $\mathbf{b}.\mathbf{c} \times \mathbf{a}$, or swapping the . and the × operations as in $\mathbf{a} \times \mathbf{b}.\mathbf{c}$, leaves the value unchanged. In each case, the result is the volume of the same parallelepiped.

Non-cyclical reorderings, such as $\mathbf{a}.\mathbf{c} \times \mathbf{b}$, give the negative of the volume, since $\mathbf{c} \times \mathbf{b}$ would point downwards in the diagram and we would have to replace **a** with $-\mathbf{a}$ to relate the product to a 'real' parallelepiped below the shaded plane.

Some authors use [**a**, **b**, **c**] for the scalar triple product, meaning $\mathbf{a}.\mathbf{b} \times \mathbf{c}$, $\mathbf{a} \times \mathbf{b}.\mathbf{c}$ and any cyclical reordering of these.

Volume of a tetrahedron

The volume of the tetrahedron generated by vectors **a**, **b** and **c**, as shown, is one sixth of the volume of the corresponding parallelepiped.

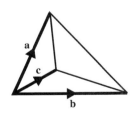

$$\text{Volume of tetrahedron} = \tfrac{1}{6}\mathbf{a}.\mathbf{b} \times \mathbf{c}$$

Other geometrical uses of the vector product and the scalar triple product

Distance from a point to a line

Suppose we have a line passing through a point A, position vector **a**, and with direction vector **b**, as shown. We wish to find the distance d from a point P, position vector **p**, to the line.

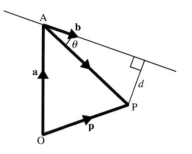

We have
$$d = \text{AP} \sin \theta$$
But $\overrightarrow{\text{AP}} = \mathbf{p} - \mathbf{a}$, which gives
$$d = |\mathbf{p} - \mathbf{a}| \sin \theta$$
We know that
$$|\mathbf{b} \times (\mathbf{p} - \mathbf{a})| = |\mathbf{b}|\,|\mathbf{p} - \mathbf{a}| \sin \theta$$
So, substituting for $|\mathbf{p} - \mathbf{a}| \sin \theta$, we get
$$d = \frac{|\mathbf{b} \times (\mathbf{p} - \mathbf{a})|}{|\mathbf{b}|}$$

Example 3 Find the distance from the point $P(2, 3, -1)$ to the line $\mathbf{r} = (3 - \lambda)\mathbf{i} + (2\lambda - 1)\mathbf{j} + (\lambda + 4)\mathbf{k}$

SOLUTION

We rewrite the equation as
$$\mathbf{r} = (3\mathbf{i} - \mathbf{j} + 4\mathbf{k}) + \lambda(-\mathbf{i} + 2\mathbf{j} + \mathbf{k})$$
From this we see that the line passes through the point with position vector $\mathbf{a} = 3\mathbf{i} - \mathbf{j} + 4\mathbf{k}$ and has direction vector $\mathbf{b} = -\mathbf{i} + 2\mathbf{j} + \mathbf{k}$.

The point P has position vector $\mathbf{p} = 2\mathbf{i} + 3\mathbf{j} - \mathbf{k}$, giving
$$\mathbf{p} - \mathbf{a} = -\mathbf{i} + 4\mathbf{j} - 5\mathbf{k}$$
The required distance d is given by
$$d = \frac{|(-\mathbf{i} + 2\mathbf{j} + \mathbf{k}) \times (-\mathbf{i} + 4\mathbf{j} - 5\mathbf{k})|}{|-\mathbf{i} + 2\mathbf{j} + \mathbf{k}|}$$
Simplifying the numerator, we get
$$(-\mathbf{i} + 2\mathbf{j} + \mathbf{k}) \times (-\mathbf{i} + 4\mathbf{j} - 5\mathbf{k}) = -14\mathbf{i} - 6\mathbf{j} - 2\mathbf{k}$$
which gives
$$d = \frac{|-14\mathbf{i} - 6\mathbf{j} - 2\mathbf{k}|}{|-\mathbf{i} + 2\mathbf{j} + \mathbf{k}|}$$
$$\Rightarrow \quad d = \frac{\sqrt{(-14)^2 + (-6)^2 + (-2)^2}}{\sqrt{(-1)^2 + 2^2 + 1^2}} = \frac{\sqrt{236}}{\sqrt{6}} = 11.7 \text{ units}$$

GEOMETRICAL USES OF THE VECTOR PRODUCT AND SCALAR TRIPLE PRODUCT

Shortest distance between a pair of skew lines

Suppose we have lines l_1 and l_2, which have equations $\mathbf{r} = \mathbf{a}_1 + \lambda\mathbf{b}_1$ and $\mathbf{r} = \mathbf{a}_2 + \mu\mathbf{b}_2$ respectively. Points A_1 and A_2, having position vectors $\mathbf{a}_1$ and $\mathbf{a}_2$, lie on l_1 and l_2 as shown.

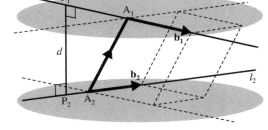

If we draw lines parallel to l_1 and l_2 through A_2 and A_1 respectively, we define two parallel planes, shaded in the diagram.

The shortest distance between the lines is the length, d, of the line P_1P_2, which is perpendicular to both l_1 and l_2. This is the perpendicular distance between the two shaded planes.

The vector $\overrightarrow{A_2A_1} = \mathbf{a}_1 - \mathbf{a}_2$, and this together with $\mathbf{b}_1$ and $\mathbf{b}_2$ define a parallelepiped, as shown in the diagram. The perpendicular height of this parallelepiped is d. If its base area is A and its volume is V, we know that

$$A = |\mathbf{b}_1 \times \mathbf{b}_2| \quad \text{and} \quad V = |(\mathbf{a}_1 - \mathbf{a}_2).\mathbf{b}_1 \times \mathbf{b}_2|$$

where the modulus bars are needed for V to ensure a positive volume.

But as $V = Ad$, we have

$$d = \frac{|(\mathbf{a}_1 - \mathbf{a}_2).\mathbf{b}_1 \times \mathbf{b}_2|}{|\mathbf{b}_1 \times \mathbf{b}_2|}$$

Intersecting lines

If the lines intersect, the shortest distance between them is zero. The condition for intersecting lines is therefore

$$(\mathbf{a}_1 - \mathbf{a}_2).\mathbf{b}_1 \times \mathbf{b}_2 = 0$$

Example 4 Find the shortest distance between the lines

$$\mathbf{r} = \begin{pmatrix} 1 \\ 0 \\ 2 \end{pmatrix} + \lambda \begin{pmatrix} 2 \\ -1 \\ 1 \end{pmatrix} \quad \text{and} \quad \mathbf{r} = \begin{pmatrix} -2 \\ 2 \\ 1 \end{pmatrix} + \mu \begin{pmatrix} 1 \\ 1 \\ -3 \end{pmatrix}$$

SOLUTION

The points having position vectors $\mathbf{a}_1 = \begin{pmatrix} 1 \\ 0 \\ 2 \end{pmatrix}$ and $\mathbf{a}_2 = \begin{pmatrix} -2 \\ 2 \\ 1 \end{pmatrix}$ are on the respective lines.

The lines have direction vectors $\mathbf{b}_1 = \begin{pmatrix} 2 \\ -1 \\ 1 \end{pmatrix}$ and $\mathbf{b}_2 = \begin{pmatrix} 1 \\ 1 \\ -3 \end{pmatrix}$.

Hence, we have

$$\mathbf{a}_1 - \mathbf{a}_2 = \begin{pmatrix} 3 \\ -2 \\ 1 \end{pmatrix} \quad \Rightarrow \quad (\mathbf{a}_1 - \mathbf{a}_2).\mathbf{b}_1 \times \mathbf{b}_2 = \begin{vmatrix} 3 & -2 & 1 \\ 2 & -1 & 1 \\ 1 & 1 & -3 \end{vmatrix} = -5$$

We also have

$$\mathbf{b}_1 \times \mathbf{b}_2 = \begin{vmatrix} \mathbf{i} & \mathbf{j} & \mathbf{k} \\ 2 & -1 & 1 \\ 1 & 1 & -3 \end{vmatrix} = 2\mathbf{i} + 7\mathbf{j} + 3\mathbf{k} \quad \Rightarrow \quad |\mathbf{b}_1 \times \mathbf{b}_2| = \sqrt{62}$$

The shortest distance between the lines is, therefore,

$$d = \frac{|-5|}{\sqrt{62}} = 0.635 \text{ units}$$

Exercise 2B

1 Find the volume of the parallelepiped generated by the vectors $\mathbf{i}+\mathbf{j}+2\mathbf{k}$, $2\mathbf{i}-\mathbf{j}-\mathbf{k}$ and $3\mathbf{i}+2\mathbf{j}-2\mathbf{k}$.

2 A tetrahedron OABC has vertices at the origin O and A$(2,1,1)$, B$(-1,-2,2)$ and C$(1,3,-2)$.
   a) Find the volume of OABC.
   b) Find the area of triangle ABC.
   c) Use your answers to parts **a** and **b** to find the distance from the origin to the plane ABC.

3 Find the volume of the tetrahedron with vertices A$(1,1,1)$, B$(2\ 1\ -1)$, C$(3,3,2)$ and D$(-3,1,0)$.

4 A tetrahedron has vertices with position vectors **a**, **b**, **c** and **d**. Show that its volume is given by
$$V = \tfrac{1}{6}|\mathbf{d}.(\mathbf{a} \times \mathbf{b} + \mathbf{b} \times \mathbf{c} + \mathbf{c} \times \mathbf{a}) - \mathbf{a}.\mathbf{b} \times \mathbf{c}|$$

5 A straight line has equation $\mathbf{r} = (1-\lambda)\mathbf{i} + \lambda\mathbf{j} + (2\lambda - 1)\mathbf{k}$. A is the point with coordinates $(3, 1, -2)$.
   a) Write down the vector AP connecting A with a general point P on the line. If AP is perpendicular to the line, find and solve an equation for λ. Hence find the distance from point A to the line.
   b) Use the vector product to confirm the value you obtained in part **a** for the distance of point A from the line.

6 a) Find the least distance between lines L_1, equation $\mathbf{r} = (\lambda + 4)\mathbf{i} + (1 - 2\lambda)\mathbf{j} - (\lambda + 1)\mathbf{k}$ and L_2, equation $\mathbf{r} = 3\mu\mathbf{i} + 2\mathbf{j} + (\mu - 1)\mathbf{k}$.
   b) A third line L_3 is parallel to L_2 and passes through the point with coordinates $(1, 2, \alpha)$. Find the value of α if L_1 and L_3 intersect.

7 Find the distance of the point P$(-2, -3, 0)$ from the lines L_1, equation $\mathbf{r} = \begin{pmatrix} 3 \\ 0 \\ 2 \end{pmatrix} + \lambda \begin{pmatrix} 1 \\ -1 \\ 0 \end{pmatrix}$, and L_2, equation $\mathbf{r} = \begin{pmatrix} -4 \\ -5 \\ -1 \end{pmatrix} + \mu \begin{pmatrix} 0 \\ -1 \\ 2 \end{pmatrix}$. Find also the distance between the lines. What can you deduce about the point P? + + + +

8 Find the shortest distance between the lines whose cartesian equations are

$$x - 1 = 2 - y = z - 1 \quad \text{and} \quad \frac{x-2}{2} = y + 1 = \frac{z+1}{2}$$

9 Examine whether the following pairs of lines intersect.

a) $\mathbf{r} = \begin{pmatrix} 3 \\ -1 \\ 1 \end{pmatrix} + \lambda \begin{pmatrix} 2 \\ 0 \\ 1 \end{pmatrix}$ and $\mathbf{r} = \begin{pmatrix} 1 \\ -2 \\ 1 \end{pmatrix} + \mu \begin{pmatrix} 1 \\ 1 \\ -1 \end{pmatrix}$

b) $\mathbf{r} = \begin{pmatrix} -2 \\ 2 \\ -1 \end{pmatrix} + \lambda \begin{pmatrix} 3 \\ -1 \\ 0 \end{pmatrix}$ and $\mathbf{r} = \begin{pmatrix} 2 \\ 0 \\ -4 \end{pmatrix} + \mu \begin{pmatrix} 2 \\ 0 \\ 3 \end{pmatrix}$

Vector equation of a curve

We can express the equations of curves, in two or three dimensions, in vector form. Essentially, these are parametric equations, since they require us to express the **i**, **j** and **k** components of the position vector **r** of a general point P on the curve in terms of a parameter, which is usually designated as t, λ or θ. The vector equation of the curve can then be written in the form

$$\mathbf{r} = \mathrm{f}(\lambda)\mathbf{i} + \mathrm{g}(\lambda)\mathbf{j} + \mathrm{h}(\lambda)\mathbf{k}$$

Example 5 Find the vector equation of a circle, centre $C(a, b)$ and radius l, lying in the **ij**-plane.

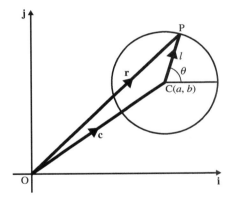

SOLUTION

In the diagram on the right, we have $\mathbf{c} = a\mathbf{i} + b\mathbf{j}$ and

$$\overrightarrow{CP} = l\cos\theta\,\mathbf{i} + l\sin\theta\,\mathbf{j}$$

The position vector of P, a general point on the circle, is given by

$$\mathbf{r} = \mathbf{c} + \overrightarrow{CP}$$
$$\Rightarrow \quad \mathbf{r} = (a\mathbf{i} + b\mathbf{j}) + (l\cos\theta\,\mathbf{i} + l\sin\theta\,\mathbf{j})$$
$$\Rightarrow \quad \mathbf{r} = (a + l\cos\theta)\mathbf{i} + (b + l\sin\theta)\mathbf{j}$$

and this is the vector equation of the circle.

Example 6 Sketch and describe the curve with vector equation

$$\mathbf{r} = a\cos\theta\,\mathbf{i} + a\sin\theta\,\mathbf{j} + b\theta\,\mathbf{k}$$

SOLUTION

The curve $\mathbf{r} = a\cos\theta\,\mathbf{i} + a\sin\theta\,\mathbf{j}$ is a circle in the **ij**-plane, with centre O and radius a. This circle (shown dashed in the diagram) is the projection of the given curve onto the **ij**-plane.

As the projected point moves at a constant rate round the circle, the point on the curve also moves at a constant rate in the **k**-direction.

The curve is a three-dimensional spiral, known as a **helix**. (This is familiar from models of the DNA molecule – the 'double helix'.)

A helix can be produced by wrapping a length of thread around a cylinder. The distance, p, between corresponding points P and Q on two neighbouring coils is constant and is called the **pitch** of the helix.

The pitch of the helix is the increase in the **k**-component of **r** as θ increases by 2π. For the given equation, the pitch is $2\pi b$.

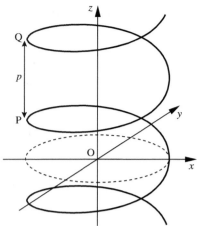

Exercise 2C

1 Find the vector equation of a circle, centre $(2, -3)$ and radius 4.

2 Find the vector equation of a circle lying in the **ik**-plane, centre the origin and radius 3.

3 Describe the curve whose vector equation is $\mathbf{r} = t\mathbf{i} + 2t^2\mathbf{j} - 4\mathbf{k}$.

4 Describe and sketch the curve whose vector equation is $\mathbf{r} = 3\cos\theta\,\mathbf{i} + 3\sin\theta\,\mathbf{j} + 2\theta\mathbf{k}$.

5 Describe and sketch the curve whose vector equation is $\mathbf{r} = \dfrac{\theta}{\pi}\mathbf{i} + \cos\theta\,\mathbf{j} + \sin\theta\,\mathbf{k}$.

6 Describe and sketch the curve whose vector equation is $\mathbf{r} = \cos\theta\,\mathbf{i} + \sin\theta\,\mathbf{j} + \cos\theta\,\mathbf{k}$.

7 Describe and sketch the curve whose vector equation is $\mathbf{r} = \cos\theta\,\mathbf{i} + \sqrt{2}\sin\theta\,\mathbf{j} + \cos\theta\,\mathbf{k}$.

8 The diagram shows a disc of radius a, which is rolling without slipping around a circular track lying in the **ij**-plane. The track is centred on the origin and has a radius of $3a$. The plane of the disc is at all times perpendicular to the **ij**-plane.

The disc starts rolling with a point P on its rim at the point $(3a, 0, 0)$. Some time later, the disc has rotated through an angle ϕ and the point of contact with the track has undergone an angular displacement θ, as shown.

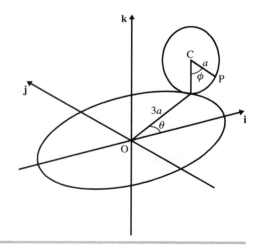

a) Find ϕ in terms of θ.
b) Find the position vector of the centre of the disc.
c) Hence find the vector equation of the path traced out by the point P.

Moments

We now examine the moment of a force **F** about an origin O.

We consider x- and y-axes in the plane containing O and **F**. Let the force be $\mathbf{F} = X\mathbf{i} + Y\mathbf{j}$ and suppose that the line of action of **F** passes through a point P with position vector $\mathbf{r} = x\mathbf{i} + y\mathbf{j}$, as shown.

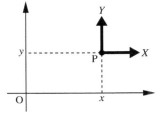

Taking the anticlockwise sense to be positive, as usual, the moment of this force about an axis through O perpendicular to the plane is

$$C = xY - yX$$

But the vector product of the position vector and the force vector gives

$$\mathbf{r} \times \mathbf{F} = (xY - yX)\mathbf{k}$$

The moment of a force through a given point can therefore be represented by the vector product of the position vector of the point and the force vector. That is, the moment, C, is given by

$$\mathbf{C} = \mathbf{r} \times \mathbf{F}$$

The moment of **F** is, therefore, a vector quantity whose direction is that of the axis through O and perpendicular to the plane containing O and **F**.

The moment of **F** is independent of the choice of the point P on the line of action of **F**. If P is chosen as above, any other point Q on the line of action of **F** has position vector $\mathbf{r} + \lambda\mathbf{F}$. If we find the moment taking Q as the point of application of **F**, we obtain

$$\begin{aligned}(\mathbf{r} + \lambda\mathbf{F}) \times \mathbf{F} &= \mathbf{r} \times \mathbf{F} + (\lambda\mathbf{F}) \times \mathbf{F} \\ &= \mathbf{r} \times \mathbf{F} + \lambda(\mathbf{F} \times \mathbf{F}) \\ &= \mathbf{r} \times \mathbf{F}\end{aligned}$$

which is the same as the moment when P is the point of application. This is called the **principle of transmissibility**.

Adding moments

We now need to establish that the total turning effect about O of a number of forces in three dimensions can be obtained as the vector sum of their moments.

Consider two forces $\mathbf{F}_1$ and $\mathbf{F}_2$ acting at points P_1 and P_2 which have position vectors $\mathbf{r}_1$ and $\mathbf{r}_2$ respectively. There are three cases to consider: the lines of action of $\mathbf{F}_1$ and $\mathbf{F}_2$ intersect, are parallel or are skew.

When the lines of action of $\mathbf{F}_1$ and $\mathbf{F}_2$ intersect at some point P or are parallel, $\mathbf{F}_1$ and $\mathbf{F}_2$ are coplanar and have a resultant $\mathbf{R} = \mathbf{F}_1 + \mathbf{F}_2$ acting through a point with position vector $\mathbf{r}$.

If the lines of action intersect at some point P, P has position vector **r** and we have

$$\begin{aligned}\text{Total turning effect} &= \mathbf{r} \times \mathbf{R} \\ &= \mathbf{r} \times (\mathbf{F}_1 + \mathbf{F}_2) \\ &= \mathbf{r} \times \mathbf{F}_1 + \mathbf{r} \times \mathbf{F}_2\end{aligned}$$

As the line of action of $\mathbf{F}_1$ passes through P and P_1, we have, by the principle of transmissibility, $\mathbf{r} \times \mathbf{F}_1 = \mathbf{r}_1 \times \mathbf{F}_1$. Similarly, we have $\mathbf{r} \times \mathbf{F}_2 = \mathbf{r}_2 \times \mathbf{F}_2$.

Hence, we have

$$\text{Total turning effect} = \mathbf{r}_1 \times \mathbf{F}_1 + \mathbf{r}_2 \times \mathbf{F}_2$$

as required.

When the lines of action of $\mathbf{F}_1$ and $\mathbf{F}_2$ are parallel, it is equivalent to having a force $\mathbf{F}_1$ acting at P_1 and a force $\mathbf{F}_2 = k\mathbf{F}_1$ acting at P_2. The resultant is, therefore, a force $(1+k)\mathbf{F}_1$ acting at the point P, position vector **r**, which divides P_1P_2 in the ratio $k:1$. That is,

$$\mathbf{r} = \frac{\mathbf{r}_1 + k\mathbf{r}_2}{1+k}$$

We therefore have

$$\begin{aligned}\text{Total turning effect} &= \frac{\mathbf{r}_1 + k\mathbf{r}_2}{1+k} \times (1+k)\mathbf{F}_1 \\ &= (\mathbf{r}_1 + k\mathbf{r}_2) \times \mathbf{F}_1 \\ &= \mathbf{r}_1 \times \mathbf{F}_1 + \mathbf{r}_2 \times (k\mathbf{F}_1) \\ &= \mathbf{r}_1 \times \mathbf{F}_1 + \mathbf{r}_2 \times \mathbf{F}_2\end{aligned}$$

as required.

When the lines of action of $\mathbf{F}_1$ and $\mathbf{F}_2$ are skew, we need to replace $\mathbf{F}_2$ with another force $\mathbf{F}_3$ acting at a point P_3 with position vector $\mathbf{r}_3$ such that the moment is unchanged but the line of action of $\mathbf{F}_3$ intersects that of $\mathbf{F}_1$.

If we have $\mathbf{r}_3 = k\mathbf{r}_2$ and $\mathbf{F}_3 = \frac{1}{k}\mathbf{F}_2$, the moment is unchanged, since

$$\mathbf{r}_3 \times \mathbf{F}_3 = k\mathbf{r}_2 \times \frac{1}{k}\mathbf{F}_2 = \mathbf{r}_2 \times \mathbf{F}_2$$

In general, there is a value of k such that $\mathbf{F}_1$ and $\mathbf{F}_3$ act along lines which intersect. We therefore have

$$\begin{aligned}\text{Total turning effect} &= \mathbf{r}_1 \times \mathbf{F}_1 + \mathbf{r}_3 \times \mathbf{F}_3 \\ &= \mathbf{r}_1 \times \mathbf{F}_1 + \mathbf{r}_2 \times \mathbf{F}_2\end{aligned}$$

as required.

We can therefore state that for any two forces $\mathbf{F}_1$ and $\mathbf{F}_2$, acting at points P_1 and P_2 which have position vectors $\mathbf{r}_1$ and $\mathbf{r}_2$,

$$\text{Total turning effect} = \mathbf{r}_1 \times \mathbf{F}_1 + \mathbf{r}_2 \times \mathbf{F}_2$$

Note The principle of moments (that the turning effect of two forces about a point is the same as that of their resultant) only applies in three dimensions if the lines of action **intersect** or are **parallel**.

Equivalent force systems and equilibrium

Two systems of forces are said to be equivalent if they have the **same resultant acting along the same line**. Another way of stating this is that they have the same resultant and produce the same turning effect about any point.

Suppose we have a set of forces $\mathbf{P}_1, \mathbf{P}_2, \ldots, \mathbf{P}_m$ acting at points with position vectors $\mathbf{r}_1, \mathbf{r}_2, \ldots, \mathbf{r}_m$, and a second set of forces $\mathbf{Q}_1, \mathbf{Q}_2, \ldots, \mathbf{Q}_n$ acting at points with position vectors $\mathbf{s}_1, \mathbf{s}_2, \ldots, \mathbf{s}_n$.

If these have the same resultant, we have

$$\sum_{i=1}^{m} \mathbf{P}_i = \sum_{j=1}^{n} \mathbf{Q}_j \qquad [1]$$

If the forces have the same turning effect about an arbitrary point A having position vector $\mathbf{a}$, we have

$$\sum_{i=1}^{m} (\mathbf{r}_i - \mathbf{a}) \times \mathbf{P}_i = \sum_{j=1}^{n} (\mathbf{s}_j - \mathbf{a}) \times \mathbf{Q}_j$$

$$\Rightarrow \sum_{i=1}^{m} [(\mathbf{r}_i \times \mathbf{P}_i) - (\mathbf{a} \times \mathbf{P}_i)] = \sum_{j=1}^{n} [(\mathbf{s}_j \times \mathbf{Q}_j) - (\mathbf{a} \times \mathbf{Q}_j)]$$

$$\Rightarrow \sum_{i=1}^{m} (\mathbf{r}_i \times \mathbf{P}_i) - \mathbf{a} \times \sum_{i=1}^{m} \mathbf{P}_i = \sum_{j=1}^{n} (\mathbf{s}_j \times \mathbf{Q}_j) - \mathbf{a} \times \sum_{j=1}^{n} \mathbf{Q}_j$$

But from equation [1], we have

$$\mathbf{a} \times \sum_{i=1}^{m} \mathbf{P}_i = \mathbf{a} \times \sum_{j=1}^{n} \mathbf{Q}_j$$

$$\Rightarrow \sum_{i=1}^{m} (\mathbf{r}_i \times \mathbf{P}_i) = \sum_{j=1}^{n} (\mathbf{s}_j \times \mathbf{Q}_j)$$

That is, having the same total moment about any arbitrary point implies they have the same total moment about the origin and hence about any other chosen point.

We can therefore state the conditions for the equivalence of two systems of forces as follows.

> Two systems of forces are equivalent if their **vector sums are equal** and the **vector sums of their moments about any one chosen point are equal**.

This leads us to the conditions for equilibrium, since a system of forces in equilibrium is equivalent to a system containing no forces. We can therefore state these conditions as follows.

> A system of forces is in equilibrium if its **vector sum is zero** and the **vector sum of its moments about any one chosen point is zero**.

CHAPTER 2 VECTORS II

Forces and couples

It is reasonable to ask what equivalent systems are possible for a general system of forces. In two dimensions, it is always possible to reduce a system of forces to a single force at some point, unless the system forms a couple. The same is **not true** in three dimensions.

Suppose we have a system of forces $\mathbf{F}_i$ acting at points with position vectors $\mathbf{r}_i$, for $i = 1, \ldots, n$. Let the resultant be $\mathbf{R} = \sum_{i=1}^{n} \mathbf{F}_i$ and the total moment be $\mathbf{C} = \sum_{i=1}^{n} (\mathbf{r}_i \times \mathbf{F}_i)$. Also, let A be an arbitrary point with position vector $\mathbf{a}$.

We can transpose each force $\mathbf{F}_i$ so that it acts through A provided we introduce a couple $(\mathbf{r}_i - \mathbf{a}) \times \mathbf{F}_i$ to keep the total moment constant. This leads to an equivalent system comprising a single force $\mathbf{R}$ acting through A, together with a couple

$$\mathbf{C}_A = \sum_{i=1}^{n} [(\mathbf{r}_i - \mathbf{a}) \times \mathbf{F}_i]$$

It follows that:

> Any system of forces is equivalent to a **single force acting through any chosen point, together with a couple**.

There are three particular cases to consider.

- Clearly, if $\mathbf{R} = \mathbf{0}$ and $\mathbf{C}_A = \mathbf{0}$, the system is in equilibrium.
- If $\mathbf{R} = \mathbf{0}$ and $\mathbf{C}_A \neq \mathbf{0}$, the system is equivalent to a couple, whose moment is

$$\sum_{i=1}^{n} [(\mathbf{r}_i - \mathbf{a}) \times \mathbf{F}_i] = \sum_{i=1}^{n} (\mathbf{r}_i \times \mathbf{F}_i - \mathbf{a} \times \mathbf{F}_i)$$
$$= \sum_{i=1}^{n} (\mathbf{r}_i \times \mathbf{F}_i) - \mathbf{a} \times \sum_{i=1}^{n} \mathbf{F}_i$$

But $\sum_{i=1}^{n} \mathbf{F}_i = \mathbf{R} = \mathbf{0}$, so we have

$$\sum_{i=1}^{n} [(\mathbf{r}_i - \mathbf{a}) \times \mathbf{F}_i] = \sum_{i=1}^{n} (\mathbf{r}_i \times \mathbf{F}_i)$$

That is, the moment of the couple is the same about any point.

- If $\mathbf{R} \neq \mathbf{0}$ and $\mathbf{C}_A = \mathbf{0}$, the system is equivalent to a single force through the point A. So, we have

$$\mathbf{C}_A = \sum_{i=1}^{n} [(\mathbf{r}_i - \mathbf{a}) \times \mathbf{F}_i] = \mathbf{0}$$
$$\Rightarrow \sum_{i=1}^{n} (\mathbf{r}_i \times \mathbf{F}_i) - \mathbf{a} \times \sum_{i=1}^{n} \mathbf{F}_i = \mathbf{0}$$
$$\Rightarrow \sum_{i=1}^{n} (\mathbf{r}_i \times \mathbf{F}_i) = \mathbf{a} \times \sum_{i=1}^{n} \mathbf{F}_i$$
$$\Rightarrow \mathbf{C} = \mathbf{a} \times \mathbf{R}$$

This means that either $\mathbf{a} = \mathbf{0}$ and the resultant force passes through the origin, or $\mathbf{C}$ is perpendicular to $\mathbf{R}$.

We can therefore state:

A system of forces is equivalent to a single force provided that the **line of action of the resultant passes through the origin or is perpendicular to the total moment of the system.**

We have seen that it is always possible to choose a point A and reduce the system of forces to a resultant $\mathbf{R}$, acting through A, together with a couple $\mathbf{C}_A$. If $\mathbf{R} \neq \mathbf{0}$ and $\mathbf{C}_A \neq \mathbf{0}$, it is not in general possible to reduce $\mathbf{C}_A$ to zero by moving the point A. We can, however, choose that $\mathbf{C}_A$ has the same direction as $\mathbf{R}$. This can be seen as follows.

Suppose we choose the axes so that $\mathbf{R} = R\mathbf{i}$ and the line of action of $\mathbf{R}$ passes through the origin; and suppose further that the total moment about the origin is

$$\mathbf{C} = C_x\mathbf{i} + C_y\mathbf{j} + C_z\mathbf{k}$$

It can be seen that, by changing the line of action of $\mathbf{R}$ so that it passes through some other point in the plane $x = 0$, we can change the values of C_y and C_z, but not that of C_x.

In particular, if we choose its point of application to be the point A having position vector

$$\mathbf{a} = -\frac{C_z}{R}\mathbf{j} + \frac{C_y}{R}\mathbf{k}$$

then we have

$$\mathbf{a} \times \mathbf{R} = C_y\mathbf{j} + C_z\mathbf{k}$$

The system therefore consists of the force $\mathbf{R}$ acting at A, together with a couple $C_x\mathbf{i}$.

We can therefore state:

Any system of forces is equivalent to a **resultant R and a couple λR**. Such a system is called a **wrench**.

Example 7 The diagram shows a uniform rectangular block OABCDEFG, with a vertex at the origin O and with $\overrightarrow{OA} = 6\mathbf{i}$, $\overrightarrow{OC} = 4\mathbf{j}$ and $\overrightarrow{OD} = 2\mathbf{k}$. The weight of the block is 10 N. Forces $\mathbf{F}_1 = \mathbf{i} - 2\mathbf{j} + 3\mathbf{k}$, $\mathbf{F}_2 = 3\mathbf{i} + \mathbf{j} + 2\mathbf{k}$ and $\mathbf{F}_3 = -2\mathbf{i} + \mathbf{j} + 3\mathbf{k}$ act at points E, F and G respectively. A fourth force, **P**, is applied in an attempt to achieve equilibrium.

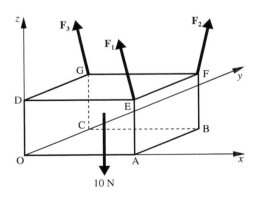

a) Show that it is not possible to achieve equilibrium with this single force, wherever it is applied.

b) If **P** is applied at D, calculate the additional couple needed to maintain equilibrium.

SOLUTION

a) The sum of the forces is

$$\begin{pmatrix} 0 \\ 0 \\ -10 \end{pmatrix} + \begin{pmatrix} 3 \\ 1 \\ 2 \end{pmatrix} + \begin{pmatrix} 1 \\ -2 \\ 3 \end{pmatrix} + \begin{pmatrix} -2 \\ 1 \\ 3 \end{pmatrix} = \begin{pmatrix} 2 \\ 0 \\ -2 \end{pmatrix}$$

The total moment of the forces is

$$\begin{pmatrix} 3 \\ 2 \\ 1 \end{pmatrix} \times \begin{pmatrix} 0 \\ 0 \\ -10 \end{pmatrix} + \begin{pmatrix} 6 \\ 0 \\ 2 \end{pmatrix} \times \begin{pmatrix} 1 \\ -2 \\ 3 \end{pmatrix} + \begin{pmatrix} 6 \\ 4 \\ 2 \end{pmatrix} \times \begin{pmatrix} 3 \\ 1 \\ 2 \end{pmatrix} +$$

$$+ \begin{pmatrix} 0 \\ 4 \\ 2 \end{pmatrix} \times \begin{pmatrix} -2 \\ 1 \\ 3 \end{pmatrix} = \begin{pmatrix} 0 \\ 4 \\ -10 \end{pmatrix}$$

In order that the system be equivalent to a single force, the resultant must be perpendicular to the total moment. But we have

$$\begin{pmatrix} 2 \\ 0 \\ -2 \end{pmatrix} \cdot \begin{pmatrix} 0 \\ 4 \\ -10 \end{pmatrix} = 20 \neq 0$$

So, we cannot replace the system by a single force. We cannot, therefore, maintain equilibrium by the addition of a single force.

b) To make the total force zero, we need to add a force $\mathbf{P} = -2\mathbf{i} + 2\mathbf{k}$. If we apply this at D, the total moment of the forces is

$$\begin{pmatrix} 0 \\ 4 \\ -10 \end{pmatrix} + \begin{pmatrix} 0 \\ 0 \\ 2 \end{pmatrix} \times \begin{pmatrix} -2 \\ 0 \\ 2 \end{pmatrix} = \begin{pmatrix} 0 \\ 0 \\ -10 \end{pmatrix}$$

We therefore need to add a couple of $10\mathbf{k}$ to maintain equilibrium.

Example 8 A system consists of forces $2\mathbf{j} + \mathbf{k}$, $3\mathbf{i} - \mathbf{j} - 2\mathbf{k}$ and $\mathbf{i} - 3\mathbf{k}$ acting at the points $(1, 0, 1)$, $(1, 2, 0)$ and $(0, 0, 3)$ respectively. Reduce this system to

a) a force acting at the point $(1, 2, 1)$ plus a couple

b) a single force acting at a point to be found.

SOLUTION

a) The sum of the forces is

$$\mathbf{R} = \begin{pmatrix} 0 \\ 2 \\ 1 \end{pmatrix} + \begin{pmatrix} 3 \\ -1 \\ -2 \end{pmatrix} + \begin{pmatrix} 1 \\ 0 \\ -3 \end{pmatrix} = \begin{pmatrix} 4 \\ 1 \\ -4 \end{pmatrix}$$

The total moment of the forces is

$$\mathbf{C} = \begin{pmatrix} 1 \\ 0 \\ 1 \end{pmatrix} \times \begin{pmatrix} 0 \\ 2 \\ 1 \end{pmatrix} + \begin{pmatrix} 1 \\ 2 \\ 0 \end{pmatrix} \times \begin{pmatrix} 3 \\ -1 \\ -2 \end{pmatrix} +$$

$$+ \begin{pmatrix} 0 \\ 0 \\ 3 \end{pmatrix} \times \begin{pmatrix} 1 \\ 0 \\ -3 \end{pmatrix} = \begin{pmatrix} -6 \\ 4 \\ -5 \end{pmatrix}$$

If the system is equivalent to **R** acting through the point $(1, 2, 1)$ together with a couple **L**, then we have

$$\begin{pmatrix} 1 \\ 2 \\ 1 \end{pmatrix} \times \mathbf{R} + \mathbf{L} = \mathbf{C} \quad \Rightarrow \quad \begin{pmatrix} 1 \\ 2 \\ 1 \end{pmatrix} \times \begin{pmatrix} 4 \\ 1 \\ -4 \end{pmatrix} + \mathbf{L} = \begin{pmatrix} -6 \\ 4 \\ -5 \end{pmatrix}$$

$$\Rightarrow \quad \mathbf{L} = \begin{pmatrix} 3 \\ -4 \\ 2 \end{pmatrix}$$

b) The system can be reduced to a single force, since $\mathbf{R} \cdot \mathbf{C} = 0$. Suppose that **R** then acts through the point (x, y, z). We have

$$\begin{pmatrix} x \\ y \\ z \end{pmatrix} \times \begin{pmatrix} 4 \\ 1 \\ -4 \end{pmatrix} = \begin{pmatrix} -6 \\ 4 \\ -5 \end{pmatrix}$$

which gives the equations $4y + z = 6$, $x + z = 1$ and $x - 4y = -5$.

Solving these, we obtain the straight line equations

$$x = 4y - 5 = 1 - z$$

This is the line of action of **R**.

Exercise 2D

1 Forces of $\mathbf{i} - \mathbf{j} + \mathbf{k}$, $2\mathbf{i} + \mathbf{j} + 3\mathbf{k}$ and $3\mathbf{j} - \mathbf{k}$ act through the points $(2, -3, 4)$, $(2, 4, 6)$ and $(-2, 2, 0)$ respectively. Show that the system can be held in equilibrium by the addition of a fourth force. Find this force and the equation of its line of action.

2 Forces of $3\mathbf{i} - 2\mathbf{j} - 4\mathbf{k}$, $-\mathbf{i} + \mathbf{j}$ and $-\mathbf{i} + 4\mathbf{k}$ act through the points $(1, 0, 1)$, $(0, 1, 1)$ and $(1, 1, 1)$ respectively.

   a) Show that the system does not reduce to a single force.
   b) Reduce the system to a single force through the point $(1, 0, 0)$, together with a couple.

3 Forces of $\mathbf{i}$, $\mathbf{j}$ and $\mathbf{k}$ act through the points $(1, 0, 0)$, $(1, 0, 0)$ and $(1, 1, 0)$ respectively. Reduce the system to a single force, whose line of action is to be found, and a couple in the same direction as the force. (That is, reduce the system to a wrench).

4 A system consists of a force $\mathbf{i} - \mathbf{k}$, acting at the origin, and a couple $2\mathbf{i} + \mathbf{j}$. Find an equivalent system consisting of a single force and a couple in the same direction (a wrench).

5 A tetrahedron has vertices at the origin and at $A(1,0,0)$, $B(1,2,0)$ and $C(1,2,2)$. Forces of magnitude 5 N, 9 N, 4 N and 6 N act along OA, OC, AB and BC respectively, in the directions indicated by the order of letters. Reduce the system to a single force through A and a couple.

6 A vertical post, 6 m high and of weight 40 N, is positioned in the direction of the z-axis and with its base at the origin. Taut cables attach the top of the post to the points $(-8,0,0)$, $(2,3,0)$ and $(2,-3,0)$.

If the tension in the first cable is 50 N, find the (equal) tensions in the other cables and the reaction of the ground on the post.

3 Differential equations

There is nothing in this world constant, but inconstancy.
JONATHAN SWIFT

Some aspects of this topic are covered in *Introducing Mechanics*, pages 420–40. However, some syllabuses do not include differential equations in the single-subject A-level, so for the sake of coherence and completeness, we repeat some of the text, examples and exercises in the present chapter.

Definitions and classification

A differential equation is any equation containing a derivative. Here are four examples:

i) $\dfrac{dy}{dx} + 2y = \sin x$ ii) $3\dfrac{d^2x}{dt^2} - 3\dfrac{dx}{dt} + 6x = 4\cos 3t$

iii) $\left(\dfrac{dy}{dt}\right)^2 - y\dfrac{d^2y}{dt^2} = \dfrac{dy}{dt}$ iv) $\dfrac{dy}{dx} - x^2y = x^3 - 3x + 4$

When we speak of 'solving' a differential equation, we mean obtaining a relationship connecting the two variables which does not involve a derivative.

A derivative corresponds to a rate of change. Many real-life problems involve quantities which are changing continuously, and the associated mathematical models will be expressed in the form of differential equations. The solution of such equations is therefore an important branch of mathematics. In this chapter, we will encounter applications not only in mechanics but in a range of other fields.

You have already met simple differential equations such as $\dfrac{dy}{dx} = 2x$ and can solve these by integration. Some other types of differential equation can be solved using analytical techniques, but many differential equations can only be solved by numerical methods.

Differential equations are categorised by their style and complexity.

- The **order** of a differential equation is the **order of the highest derivative** appearing in the equation. Of the examples above, **i** and **iv** are first-order equations, while **ii** and **iii** are second-order equations.
- The **degree** of a differential equation in y is the **highest power of y or its derivatives** appearing in the equation. Of the examples above, **i**, **ii** and **iv** are first-degree equations, while **iii** is a second-degree equation because it contains $\left(\dfrac{dy}{dt}\right)^2$. An equation of the first degree is called **linear**.

CHAPTER 3 DIFFERENTIAL EQUATIONS

In this chapter, we will concentrate on linear equations.

Forming differential equations

Before we consider how to solve the various types of differential equation, we should look at some situations in which they arise.

Any situation involving a continuous rate of change can be modelled using differential equations. Often, the rate of change is with respect to time. For example, we might consider the rate at which the temperature, T, of an object is increasing as it is heated. This appears in the equation as $\dfrac{dT}{dt}$.

When time is the independent variable, it is not usual to say so explicitly – we just refer to the **rate of change**. In other cases, we have to be explicit. For example, we might consider the rate at which temperature changes as we bore down into the surface of the Earth. This would appear as $\dfrac{dT}{dx}$, say, where x is the depth below the surface. We would refer to this as the **rate of change of temperature with respect to depth**.

Example 1 The rate at which the population of mice on an island is increasing is modelled as being proportional to the size of the population. It is estimated that when the population reaches 3000, it is increasing at a rate of 900 per week. Express this model in the form of a differential equation.

SOLUTION

Let t be the time in weeks and p be the population. The rate of increase of the population is then expressed as $\dfrac{dp}{dt}$. As this is proportional to p, we have

$$\frac{dp}{dt} = kp$$

where k is a positive constant.

We know that $\dfrac{dp}{dt} = 900$ when $p = 3000$. Substituting these, we have

$$900 = 3000k \quad \Rightarrow \quad k = 0.3$$

So, the required differential equation is

$$\frac{dp}{dt} = 0.3p$$

Example 2 A spherical air freshener is evaporating at a rate proportional to its surface area. Find an expression for the rate of change of its radius.

SOLUTION

Let the volume be V, the surface area be A and the radius be r at time t.

The information given corresponds to the differential equation

$$\frac{dV}{dt} = -kA$$

where k is a positive constant. (The negative sign indicates that the volume is decreasing.)

We know that $A = 4\pi r^2$, giving

$$\frac{dV}{dt} = -4\pi k r^2$$

We also know that $V = \frac{4}{3}\pi r^3$, giving

$$\frac{dV}{dr} = 4\pi r^2$$

We require an expression for $\frac{dr}{dt}$. These three rates of change are related by the chain rule

$$\frac{dV}{dt} = \frac{dV}{dr} \times \frac{dr}{dt}$$

$$\Rightarrow \quad -4\pi k r^2 = 4\pi r^2 \times \frac{dr}{dt}$$

which gives

$$\frac{dr}{dt} = -k$$

The radius is therefore decreasing at a constant rate.

Exercise 3A

1. An object of mass $5\,\text{kg}$ is dropped from a hot-air balloon. As it falls, it is subjected to air resistance $R\,\text{N}$ proportional to its speed $v\,\text{m s}^{-1}$. By considering Newton's second law, express this as a differential equation connecting v, the time $t\,\text{s}$ and a constant k. As it falls, the object's acceleration decreases and it approaches terminal velocity. If the terminal velocity of the object is $60\,\text{m s}^{-1}$, find the value of k.

2. If a capacitor is subjected to a potential difference V, it gains charge up to a maximum, Q_{MAX}. The rate at which it gains charge is proportional to the difference between its charge and the maximum charge. Express this as a differential equation connecting the time, t, the charge, Q, on the capacitor at that time, and a constant k.

3. The radius of a sphere is increasing at a constant rate, k. Find a differential equation connecting the time, t, and the volume, V, of the sphere.

4. Salt is being poured at the rate of $2\,\text{cm}^3\,\text{s}^{-1}$ onto a conical pile. Assuming that the cone always has equal height and base diameter, find a differential equation connecting the time, $t\,\text{s}$, and the height, $h\,\text{cm}$, of the cone.

CHAPTER 3 DIFFERENTIAL EQUATIONS

5 In a population of size N, the number of people, n, who have heard a particular joke increases at a rate proportional to the number of people who have heard it, but also to the number of people who have yet to hear it. Express this as a differential equation.

6 A moorland fire, burning in windless conditions, advances outwards from the point at which it started at a constant rate, $k\,\mathrm{m\,s^{-1}}$. Find a differential equation connecting the time, t s, and the area $A\,\mathrm{m}^2$, which has been burned.

First-order differential equations

Tangent fields

A first-order differential equation contains the derivative $\dfrac{dy}{dx}$ only.

If the solution to the equation is $y = f(x)$, $\dfrac{dy}{dx}$ represents the gradient of the graph of $f(x)$. We can therefore calculate the gradient of the solution at various points in the xy-plane. Plotting these we obtain a diagram called a **tangent field**. The locus of points where the solution has the same gradient is called an **isocline**.

Example 3 Plot the tangent field representing the differential equation $\dfrac{dy}{dx} = 2x$ at integer points (x, y), where $-3 \leqslant x \leqslant 3$ and $-5 \leqslant y \leqslant 5$.

Describe the isocline $\dfrac{dy}{dx} = m$.

SOLUTION

We first calculate the gradients at the integer point required.

	x						
y	-3	-2	-1	0	1	2	3
5	-6	-4	-2	0	2	4	6
4	-6	-4	-2	0	2	4	6
3	-6	-4	-2	0	2	4	6
2	-6	-4	-2	0	2	4	6
1	-6	-4	-2	0	2	4	6
0	-6	-4	-2	0	2	4	6
-1	-6	-4	-2	0	2	4	6
-2	-6	-4	-2	0	2	4	6
-3	-6	-4	-2	0	2	4	6
-4	-6	-4	-2	0	2	4	6
-5	-6	-4	-2	0	2	4	6

Notice that, since $\dfrac{dy}{dx}$ is independent of y, we did not need to produce this large table – calculating the values once would suffice.

A large table is only necessary when $\dfrac{dy}{dx}$ is dependent on both x and y.

We then draw a short line with the calculated gradient at each integer point, as shown on the right.

The isocline $\dfrac{\mathrm{d}y}{\mathrm{d}x} = m$ is the line $x = \tfrac{1}{2}m$.

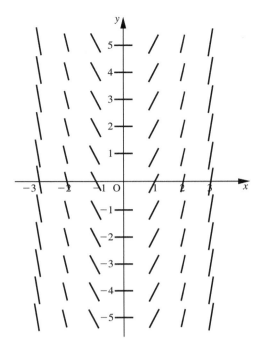

Note Some computer graphing packages are able to plot tangent fields, and if you have access to such a package you should explore these examples and those in Exercise 3B more fully.

Example 4 Plot the tangent field representing the differential equation $\dfrac{\mathrm{d}y}{\mathrm{d}x} = x + y$ at integer points (x, y), where $-3 \leqslant x \leqslant 3$ and $-5 \leqslant y \leqslant 5$.

SOLUTION

We first calculate the gradients at the integer points required.

		x						
		-3	-2	-1	0	1	2	3
	5	2	3	4	5	6	7	8
	4	1	2	3	4	5	6	7
	3	0	1	2	3	4	5	6
	2	-1	0	1	2	3	4	5
	1	-2	-1	0	1	2	3	4
y	0	-3	-2	-1	0	1	2	3
	-1	-4	-3	-2	-1	0	1	2
	-2	-5	-4	-3	-2	-1	0	1
	-3	-6	-5	-4	-3	-2	-1	0
	-4	-7	-6	-5	-4	-3	-2	-1
	-5	-8	-7	-6	-5	-4	-3	-2

We then plot the tangent field, as shown at the top of the next page.

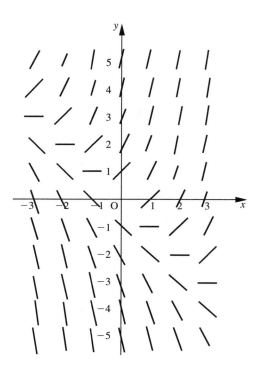

We note that some differential equations, such as $\dfrac{dy}{dx} = \dfrac{x+3}{y-2}$, can produce a 'division by zero' error at some points (in this case, those at which $y = 2$). This can usually be interpreted as an 'infinite gradient', giving a vertical line in the tangent field, unless there are further complications. The point $(-3, 2)$ causes problems in this case, as the gradient is completely undefined and no line segment can be drawn. In fact, the solution containing this point is the pair of straight lines $x + y = -1$ and $x - y = -5$, and not a curve at all.

Exercise 3B

1 Sketch each of the following tangent fields for $\dfrac{dy}{dx} = f(x, y)$, taking integer values of x and y in the ranges indicated, where $f(x, y)$ is

a) x^2 for $-2 \leqslant x \leqslant 2$, $-2 \leqslant y \leqslant 2$

b) $x - y$ for $-3 \leqslant x \leqslant 3$, $-3 \leqslant y \leqslant 3$

c) xy for $-2 \leqslant x \leqslant 2$, $-2 \leqslant y \leqslant 2$

d) $\dfrac{x}{y}$ for $-3 \leqslant x \leqslant 3$, $-3 \leqslant y \leqslant 3$

e) $x^2 + y^2$ for $-2 \leqslant x \leqslant 2$, $-2 \leqslant y \leqslant 2$

2 A cup of coffee is left standing in a room where the temperature is $30\,°C$. The rate at which the coffee temperature changes depends on the difference between the temperature of the coffee and the temperature of its surroundings. This leads to the differential equation

$$\dfrac{d\theta}{dt} = 0.6(30 - \theta)$$

where θ is the temperature measured in °C and t is the time measured in minutes. Sketch a tangent field for the intervals $0 \leqslant t \leqslant 5$ and $0 \leqslant \theta \leqslant 100$, with values of t taken every minute, and values of θ taken every 20 °C.

3 The growth of a population of bacteria in a dish can be modelled by the differential equation

$$\frac{dP}{dt} = 0.4P(6 - P)$$

where the population, P, is measured as an area in cm² and the time t is measured in days.

Sketch a tangent field representing the growth of the bacteria for the intervals $0 \leqslant P \leqslant 9$ and $0 \leqslant t \leqslant 10$, taking values of t every 2 days.

Families of solution curves

The tangent fields you drew in Exercise 3B show the direction of the solution curves going through the points chosen for the grid. There will be an infinite number of solution curves, none of which overlaps.

For example, if we sketch some solution curves for Example 3 we get the family shown below. On each of these curves, the gradient is given at each point by $\dfrac{dy}{dx} = 2x$.

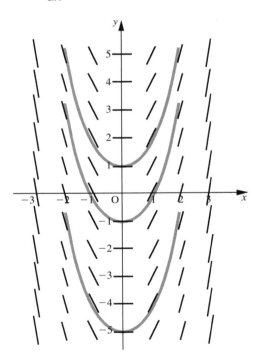

Similarly, we can show some solution curves for the differential equation in Example 4:

$$\frac{dy}{dx} = x + y$$

CHAPTER 3 DIFFERENTIAL EQUATIONS

Notice that it is not possible to draw the solution curves accurately from the tangent field, but we can sketch a good approximation to them.

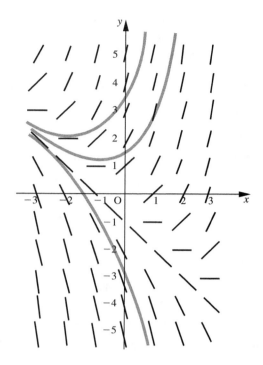

Particular solutions

As we have seen, the solution of a differential equation gives us a family of curves. We will often need to find the solution curve which passes through a particular point. This happens when we have additional information about the problem.

For example, if in Question **2** of Exercise 3B we knew that the cup of coffee had a starting temperature of 80 °C, we would need the curve which passed through the point (0, 80). Similarly, in Question **3**, the bacteria colony may have had an original size of 0.1 cm^2, corresponding to the solution curve passing through the point (0, 0.1).

In such cases, we are looking for a **particular solution**. This will be unique.

On other occasions, we do not need a particular solution but we do need an equation governing the whole family of solution curves. This is called the **general solution**.

When we are solving differential equations analytically, we first find the general solution and the particular solution is found from this.

Example 5 Sketch the particular solution of $\dfrac{dy}{dx} = x + y$ passing through the point (1, 2).

SOLUTION

We plot the tangent field and sketch the curve as shown on the right.

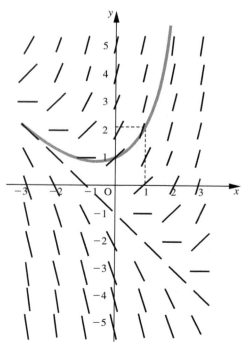

42

Exercise 3C

1 On your tangent-field diagrams from Exercise 3B, sketch the particular solutions for $\dfrac{dy}{dx} = f(x, y)$ passing through the points indicated.

	$f(x, y)$	i	ii
a)	x^2	(0, 1)	(1, 0)
b)	$x - y$	(−2, 2)	(2, −2)
c)	xy	(−2, 1)	(1, −2)
d)	$\dfrac{x}{y}$	(0, 1)	(1, 0)
e)	$x^2 + y^2$	(0, 0)	(0, 2)

2 In Question **2** of Exercise 3B, the coffee cup had an initial temperature of $T\,°C$. Sketch the solution curves corresponding to

a) $T = 80\,°C$ b) $T = 10\,°C$

3 The population of bacteria in Question **3** of Exercise 3B had an initial size of A. You can see that the form of the equation would indicate a low but increasing rate of growth if A is small, and gives a decreasing population if A exceeds 6 (to model the effects of a limited supply of nutrients). Sketch the solution curves corresponding to

a) $A = 0.5\,cm^2$ b) $A = 9\,cm^2$

Separation of variables

We now look at the method of solution for a particular group of first-order differential equations.

The term **separation of variables** is used to cover equations which are given or can be rewritten in the form

$$g(y)\frac{dy}{dx} = f(x)$$

We then integrate both sides with respect to x, which gives

$$\int g(y)\frac{dy}{dx}\,dx = \int f(x)\,dx$$

We can simplify the left-hand side of this to give

$$\int g(y)\,dy = \int f(x)\,dx$$

and in practice we move directly to this without the intervening statement.

Once the equation is written in this form, we can obtain the general solution provided that we can integrate the functions f and g.

Example 6 Find the general solution of the equation $\dfrac{dy}{dx} = xy$.

SOLUTION

We separate the variables to give

$$\frac{1}{y}\frac{dy}{dx} = x$$

Integrating, we obtain

$$\int \frac{1}{y}\,dy = \int x\,dx$$

$$\Rightarrow \quad \ln y = \tfrac{1}{2}x^2 + C$$

Note We need only one arbitrary constant of integration. Had we introduced one for each integral, we could have subtracted one of them from both sides and simplified to give the above result.

We have

$$\ln y = \tfrac{1}{2}x^2 + C \quad \Rightarrow \quad y = e^{\frac{1}{2}x^2 + C}$$

which is commonly written as

$$y = e^{\frac{1}{2}x^2} e^C$$

giving the general solution as

$$y = A\,e^{\frac{1}{2}x^2}$$

where $A = e^C$.

The particular solution

The general solution of a differential equation contains arbitrary constants. By altering the values of the constants, we can obtain any one of the family of solutions. If we have further information about the problem, we can find the values of the constants. We then have the **particular solution** to the problem.

When we looked at tangent fields, we asked for the curve 'passing through the point...'. This is equivalent to finding the value of the arbitrary constant which is consistent with a known pair of values of the variables. Often the information is given in the form 'the value of x when $t = 0$ is ...' or 'the value of y when $x = 0$ is ...'.

These are usually called **initial conditions** if the stated value of the independent variable is zero, and more generally are called **boundary conditions**.

Initial conditions such as '$x = 2$ when $t = 0$' are sometimes given in the form '$x(0) = 2$'.

Example 7 Find the general solution of the equation

$$\frac{dy}{dx} = \frac{\cos x}{y}$$

Hence find the particular solution if the boundary conditions are $y = 2$ when $x = \frac{\pi}{6}$.

SOLUTION

First, we separate the variables to give

$$y \frac{dy}{dx} = \cos x$$

Then we integrate, getting

$$\int y \, dy = \int \cos x \, dx$$

$$\Rightarrow \quad \tfrac{1}{2} y^2 = \sin x + C$$
$$\Rightarrow \quad y^2 = 2 \sin x + K \quad \text{where } K = 2C$$
$$\Rightarrow \quad y = \sqrt{2 \sin x + K}$$

This is the general solution.

For the particular solution, we substitute $y = 2$ and $x = \frac{\pi}{6}$, which gives

$$2 = \sqrt{1 + K}$$
$$\Rightarrow \quad K = 3$$

The particular solution is, therefore, $y = \sqrt{2 \sin x + 3}$.

Homogeneous equations

These are the equations of the form

$$\frac{dy}{dx} = f\left(\frac{y}{x}\right)$$

Such an equation can be converted to one in which the variables are separable by replacing $\frac{y}{x}$ by v – that is, by making the substitution $y = vx$.

CHAPTER 3 DIFFERENTIAL EQUATIONS

Example 8 Find the general solution of the equation

$$\frac{dy}{dx} = \frac{x^2 + y^2}{xy}$$

Given that $y = 2$ when $x = 1$, find the value of y when $x = 3$.

SOLUTION

The equation is homogeneous because it can be written in the form

$$\frac{dy}{dx} = \frac{1 + \left(\frac{y}{x}\right)^2}{\left(\frac{y}{x}\right)}$$

Putting $y = vx$, we have

$$\frac{dy}{dx} = v + x\frac{dv}{dx}$$

and the equation becomes

$$v + x\frac{dv}{dx} = \frac{1 + v^2}{v}$$

$$\Rightarrow \quad x\frac{dv}{dx} = \frac{1 + v^2}{v} - v = \frac{1}{v}$$

We can now separate the variables, giving

$$v\frac{dv}{dx} = \frac{1}{x}$$

$$\Rightarrow \quad \int v\, dv = \int \frac{1}{x}\, dx$$

$$\Rightarrow \quad \tfrac{1}{2}v^2 = \ln x + C$$

$$\Rightarrow \quad v = \sqrt{2\ln x + K} \quad \text{where } K = 2C$$

Putting $v = \frac{y}{x}$, we obtain

$$y = x\sqrt{2\ln x + K}$$

which is the general solution.

For the particular solution, we substitute $y = 2$ and $x = 1$, which gives

$$2 = \sqrt{K} \quad \Rightarrow \quad K = 4$$

So, the particular solution is

$$y = x\sqrt{2\ln x + 4}$$

Hence, when $x = 3$, we have

$$y = 3\sqrt{2\ln 3 + 4} = 7.47 \quad \text{(to 3 sf)}$$

Exercise 3D

1 Find the general solution of $\dfrac{dy}{dx} = f(x, y)$ by separating the variables, where $f(x, y)$ is

a) $\dfrac{x}{y}$ b) $\dfrac{x}{y^2 - 2y}$ c) $(x-2)(y-3)$ d) $x(y^2 - 4)$

2 Find the particular solution of the equation $\dfrac{dy}{dx} = \dfrac{\sin x}{\cos y}$, given that $y = 0$ when $x = 0$.

3 Given that $\dfrac{dy}{dx} = e^{x+y}$ and that $y = 0$ when $x = 0$, show that $y = \ln\left(\dfrac{1}{2 - e^x}\right)$.

4 Find the general solution of each of the following differential equations by separating the variables.

a) $x\dfrac{dy}{dx} + y^2 = 1$ b) $\dfrac{dy}{dx} - 2y = xy$ c) $x \tan y \dfrac{dy}{dx} = 1$ d) $(1 - \cos x)\dfrac{dy}{dx} = \sin x \cot y$

5 Use the substitution $z = y - x$ to rewrite $\dfrac{dy}{dx} = (y - x)^2$ as a differential equation in z and x.
Hence

a) find the general solution of the original equation in the form $y = f(x)$
b) find the particular solution, given that $y = 0$ when $x = 0$
c) show that, in this case, $y = \dfrac{2}{1 + e^2}$ when $x = 1$.

6 Find the general solution of each of the following homogeneous differential equations.

a) $\dfrac{dy}{dx} = \dfrac{x^2 + y^2}{2x^2}$ b) $\dfrac{dy}{dx} = \dfrac{y + x}{y - x}$ c) $\dfrac{dy}{dx} = \dfrac{xy}{x^2 - y^2}$ d) $x^2 \dfrac{dy}{dx} = 4x^2 - 3xy + y^2$

7 An object of mass 1 kg falls from rest. Air resistance is of magnitude $0.1v$ N, where v is the speed of the object. Write a differential equation connecting v and t, and solve it to find v in terms of t. Hence show that v cannot exceed $98 \, \text{m s}^{-1}$.

8 The motion of a go-kart with a laden mass of 100 kg is modelled by assuming that its engine exerts a constant power of 4 kW, and that when travelling at a speed of $v \, \text{m s}^{-1}$ it is subject to a resistance force of $10v$ N.

a) Use this model to write a differential equation connecting v and t.
b) Find the particular solution if the go-kart starts from rest.
c) How long does the go-kart take to reach half its maximum speed?

CHAPTER 3 DIFFERENTIAL EQUATIONS

First-order linear equations

A first-order linear equation can be written in the form

$$\frac{dy}{dx} + P(x)y = Q(x) \qquad [1]$$

We can solve such equations by using an **integrating factor**.

Consider the expression $e^{\int P(x)\,dx}y$. Differentiating this with respect to x, we have

$$\frac{d}{dx}(e^{\int P(x)\,dx}y) = e^{\int P(x)\,dx}\frac{dy}{dx} + P(x)e^{\int P(x)\,dx}y \qquad [2]$$

Multiplying through equation [1] by $e^{\int P(x)\,dx}$, we get

$$e^{\int P(x)\,dx}\frac{dy}{dx} + P(x)e^{\int P(x)\,dx}y = e^{\int P(x)\,dx}Q(x)$$

which from [2] gives

$$\frac{d}{dx}(e^{\int P(x)\,dx}y) = e^{\int P(x)\,dx}Q(x)$$

Integrating, we have

$$e^{\int P(x)\,dx}y = \int e^{\int P(x)\,dx}Q(x)\,dx$$

and, provided we can integrate the right-hand side, the equation is solved.

The expression $e^{\int P(x)\,dx}$ is called the **integrating factor**.

Example 9 Find the general solution of the differential equation

$$\frac{dy}{dx} + 2xy = x$$

SOLUTION

Comparing this with equation [1], we have $P(x) = 2x$, which gives

$$\int P(x)\,dx = \int 2x\,dx = x^2$$

The integrating factor $e^{\int P(x)\,dx}$ is, therefore, e^{x^2}. Multiplying through by this, we get

$$e^{x^2}\frac{dy}{dx} + 2xe^{x^2}y = xe^{x^2}$$

$$\Rightarrow \quad \frac{d}{dx}(e^{x^2}y) = xe^{x^2}$$

$$\Rightarrow \quad e^{x^2} y = \int x e^{x^2} \, dx$$

$$\Rightarrow \quad e^{x^2} y = \tfrac{1}{2} e^{x^2} + C$$

which gives the general solution as

$$y = \tfrac{1}{2} + C e^{-x^2}$$

Example 10 Find the general solution of the differential equation

$$(x^2 + 1) \frac{dy}{dx} + xy = 3x$$

and hence find the particular solution for which $y = 2$ when $x = 0$.

SOLUTION

The equation is not yet in the correct format. To correct this, we divide by $(x^2 + 1)$ to get

$$\frac{dy}{dx} + \frac{xy}{x^2 + 1} = \frac{3x}{x^2 + 1}$$

Comparing this with equation [1], we have $P(x) = \dfrac{x}{x^2 + 1}$, which gives

$$\int P(x) \, dx = \int \frac{x}{x^2 + 1} \, dx = \tfrac{1}{2} \ln(x^2 + 1) = \ln \sqrt{x^2 + 1}$$

The integrating factor is, therefore,

$$e^{\int P(x) \, dx} = e^{\ln \sqrt{x^2 + 1}} = \sqrt{x^2 + 1}$$

Multiplying through by this, we get

$$\sqrt{x^2 + 1} \, \frac{dy}{dx} + \frac{xy}{\sqrt{x^2 + 1}} = \frac{3x}{\sqrt{x^2 + 1}}$$

$$\Rightarrow \quad \frac{d}{dx} \left(y \sqrt{x^2 + 1} \right) = \frac{3x}{\sqrt{x^2 + 1}}$$

$$\Rightarrow \quad y \sqrt{x^2 + 1} = \int \frac{3x}{\sqrt{x^2 + 1}} \, dx = 3 \sqrt{x^2 + 1} + C$$

$$\Rightarrow \quad y = 3 + \frac{C}{\sqrt{x^2 + 1}}$$

which is the general solution.

To find the particular solution, we substitute $x = 0$ and $y = 2$, giving

$$2 = 3 + C \quad \Rightarrow \quad C = -1$$

So, the particular solution is

$$y = 3 - \frac{1}{\sqrt{x^2 + 1}}$$

CHAPTER 3 DIFFERENTIAL EQUATIONS

Exercise 3E

1 Find the general solution of each of the following first-order linear differential equations by using an integrating factor.

a) $\dfrac{dy}{dx} + 2y = 4$

b) $\dfrac{dy}{dx} + \dfrac{y}{x} = (x+2)$

c) $\dfrac{dy}{dx} + y = x$

d) $\dfrac{dy}{dx} + 3y = e^{-x}$

e) $\dfrac{dy}{dx} + 2xy = 4x$

f) $\dfrac{dy}{dx} - y\sin x = \sin x$

g) $\dfrac{dy}{dx} + 3x^2 y = 18x^3$

h) $x\dfrac{dy}{dx} + y = x^2$

i) $x^2 \dfrac{dy}{dx} - 2xy = x^2 \ln x$

j) $\cos x \dfrac{dy}{dx} + y \sin x = \cos^2 x$

2 A particle has a mass of 1 kg and starts moving from rest. For the first 10 s of its motion it is subject at time t s to a force of magnitude $(10 - t)$ N in a constant direction, and to a resistance force of magnitude $0.2v$ N, where v is the speed of the particle. Express this information as a differential equation connecting v and t, and, by solving it, find the speed at which the particle is moving at the end of the 10 s period.

Linear equations with constant coefficients

The integrating factor technique is applicable to any first-order linear differential equation (provided the right-hand side can be integrated), but if the equation has constant coefficients there is an alternative technique.

A constant-coefficient linear differential equation has the form

$$a_0 y + a_1 \frac{dy}{dx} + a_2 \frac{d^2 y}{dx^2} + a_3 \frac{d^3 y}{dx^3} + \ldots = f(x)$$

where $a_0, a_1, a_2, a_3, \ldots$ are constants.

If the function $f(x) = 0$, the equation is called a **homogeneous linear equation** (not to be confused with the homogeneous equations we met on pages 45–6).

Initially, we will consider a homogeneous first-order linear equation with constant coefficients, such as

$$\frac{dy}{dx} + 5y = 0$$

We already know how to solve this using the integrating factor technique.

The integrating factor is e^{5x}. Multiplying through by this, we get

$$e^{5x} \frac{dy}{dx} + 5e^{5x} y = 0$$

$$\Rightarrow \quad \frac{d}{dx}(e^{5x}y) = 0$$
$$\Rightarrow \quad e^{5x}y = C$$
$$\Rightarrow \quad y = Ce^{-5x}$$

However, it is easy to see that **all** such equations will yield an exponential function, and knowing this, we can solve such equations as follows.

Consider again the equation

$$\frac{dy}{dx} + 5y = 0$$

We assume that the solution to this equation is of the form $y = Ae^{mx}$.

Differentiating this, we get

$$\frac{dy}{dx} = Ame^{mx} = my$$

Substituting in the differential equation, we have

$$my + 5y = 0$$

Now, y cannot be zero (or we have the trivial particular solution $y = 0$) and so

$$m + 5 = 0 \qquad [3]$$
$$\Rightarrow \quad m = -5$$

The general solution is, therefore, $y = Ae^{-5x}$.

Equation [3] is called the **auxiliary equation**.

In general, for the equation $\frac{dy}{dx} + ky = 0$, the auxiliary equation is $m + k = 0$.

Example 11 Find the general solution of the homogeneous linear differential equation $\frac{dy}{dx} + 4y = 0$.

SOLUTION

The auxiliary equation is $m + 4 = 0$, from which $m = -4$.

Thus the general solution is $y = Ae^{-4x}$.

Exercise 3F

1 Find the general solution of the differential equation $\frac{dy}{dx} - 4y = 0$

   a) by separating the variables
   b) by using an integrating factor
   c) by using the auxiliary equation.

CHAPTER 3 DIFFERENTIAL EQUATIONS

2 Find the general solution of each of the following first-order homogeneous linear differential equations by using the auxiliary equation.

a) $\dfrac{dy}{dx} + y = 0$ b) $\dfrac{dx}{dt} - 5x = 0$ c) $\dfrac{dp}{dt} + 2p = 0$

d) $\dfrac{dx}{dt} - 0.5x = 0$ e) $\dfrac{dy}{dx} + 3y = 0$ f) $\dfrac{dz}{dy} = 3z$

The complementary function and the particular integral

We now have to find the solution to the non-homogeneous linear equation

$$\frac{dy}{dx} + py = f(x)$$

To see what form our solution should take, let us use the integrating factor method to solve a simple example in which $f(x)$ is a constant:

$$\frac{dy}{dx} + 5y = 3$$

The integrating factor is e^{5x}. Multiplying through by this, we get

$$e^{5x}\frac{dy}{dx} + 5e^{5x}y = 3e^{5x}$$

$$\Rightarrow \quad \frac{d}{dx}(e^{5x}y) = 3e^{5x}$$

$$\Rightarrow \quad e^{5x}y = \frac{3e^{5x}}{5} + C$$

$$\Rightarrow \quad y = Ce^{-5x} + \tfrac{3}{5}$$

which is the general solution.

The solution we have found consists of two parts:

- $y = Ce^{-5x}$ is the solution to the **associated homogeneous equation**
$\dfrac{dy}{dx} + 5y = 0$.

  Ce^{-5x} is called the **complementary function**.

- $y = \tfrac{3}{5}$ is one possible solution to the original equation $\dfrac{dy}{dx} + 5y = 3$.

  (This can be checked by differentiating it.)

  $\tfrac{3}{5}$ is called a **particular integral**.

If we can find these two parts for a given equation, our task is complete. Our approach is, therefore, as follows.

THE COMPLEMENTARY FUNCTION AND THE PARTICULAR INTEGRAL

- First, solve the associated homogeneous equation to find the complementary function (CF).
- Then find a particular integral (PI) which satisfies the original equation.

The general solution is then

$$y = \text{CF} + \text{PI}$$

Note You need to be clear as to the difference between the terms **particular integral**, as used above, and **particular solution**, used to indicate the solution consistent with a given set of boundary conditions.

We find the particular integral by deciding on its likely form and substituting our trial solution into the equation.

Example 12 Find the general solution to the differential equation

$$\frac{dy}{dx} - 4y = 2x - 8x^2$$

SOLUTION

The complementary function is found from the associated homogenous equation

$$\frac{dy}{dx} - 4y = 0$$

The auxiliary equation $m - 4 = 0$ gives $m = 4$, and so the CF is $y = A\,e^{4x}$.

We now need a particular integral which will generate the $(2x - 8x^2)$ term. Because this is a quadratic polynomial, we try the general quadratic polynomial

$$y = Px^2 + Qx + R$$

Differentiating this, we get

$$\frac{dy}{dx} = 2Px + Q$$

Substituting in the differential equation, we have

$$(2Px + Q) - 4(Px^2 + Qx + R) = 2x - 8x^2$$

This is an identity and so we can compare coefficients:

$$\begin{array}{lll} x^2: & -4P = -8 & \text{giving} \quad P = 2 \\ x^1: & 2P - 4Q = 2 & \text{giving} \quad Q = \tfrac{1}{2} \\ x^0: & Q - 4R = 0 & \text{giving} \quad R = \tfrac{1}{8} \end{array}$$

Hence, the PI is $y = 2x^2 + \tfrac{1}{2}x + \tfrac{1}{8}$.

We find the general solution by combining the CF and the PI to give

$$y = A\,e^{4x} + 2x^2 + \tfrac{1}{2}x + \tfrac{1}{8}$$

CHAPTER 3 DIFFERENTIAL EQUATIONS

Finding a particular integral

When using the above technique, our trial solution should be the most general form of the right-hand side of the original equation. Thus, we have:

When the right-hand side involves	We try
a polynomial of degree n	a general polynomial of degree n
e^{nx}	a general exponential function, Pe^{nx}
$\sin nx$, $\cos nx$ or a combination of these	a general trigonometric function $P \sin nx + Q \cos nx$
a combination of the above	a combination of the general functions above

Exercise 3G

In each of the following, use the auxiliary equation method to find the general solution.

1. $\dfrac{dy}{dx} + 3y = 2x - 4$

2. $\dfrac{dx}{dt} - 2x = \cos t$

3. $\dfrac{dy}{dx} + y = e^x$

4. $\dfrac{dp}{dt} - 5p = t^3$

5. $\dfrac{dz}{dx} + 2z = 2 \sin 3x - \cos 3x$

6. $\dfrac{dx}{dt} - 4x = 3e^{-2t}$

7. $\dfrac{dy}{dx} + 4y = 4x - e^x$

8. $\dfrac{dx}{dt} - x = e^{2t} - \cos 2t$

Problem cases

A difficulty arises in equations where the complementary function contains an element from the right-hand side of the differential equation. For example, suppose we need to find the general solution to the equation

$$\dfrac{dy}{dx} - 2y = 4e^{2x}$$

The complementary function for this equation is $y = Ae^{2x}$.

When finding a particular integral, the right-hand side suggests a trial solution $y = Pe^{2x}$. However, when we combine the two solutions, we get

$$y = Ae^{2x} + Pe^{2x}$$
$$\Rightarrow \quad y = (A + P)e^{2x} \quad \text{or} \quad y = Be^{2x} \quad (\text{where } B = A + P)$$

This is just the solution to the homogeneous differential equation and not to the non-homogeneous one.

For an indication as to how this difficulty might be resolved, let us solve the original equation using the integrating factor technique.

The integrating factor is e^{-2x}. Multiplying through by this, we get

$$e^{-2x}\frac{dy}{dx} - 2e^{-2x}y = 4$$

$$\Rightarrow \frac{d}{dx}(e^{-2x}y) = 4$$

$$\Rightarrow e^{-2x}y = \int 4\,dx = 4x + A$$

$$\Rightarrow y = Ae^{2x} + 4xe^{2x}$$

In this solution, we have our complementary function Ae^{2x} as before, but the particular integral is not of the form Pe^{2x} as expected, but Pxe^{2x}.

This gives us the required strategy. If the complementary function is Ae^{kx} and our first choice of trial solution contains Pe^{kx}, we replace it by Pxe^{kx} to get an alternative trial solution. If Pxe^{kx} is already part of our trial solution, we use Px^2e^{kx}, and so on.

Note We do not know whether there will be a problem until we have found the complementary function, and so we always find this first.

Example 13 Find the general solution to the differential equation

$$\frac{dx}{dt} - 3x = 2e^{3t} + 9t$$

SOLUTION

The auxiliary equation is $m - 3 = 0$, giving $m = 3$.

Thus, the complementary function is $x = Ae^{3t}$.

This is contained within the right-hand side and so our trial solution will be

$$x = Pte^{3t} + Qt + R$$

Differentiating, we have

$$\frac{dx}{dt} = 3Pte^{3t} + Pe^{3t} + Q$$

Substituting into the differential equation, we get

$$(3Pte^{3t} + Pe^{3t} + Q) - 3(Pte^{3t} + Qt + R) = 2e^{3t} + 9t$$
$$\Rightarrow Pe^{3t} - 3Qt + (Q - 3R) = 2e^{3t} + 9t$$

This is an identity and so we can compare coefficients:

$$t^0: \quad Q - 3R = 0$$
$$t^1: \quad -3Q = 9 \quad \text{giving} \quad Q = -3 \text{ and } R = -1$$
$$e^{3t}: \quad P = 2$$

Thus, the PI is
$$x = 2te^{3t} - 3t - 1$$
The general solution is, therefore,
$$x = Ae^{3t} + 2te^{3t} - 3t - 1$$

If, in Example 13, the right-hand side had been $2e^{3t} + 4te^{3t} + 9t$, which contains both the terms e^{3t} and te^{3t}, our trial solution would have been
$$Pt^2 e^{3t} + Qte^{3t} + Rt + S$$
As an exercise, you might try to show that in this case $P = Q = 2$ and $R = S = -3$.

Exercise 3H

Find the general solution of each of the following differential equations by finding the complementary function and the particular integral. Hence find in each case the particular solution consistent with the given initial conditions.

1 $\dfrac{dy}{dx} - 4y = 8e^{4x}$, given that $y = 4$ when $x = 0$.

2 $\dfrac{dx}{dt} + 5x = 10e^{-5t} + 2t$, given that $x = 2$ when $t = 0$.

3 $\dfrac{dy}{dx} = 2y - 6e^{2x}$, given that $y = 2$ when $x = 0$.

4 $\dfrac{dx}{dt} = 3t - 3x + 12e^{-3t}$, with initial condition $x(0) = 6$.

5 $\dfrac{dy}{dx} - 2y = 4e^{2x} - 8xe^{2x}$, with initial condition $y(0) = 8$.

6 $\dfrac{dx}{dt} + 3x = 6t + 9e^{-3t} - 15te^{-3t}$, with initial condition $x(0) = 9$.

Second-order linear equations with constant coefficients

With first-order linear differential equations with constant coefficients, it is a matter of individual choice whether to use the auxiliary equation technique or the integrating factor technique. However, when we progress to second-order equations, we cannot use the integrating factor technique but we can still use the auxiliary equation technique.

The complementary function

Consider the non-homogeneous second-order differential equation with constant coefficients

$$a\frac{d^2y}{dx^2} + b\frac{dy}{dx} + cy = f(x)$$

The associated homogeneous equation is

$$a\frac{d^2y}{dx^2} + b\frac{dy}{dx} + cy = 0 \qquad [4]$$

Let us assume that this has a solution of the form $y = A e^{mx}$. Differentiating this, we have

$$\frac{dy}{dx} = Am e^{mx} = my \quad \text{and} \quad \frac{d^2y}{dx^2} = Am^2 e^{mx} = m^2 y$$

Substituting into equation [4], we obtain

$$am^2 y + bmy + cy = 0$$

Since $y \neq 0$ (this would be a trivial particular solution), we have

$$am^2 + bm + c = 0$$

This is the quadratic auxiliary equation derived from the second-order homogeneous differential equation. If its roots are m_1 and m_2, then we have two solutions to the differential equation:

$$y_1 = A e^{m_1 x} \quad \text{and} \quad y_2 = B e^{m_2 x}$$

If y_1 and y_2 are two solutions to a differential equation, then $y_1 + y_2$ is also a solution. (This is called the **principle of superposition of solutions.**)

Hence, the **most general form** of the complementary function is

$$y = A e^{m_1 x} + B e^{m_2 x}$$

Note that this equation contains two arbitrary constants. Solving a second-order differential equation is equivalent to integrating twice, generating an arbitrary constant each time we integrate. (In general, the solution to an nth order differential equation has n arbitrary constants.)

However, the quadratic auxiliary equation $am^2 + bm + c = 0$ has three types of solution:

- Two real, distinct roots.
- A real, repeated root.
- Two complex roots (which are conjugates of each other).

We will deal with each case separately.

CHAPTER 3 DIFFERENTIAL EQUATIONS

Two real, distinct roots

Essentially, this is the case dealt with above. If the two roots are m_1 and m_2, then the complementary function takes the form

$$y = A e^{m_1 x} + B e^{m_2 x}$$

A real, repeated root

The problem here is that if the repeated root is m, say, then the complementary function would be

$$y = A e^{mx} + B e^{mx}$$
$$\Rightarrow \quad y = (A + B) e^{mx}$$
$$\Rightarrow \quad y = C e^{mx}$$

This cannot be the complete complementary function, since it has only one arbitrary constant.

If the auxiliary equation has a repeated root m, the associated homogeneous equation is

$$\frac{d^2 y}{dx^2} - 2m \frac{dy}{dx} + m^2 y = 0$$

Consider $y = Bx e^{mx}$. Differentiating this, we obtain

$$\frac{dy}{dx} = B(mx e^{mx} + e^{mx}) \quad \text{and} \quad \frac{d^2 y}{dx^2} = B(m^2 x e^{mx} + 2m e^{mx})$$

We can establish by substitution that these satisfy the equation. (You should check that this is so.) Hence, using the principle of superposition, the most general form of solution is

$$y = A e^{mx} + Bx e^{mx} \quad \text{or} \quad y = (A + Bx) e^{mx}$$

Two complex roots

The roots of the auxiliary equation will be complex conjugates:

$$m_1 = \lambda + i\mu \quad \text{and} \quad m_2 = \lambda - i\mu$$

(Some examination boards use j rather than i to represent $\sqrt{-1}$.)

The complementary function is, therefore,

$$y = C e^{(\lambda + i\mu)x} + D e^{(\lambda - i\mu)x} = e^{\lambda x}(C e^{i\mu x} + D e^{-i\mu x})$$

However, because in general $e^{i\theta} = \cos\theta + i\sin\theta$, we have

$$y = e^{\lambda x}[C(\cos\mu x + i\sin\mu x) + D(\cos\mu x - i\sin\mu x)]$$
$$\Rightarrow \quad y = e^{\lambda x}[(C + D)\cos\mu x + (iC - iD)\sin\mu x]$$

which we can write as

$$y = e^{\lambda x}(A \cos\mu x + B \sin\mu x)$$

SECOND-ORDER LINEAR EQUATIONS WITH CONSTANT COEFFICIENTS

Note A and B can both be real. If we choose $C = p + iq$ and $D = p - iq$, then we have

$$A = C + D = 2p \quad \text{which is real}$$

and

$$B = i(C - D) = i(2iq) = -2q \quad \text{which is also real}$$

Hence, the complementary function is

$$y = e^{\lambda x}(A \cos \mu x + B \sin \mu x)$$

This can also be written in the alternative form

$$y = R e^{\lambda x} \cos(\mu x + \phi)$$

where the arbitrary constants are now R and ϕ.

Example 14 Find the general solution to the equation

$$\frac{d^2 y}{dx^2} - 5\frac{dy}{dx} + 6y = 0$$

SOLUTION

The auxiliary equation is

$$m^2 - 5m + 6 = 0$$
$$\Rightarrow (m - 3)(m - 2) = 0$$
$$\Rightarrow m = 3 \quad \text{or} \quad m = 2$$

Thus, the general solution is $y = A e^{3x} + B e^{2x}$.

Example 15 Find the general solution to the equation

$$\frac{d^2 y}{dx^2} - 6\frac{dy}{dx} + 9y = 0$$

SOLUTION

The auxiliary equation is

$$m^2 - 6m + 9 = 0$$
$$\Rightarrow (m - 3)^2 = 0$$
$$\Rightarrow m = 3 \quad \text{(repeated root)}$$

Thus, the general solution is $y = e^{3x}(Ax + B)$.

Example 16 Find the general solution to the equation

$$\frac{d^2 y}{dx^2} - 6\frac{dy}{dx} + 13y = 0$$

SOLUTION

The auxiliary equation is
$$m^2 - 6m + 13 = 0$$
$$\Rightarrow \quad m = \frac{6 \pm \sqrt{36 - 52}}{2} = 3 \pm 2\mathrm{i}$$

Thus, the general solution is $y = \mathrm{e}^{3x}(A\cos 2x + B\sin 2x)$.

Exercise 3I

Find the general solution for each of the following equations.

1. $\dfrac{\mathrm{d}^2 y}{\mathrm{d}x^2} + 2\dfrac{\mathrm{d}y}{\mathrm{d}x} - 3y = 0$
2. $\dfrac{\mathrm{d}^2 y}{\mathrm{d}x^2} - 4\dfrac{\mathrm{d}y}{\mathrm{d}x} + 4y = 0$
3. $\dfrac{\mathrm{d}^2 x}{\mathrm{d}t^2} + 4\dfrac{\mathrm{d}x}{\mathrm{d}t} + 8x = 0$
4. $\dfrac{\mathrm{d}^2 y}{\mathrm{d}x^2} + 6\dfrac{\mathrm{d}y}{\mathrm{d}x} + 9y = 0$
5. $\dfrac{\mathrm{d}^2 x}{\mathrm{d}t^2} - 2\dfrac{\mathrm{d}x}{\mathrm{d}t} + 5x = 0$
6. $\dfrac{\mathrm{d}^2 y}{\mathrm{d}x^2} + 6\dfrac{\mathrm{d}y}{\mathrm{d}x} + 8y = 0$
7. $\dfrac{\mathrm{d}^2 x}{\mathrm{d}t^2} + 2\dfrac{\mathrm{d}x}{\mathrm{d}t} + 2x = 0$
8. $\dfrac{\mathrm{d}^2 y}{\mathrm{d}x^2} - 3\dfrac{\mathrm{d}y}{\mathrm{d}x} - 4y = 0$
9. $\dfrac{\mathrm{d}^2 y}{\mathrm{d}x^2} - 8\dfrac{\mathrm{d}y}{\mathrm{d}x} + 16y = 0$
10. $\dfrac{\mathrm{d}^2 y}{\mathrm{d}x^2} + 2\dfrac{\mathrm{d}y}{\mathrm{d}x} - 8y = 0$
11. $\dfrac{\mathrm{d}^2 y}{\mathrm{d}x^2} + 2\dfrac{\mathrm{d}y}{\mathrm{d}x} + y = 0$
12. $\dfrac{\mathrm{d}^2 x}{\mathrm{d}t^2} + 4\dfrac{\mathrm{d}x}{\mathrm{d}t} + 13x = 0$
13. $\dfrac{\mathrm{d}^2 y}{\mathrm{d}x^2} - 9y = 0$
14. $\dfrac{\mathrm{d}^2 x}{\mathrm{d}t^2} + 4x = 0$
15. $\dfrac{\mathrm{d}^2 y}{\mathrm{d}x^2} - 3\dfrac{\mathrm{d}y}{\mathrm{d}x} = 0$
16. $\dfrac{\mathrm{d}^2 y}{\mathrm{d}x^2} + 4\dfrac{\mathrm{d}y}{\mathrm{d}x} = 0$

The particular integral and problem cases

The approach to the particular integral is the same for second-order equations as for first. The trial solution is a general form of the right-hand side of the equation. If part of the complementary function is present in the first-choice trial solution, we multiply the trial solution by x, x^2 and so on, as before.

Example 17 Find the general solution of the differential equation

$$\frac{d^2x}{dt^2} + 6\frac{dx}{dt} + 25x = 3\cos 2t$$

SOLUTION

The auxiliary equation is

$$m^2 + 6m + 25 = 0$$
$$\Rightarrow \quad m = -3 \pm 4i$$

The complementary function is

$$x = e^{-3t}(A\cos 4t + B\sin 4t)$$

The trial solution is

$$x = P\cos 2t + Q\sin 2t$$

which gives

$$\frac{dx}{dt} = -2P\sin 2t + 2Q\cos 2t$$

and

$$\frac{d^2x}{dt^2} = -4P\cos 2t - 4Q\sin 2t$$

Substituting these into the original equation, we obtain

$$(-4P\cos 2t - 4Q\sin 2t) + 6(-2P\sin 2t + 2Q\cos 2t) +$$
$$+ 25(P\cos 2t + Q\sin 2t) = 3\cos 2t$$

Comparing coefficients, we have

$$\cos 2t: \quad -4P + 12Q + 25P = 3 \quad \Rightarrow \quad 7P + 4Q = 1$$
$$\sin 2t: \quad -4Q - 12P + 25Q = 0 \quad \Rightarrow \quad -4P + 7Q = 0$$

Solving these, we obtain $P = \frac{7}{65}$ and $Q = \frac{4}{65}$.

So, the general solution is

$$x = e^{-3t}(A\cos 4t + B\sin 4t) + \frac{1}{65}(7\cos 2t + 4\sin 2t)$$

This solution is typical of many mechanical systems where we have damped, forced vibrations. As time increases, the first part of the solution containing the term in e^{-3t} will diminish and eventually the second part will dominate.

The first part is called a **transient** and represents the way that the damped system will oscillate naturally. The second part is the **steady-state** solution produced by whatever is driving the oscillations. This will be covered in more detail on pages 126–34.

CHAPTER 3 DIFFERENTIAL EQUATIONS

Example 18 Find the general solution of the differential equation

$$\frac{d^2y}{dx^2} - 5\frac{dy}{dx} + 6y = 4e^{3x} - 2e^x$$

SOLUTION

We already have the complementary function $y = Ae^{3x} + Be^{2x}$ (see Example 14).

Our first choice for a trial solution would be $y = Pe^{3x} + Qe^x$. However, as the term in e^{3x} occurs in the complementary function, we use a revised trial solution $y = Pxe^{3x} + Qe^x$, from which

$$\frac{dy}{dx} = Pe^{3x} + 3Pxe^{3x} + Qe^x$$

and

$$\frac{d^2y}{dx^2} = 6Pe^{3x} + 9Pxe^{3x} + Qe^x$$

Substituting into the original equation, we obtain

$$(6Pe^{3x} + 9Pxe^{3x} + Qe^x) - 5(Pe^{3x} + 3Pxe^{3x} + Qe^x) + \\ + 6(Pxe^{3x} + Qe^x) = 4e^{3x} - 2e^x$$

$$\Rightarrow \quad Pe^{3x} + 2Qe^x = 4e^{3x} - 2e^x$$

Comparing coefficients, we have

$$e^{3x}: \quad P = 4$$
$$e^x: \quad 2Q = -2 \quad \text{giving} \quad Q = -1$$

Hence, the particular integral is $y = 4xe^{3x} - e^x$.

The general solution is CF + PI. That is,

$$y = Ae^{3x} + Be^{2x} + 4xe^{3x} - e^x$$

Example 19 Find the general solution of the differential equation

$$\frac{d^2x}{dt^2} + 4x = \cos 2t$$

SOLUTION

The auxiliary equation is $m^2 + 4 = 0$, which gives $m = \pm 2i$.

Thus, the complementary function is $x = A\cos 2t + B\sin 2t$.

Our first-choice trial solution would be $x = P\cos 2t + Q\sin 2t$. However, the $\cos 2t$ term occurs in the complementary function. Our trial solution will therefore be

$$x = t(P\cos 2t + Q\sin 2t)$$

which gives
$$\frac{dx}{dt} = P\cos 2t + Q\sin 2t + 2t(-P\sin 2t + Q\cos 2t)$$
and
$$\frac{d^2x}{dt^2} = -4P\sin 2t + 4Q\cos 2t - 4t(P\cos 2t + Q\sin 2t)$$

Substituting into the original equation, we obtain
$$[-4P\sin 2t + 4Q\cos 2t - 4t(P\cos 2t + Q\sin 2t)] +$$
$$+ 4t(P\cos 2t + Q\sin 2t) = \cos 2t$$
$$\Rightarrow \quad -4P\sin 2t + 4Q\cos 2t = \cos 2t$$

Comparing coefficients, we have
$$\sin 2t: \quad -4P = 0 \quad \text{giving} \quad P = 0$$
$$\cos 2t: \quad 4Q = 1 \quad \text{giving} \quad Q = \tfrac{1}{4}$$

Thus, the particular integral is $x = \tfrac{1}{4}t\sin 2t$.

The general solution is CF + PI. That is,
$$x = A\cos 2t + B\sin 2t + \tfrac{1}{4}t\sin 2t$$

Example 20 Find the general solution of the differential equation
$$\frac{d^2y}{dx^2} + 4\frac{dy}{dx} = 4x - 7$$

SOLUTION

The auxiliary equation is $m^2 + 4m = 0$, which gives $m = 0$ or $m = -4$.

Thus, the complementary function is $y = A + Be^{-4x}$.

We note that this contains a constant term which will cause problems given the format of the right-hand side of the differential equation. Our first-choice trial solution of $y = Px + Q$ must therefore be amended to
$$y = Px^2 + Qx$$
which gives
$$\frac{dy}{dx} = 2Px + Q$$
and
$$\frac{d^2y}{dx^2} = 2P$$

Substituting into the original equation, we obtain
$$2P + 4(2Px + Q) = 4x - 7$$

CHAPTER 3 DIFFERENTIAL EQUATIONS

Comparing coefficients, we have

$$x^1: \quad 8P = 4 \quad \text{giving} \quad P = \tfrac{1}{2}$$
$$x^0: \quad 2P + 4Q = -7 \quad \text{giving} \quad Q = -2$$

Hence, the particular integral is $y = \tfrac{1}{2}x^2 - 2x$.

The general solution is CF + PI. That is,

$$y = A + Be^{-4x} + \tfrac{1}{2}x^2 - 2x$$

Exercise 3J

Use your results from Exercise 3I to find the general solution for each of the following differential equations.

1. $\dfrac{d^2y}{dx^2} + 2\dfrac{dy}{dx} - 3y = x^2 - 3x$

2. $\dfrac{d^2y}{dx^2} - 4\dfrac{dy}{dx} + 4y = 4e^{3x}$

3. $\dfrac{d^2x}{dt^2} + 4\dfrac{dx}{dt} + 8x = 4t - 3$

4. $\dfrac{d^2y}{dx^2} + 6\dfrac{dy}{dx} + 9y = 4\sin 2x$

5. $\dfrac{d^2x}{dt^2} - 2\dfrac{dx}{dt} + 5x = 5e^{-t}$

6. $\dfrac{d^2y}{dx^2} + 6\dfrac{dy}{dx} + 8y = 3\cos 3x$

7. $\dfrac{d^2x}{dt^2} + 2\dfrac{dx}{dt} + 2x = 3e^{-2t} - 2t$

8. $\dfrac{d^2y}{dx^2} - 3\dfrac{dy}{dx} - 4y = 5x^2 + 4$

9. $\dfrac{d^2y}{dx^2} - 8\dfrac{dy}{dx} + 16y = 2e^{2x} + 2\cos x$

10. $\dfrac{d^2y}{dx^2} + 2\dfrac{dy}{dx} - 8y = e^{-2x} + 4x$

11. $\dfrac{d^2y}{dx^2} + 2\dfrac{dy}{dx} + y = 3e^{-x}$

12. $\dfrac{d^2x}{dt^2} + 4\dfrac{dx}{dt} + 13x = 3\sin 2t - 6t$

13. $\dfrac{d^2y}{dx^2} - 9y = 4e^{3x}$

14. $\dfrac{d^2x}{dt^2} + 4x = 2\sin 2t$

15. $\dfrac{d^2y}{dx^2} - 3\dfrac{dy}{dx} = 2e^{4x}$

16. $\dfrac{d^2y}{dx^2} + 4\dfrac{dy}{dx} = 2x + 3$

The particular solution

The general solution of a second-order differential equation contains two arbitrary constants, and so we need two sets of conditions to find their values. We can do this in two ways:

- Give the values of y and $\dfrac{dy}{dx}$ when x takes a particular value (usually $x = 0$).

  These are called **initial conditions**. This is the most common situation.

- Give the value of y for each of two different values of x (usually the end points of the interval being used). These are called **boundary conditions**.

Note For a given set of conditions, there will usually be a unique solution, but the simultaneous equations to find the constants may be inconsistent (no solution) or not independent (an infinite number of solutions).

Example 21 Find the particular solution of the differential equation

$$\frac{d^2y}{dx^2} + 4\frac{dy}{dx} = 4x - 7$$

given that $y = 0$ and $\frac{dy}{dx} = 6$ when $x = 0$.

SOLUTION

From Example 20, the general solution is $y = A + Be^{-4x} + \frac{1}{2}x^2 - 2x$, which gives

$$\frac{dy}{dx} = -4Be^{-4x} + x - 2$$

When $x = 0$, $y = 0$, so we have

$$0 = A + B \qquad [1]$$

When $x = 0$, $\frac{dy}{dx} = 6$, so we have

$$6 = -4B - 2 \qquad [2]$$

Solving [1] and [2], we obtain $A = 2$, $B = -2$.

So, the particular solution is

$$y = 2 - 2e^{-4x} + \frac{1}{2}x^2 - 2x$$

Example 22 Find the particular solution of the differential equation

$$\frac{d^2x}{dt^2} + 4x = \cos 2t$$

given that $x = 1$ when $t = 0$, and $x = \frac{\pi}{16}$ when $t = \frac{\pi}{4}$.

SOLUTION

From Example 19, the general solution is

$$x = A\cos 2t + B\sin 2t + \tfrac{1}{4}t\sin 2t$$

When $t = 0$, $x = 1$, so we have $A = 1$. When $t = \frac{\pi}{4}$, $x = \frac{\pi}{16}$, so we have

$$\frac{\pi}{16} = B + \frac{\pi}{16} \quad \Rightarrow \quad B = 0$$

Hence, the particular solution is

$$x = \cos 2t + \tfrac{1}{4}t\sin 2t$$

Exercise 3K

Use your results from Exercise 3J to find the particular solution for each of the following differential equations given the conditions stated.

1. $\dfrac{d^2y}{dx^2} + 2\dfrac{dy}{dx} - 3y = x^2 - 3x$

   $y = 0$ and $\dfrac{dy}{dx} = 0$ when $x = 0$

2. $\dfrac{d^2y}{dx^2} - 4\dfrac{dy}{dx} + 4y = 4e^{3x}$

   $y = 2$ and $\dfrac{dy}{dx} = 2$ when $x = 0$

3. $\dfrac{d^2x}{dt^2} + 4\dfrac{dx}{dt} + 8x = 4t - 3$

   $x = 1$ and $\dfrac{dx}{dt} = 4$ when $t = 0$

4. $\dfrac{d^2y}{dx^2} + 6\dfrac{dy}{dx} + 9y = 4\sin 2x$

   $y = 0$ and $\dfrac{dy}{dx} = 2$ when $x = 0$

5. $\dfrac{d^2x}{dt^2} - 2\dfrac{dx}{dt} + 5x = 5e^{-t}$

   $x = 0$ and $\dfrac{dx}{dt} = 2$ when $t = 0$

6. $\dfrac{d^2y}{dx^2} + 6\dfrac{dy}{dx} + 8y = 3\cos 3x$

   $y = 3$ and $\dfrac{dy}{dx} = 1$ when $x = 0$

7. $\dfrac{d^2x}{dt^2} + 2\dfrac{dx}{dt} + 2x = 3e^{-2t} - 2t$

   $x = 0$ and $\dfrac{dx}{dt} = 4$ when $t = 0$

8. $\dfrac{d^2y}{dx^2} - 3\dfrac{dy}{dx} - 4y = 5x^2 + 4$

   $y = 2$ and $\dfrac{dy}{dx} = -1$ when $x = 0$

9. $\dfrac{d^2y}{dx^2} - 8\dfrac{dy}{dx} + 16y = 2e^{2x} + 2\cos x$

   $y = 0$ and $\dfrac{dy}{dx} = 4$ when $x = 0$

10. $\dfrac{d^2y}{dx^2} + 2\dfrac{dy}{dx} - 8y = e^{-2x} + 4x$

    $y = 2$ and $\dfrac{dy}{dx} = -2$ when $x = 0$

11. $\dfrac{d^2y}{dx^2} + 2\dfrac{dy}{dx} + y = 3e^{-x}$

    $y = 0$ and $\dfrac{dy}{dx} = 0$ when $x = 0$

12. $\dfrac{d^2x}{dt^2} + 4\dfrac{dx}{dt} + 13x = 3\sin 2t - 6t$

    $x = 0$ and $\dfrac{dx}{dt} = 2$ when $t = 0$

13. $\dfrac{d^2y}{dx^2} - 9y = 4e^{3x}$

    $y = 3$ when $x = 0$ and

    $y = \dfrac{11}{3}\cosh 3$ when $x = 1$

14. $\dfrac{d^2x}{dt^2} + 4x = 2\sin 2t$

    $x = 3$ when $t = 0$ and

    $x = -1$ when $t = \dfrac{\pi}{4}$

15 $\dfrac{d^2y}{dx^2} - 3\dfrac{dy}{dx} = 2e^{4x}$

$y = 2$ and $\dfrac{dy}{dx} = 1$ when $x = 0$

16 $\dfrac{d^2y}{dx^2} + 4\dfrac{dy}{dx} = 2x + 3$

$y = 1$ when $x = 0$ and
$y = e^{10}$ when $x = -2.5$

Vector differential equations

Simple vector differential equations are, depending on dimensions, effectively a set of two or three equations of the same form, one for each component. We will consider first- and second-order linear equations in which the position or velocity vector is a function of time. We will see how the techniques we have already met – the use of an integrating factor and solutions in complementary function/particular integral form – can be applied in the vector case.

Example 23 Find the general solution of the equation

$$\frac{d\mathbf{r}}{dt} + \mathbf{r} = t\mathbf{i} + 2\mathbf{j} + e^{-t}\mathbf{k}$$

SOLUTION

We will solve this equation in two ways.

Method 1 Using an integrating factor

The integrating factor is $e^{\int dt} = e^{t}$. Multiplying through, we obtain

$$e^{t}\frac{d\mathbf{r}}{dt} + e^{t}\mathbf{r} = te^{t}\mathbf{i} + 2e^{t}\mathbf{j} + \mathbf{k}$$

$$\Rightarrow \quad \frac{d(e^{t}\mathbf{r})}{dt} = te^{t}\mathbf{i} + 2e^{t}\mathbf{j} + \mathbf{k}$$

$$\Rightarrow \quad e^{t}\mathbf{r} = \int (te^{t}\mathbf{i} + 2e^{t}\mathbf{j} + \mathbf{k})\,dt$$

$$\Rightarrow \quad e^{t}\mathbf{r} = (te^{t} - e^{t})\mathbf{i} + 2e^{t}\mathbf{j} + t\mathbf{k} + \mathbf{c}$$

where $\mathbf{c}$ is a vector constant.

Hence, we have

$$\mathbf{r} = (t - 1)\mathbf{i} + 2\mathbf{j} + te^{-t}\mathbf{k} + e^{-t}\mathbf{c}$$

Method 2 Finding the complementary function and a particular integral

The auxiliary equation is $m + 1 = 0$, giving $m = -1$.

The CF is therefore $e^{-t}\mathbf{A}$, where $\mathbf{A}$ is an arbitrary vector constant.

Our trial solution for the PI is

$$\mathbf{r} = (Pt + Q)\mathbf{i} + R\mathbf{j} + Ste^{-t}\mathbf{k}$$

(Notice the form of the **k**-component here. This is necessary because of the form of the complementary function.)

Differentiating, we obtain

$$\frac{d\mathbf{r}}{dt} = P\mathbf{i} + Se^{-t}(1-t)\mathbf{k}$$

Substituting into the given differential equation, we obtain

$$P\mathbf{i} + Se^{-t}(1-t)\mathbf{k} + (Pt+Q)\mathbf{i} + R\mathbf{j} + Ste^{-t}\mathbf{k} = t\mathbf{i} + 2\mathbf{j} + e^{-t}\mathbf{k}$$

Comparing **i**-components, we have: $Pt + P + Q = t \Rightarrow P = 1, Q = -1$

Comparing **j**-components, we have: $R = 2$

Comparing **k**-components, we have: $Se^{-t} = e^{-t} \Rightarrow S = 1$

The PI is, therefore,

$$\mathbf{r} = (t-1)\mathbf{i} + 2\mathbf{j} + te^{-t}\mathbf{k}$$

Hence, the general solution is

$$\mathbf{r} = (t-1)\mathbf{i} + 2\mathbf{j} + te^{-t}\mathbf{k} + e^{-t}\mathbf{A}$$

Example 24 Find the particular solution of the equation

$$\frac{d^2\mathbf{r}}{dt^2} + 2\frac{d\mathbf{r}}{dt} + 5\mathbf{r} = 2\cos t\,\mathbf{i} + 3\sin t\,\mathbf{j} + \mathbf{k}$$

given that when $t = 0$, $\mathbf{r} = 2\mathbf{i} + \mathbf{j}$ and $\dfrac{d\mathbf{r}}{dt} = \mathbf{k}$.

SOLUTION

The auxiliary equation is $m^2 + 2m + 5 = 0$, giving $m = -1 \pm 2i$.

The CF is, therefore,

$$\mathbf{r} = e^{-t}(\mathbf{A}\cos 2t + \mathbf{B}\sin 2t)$$

where **A** and **B** are arbitrary vector constants.

Our trial solution for the PI is

$$\mathbf{r} = (P\cos t + Q\sin t)\mathbf{i} + (R\cos t + S\sin t)\mathbf{j} + T\mathbf{k}$$

Differentiating, we obtain

$$\frac{d\mathbf{r}}{dt} = (Q\cos t - P\sin t)\mathbf{i} + (S\cos t - R\sin t)\mathbf{j}$$

and

$$\frac{d^2\mathbf{r}}{dt^2} = -(P\cos t + Q\sin t)\mathbf{i} - (R\cos t + S\sin t)\mathbf{j}$$

Substituting into the differential equation and simplifying, we have

$$[(4P + 2Q)\cos t + (4Q - 2P)\sin t]\mathbf{i} + [(4R + 2S)\cos t + (4S - 2R)\sin t]\mathbf{j} + T\mathbf{k} = 2\cos t\,\mathbf{i} + 3\sin t\,\mathbf{j} + \mathbf{k}$$

Comparing **i**-components, we have: $4P + 2Q = 2$ and $4Q - 2P = 0$
$$\Rightarrow P = 0.4, Q = 0.2$$

Comparing **j**-components, we have: $4R + 2S = 0$ and $4S - 2R = 3$
$$\Rightarrow R = -0.3, S = 0.6$$

Comparing **k**-components, we have: $T = 1$

Adding the CF and the PI, we obtain the general solution

$$\mathbf{r} = e^{-t}(\mathbf{A}\cos 2t + \mathbf{B}\sin 2t) + (0.4\cos t + 0.2\sin t)\mathbf{i} + \\ + (0.6\sin t - 0.3\cos t)\mathbf{j} + \mathbf{k}$$

Differentiating, we obtain

$$\frac{d\mathbf{r}}{dt} = e^{-t}[(2\mathbf{B} - \mathbf{A})\cos 2t - (2\mathbf{A} + \mathbf{B})\sin 2t] + \\ + (0.2\cos t - 0.4\sin t)\mathbf{i} + (0.6\cos t + 0.3\sin t)\mathbf{j}$$

Substituting the initial conditions, we have

$$\mathbf{A} + 0.4\mathbf{i} - 0.3\mathbf{j} + \mathbf{k} = 2\mathbf{i} + \mathbf{j} \quad \Rightarrow \quad \mathbf{A} = 1.6\mathbf{i} + 1.3\mathbf{j} - \mathbf{k}$$

and

$$(2\mathbf{B} - \mathbf{A}) + 0.2\mathbf{i} + 0.6\mathbf{j} = \mathbf{k} \quad \Rightarrow \quad \mathbf{B} = 0.7\mathbf{i} + 0.35\mathbf{j}$$

So, the particular solution is

$$\mathbf{r} = [e^{-t}(1.6\cos 2t + 0.7\sin 2t) + 0.4\cos t + 0.2\sin t]\mathbf{i} + \\ + [e^{-t}(1.3\cos 2t + 0.35\sin 2t) + 0.6\sin t - 0.3\cos t]\mathbf{j} + \\ + (1 - e^{-t}\cos 2t)\mathbf{k}$$

Exercise 3L

1 Use an integrating factor to find the general solution of each of the following differential equations.

a) $\dfrac{d\mathbf{r}}{dt} + 3\mathbf{r} = 3t\mathbf{j} - \mathbf{k}$

b) $\dfrac{d\mathbf{r}}{dt} + t\mathbf{r} = t\mathbf{i} + 3t\mathbf{k}$

c) $\dfrac{d\mathbf{r}}{dt} - \dfrac{\mathbf{r}}{t} = t^2\mathbf{i} + 2t\mathbf{j} + 3\mathbf{k}$

d) $(t+1)\dfrac{d\mathbf{r}}{dt} + 2\mathbf{r} = 6\mathbf{j}$

2 By finding the complementary function and a particular integral, solve each of the following first-order equations. Hence find the particular solution which satisfies the given initial conditions.

a) $\dfrac{d\mathbf{r}}{dt} - 2\mathbf{r} = (t+1)\mathbf{i} + 3\mathbf{j} + t\mathbf{k}$, given that $\mathbf{r} = 4\mathbf{i} - \mathbf{j}$ when $t = 0$.

b) $\dfrac{d\mathbf{r}}{dt} + 3\mathbf{r} = e^{-3t}\mathbf{j}$, given that $\mathbf{r} = 6\mathbf{j} + \mathbf{k}$ when $t = 0$.

c) $\dfrac{d\mathbf{r}}{dt} - \mathbf{r} = \cos t\,\mathbf{i} + \cos 2t\,\mathbf{j}$, given that $\mathbf{r} = \mathbf{0}$ when $t = 0$.

CHAPTER 3 DIFFERENTIAL EQUATIONS

3 Find the general solution of each of the following equations. Hence find the particular solution which satisfies the given initial conditions.

a) $\dfrac{d^2\mathbf{r}}{dt^2} - 3\dfrac{d\mathbf{r}}{dt} + 2\mathbf{r} = \mathbf{i} - \mathbf{j} + \mathbf{k}$, given that $\mathbf{r} = \mathbf{0}$ and $\dfrac{d\mathbf{r}}{dt} = \mathbf{0}$ when $t = 0$.

b) $\dfrac{d^2\mathbf{r}}{dt^2} + 2\dfrac{d\mathbf{r}}{dt} + \mathbf{r} = (t+1)\mathbf{i}$, given that $\mathbf{r} = \mathbf{i} + \mathbf{j} + \mathbf{k}$ and $\dfrac{d\mathbf{r}}{dt} = \mathbf{0}$ when $t = 0$.

c) $\dfrac{d^2\mathbf{r}}{dt^2} - 2\dfrac{d\mathbf{r}}{dt} + 5\mathbf{r} = \cos t\,\mathbf{j} + \sin t\,\mathbf{k}$, given that $\mathbf{r} = \mathbf{0}$ and $\dfrac{d\mathbf{r}}{dt} = \mathbf{0}$ when $t = 0$.

4 A particle of mass 2 kg is moving under the influence of a force in the **i**-direction, which at time t s has magnitude $6\cos t$ N. It is also subject to a resistance of magnitude $2v$ N, where $v\,\text{m s}^{-1}$ is the speed of the particle.

a) Form a first-order differential equation connecting its velocity **v** and the time t.
b) Given that the initial velocity of the particle is $\mathbf{v}_0 = 60\mathbf{j} - 80\mathbf{k}$, find its velocity after 5 s.

5 A particle of unit mass moves under the influence of a force, directed towards the origin, which has magnitude $0.1r$ N, where r m is its distance from the origin. The particle is also subject to a resistance force of magnitude $0.2v$ N, where $v\,\text{m s}^{-1}$ is its speed.

a) Form a second-order differential equation connecting its displacement **r** and the time t.
b) Given that the particle is initially at the point $2\mathbf{i} + \mathbf{j} + 2\mathbf{k}$ and moving with velocity $\mathbf{i} - \mathbf{j} - 5\mathbf{k}$, find its position and velocity after 2 s.

Simultaneous linear differential equations

Simultaneous linear differential equations occur when we have several variables whose rates of change are interrelated. For example, if we have two or more particles fastened to a system of springs, the motion of each particle depends on its position and those of the other particles in the system.

Similarly, suppose we have populations, P_A and P_B, of two species of animal, A and B, in an area where B preys on A. The rate of change of P_A depends on P_A (which governs the number of births) and on P_B (which governs the level of predation). The rate of change of P_B depends on P_B (which governs the number of births) and on P_A (which governs the availability of food).

Complex systems can involve several equations and variables. There are techniques for solving such systems of equations, but we will restrict ourselves to pairs of first-order equations.

Example 25 Solve the system of equations

$$\dfrac{dx}{dt} = 4x - 2y \quad [1] \quad \text{and} \quad \dfrac{dy}{dt} = x + y \quad [2]$$

given that when $t = 0$, $x = 6$ and $y = 0$.

SOLUTION

Differentiating [1] with respect to t, we obtain

$$\frac{d^2x}{dt^2} = 4\frac{dx}{dt} - 2\frac{dy}{dt}$$

Substituting from [2], we have

$$\frac{d^2x}{dt^2} = 4\frac{dx}{dt} - 2x - 2y \qquad [3]$$

From [1], we have

$$2y = 4x - \frac{dx}{dt} \qquad [4]$$

Substituting from [4] into [3], we obtain

$$\frac{d^2x}{dt^2} = 4\frac{dx}{dt} - 2x - \left(4x - \frac{dx}{dt}\right)$$

$$\Rightarrow \quad \frac{d^2x}{dt^2} - 5\frac{dx}{dt} + 6x = 0$$

Solving this second-order homogeneous differential equation, we have

$$x = Ae^{2t} + Be^{3t} \qquad [5]$$

Substituting from [5] into [4] we obtain

$$2y = 4(Ae^{2t} + Be^{3t}) - (2Ae^{2t} + 3Be^{3t})$$

$$\Rightarrow \quad y = Ae^{2t} + \tfrac{1}{2}Be^{3t} \qquad [6]$$

The pair of equations [5] and [6] is the general solution to the problem. Notice that a pair of first-order equations is equivalent to a single second-order equation and that solving them generates two arbitrary constants.

To find the particular solution, we use the initial conditions $t = 0$, $x = 6$, $y = 0$:

From [5]: $\quad 6 = A + B$
From [6]: $\quad 0 = A + \tfrac{1}{2}B$

giving $A = -6$ and $B = 12$.

The particular solution is, therefore,

$$x = -6e^{2t} + 12e^{3t} \quad \text{and} \quad y = -6e^{2t} + 6e^{3t}$$

The graph, at the top of the next page, shows the solution with both x and y plotted against t.

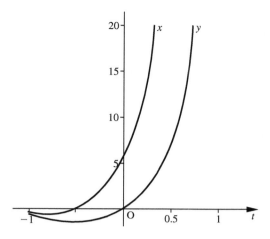

We can also investigate the relationship between x and y.

From [1] and [2], we obtain

$$\frac{dy}{dx} = \frac{dy/dt}{dx/dt} = \frac{x-y}{4x-2y}$$

The tangent field representing the family of solutions is shown in the diagram below, together with the particular solution

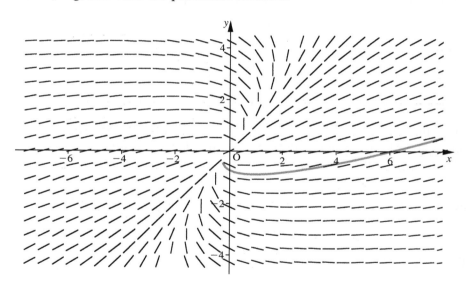

We can also obtain in a direct relationship between x and y, either by eliminating t from equations [5] and [6], which gives

$$6(x-y)^2 = (x-2y)^3$$

or by solving the equation

$$\frac{dy}{dx} = \frac{x-y}{4x-2y}$$

(though this leads to some awkward integration in this case).

SIMULTANEOUS LINEAR DIFFERENTIAL EQUATIONS

Example 26

a) Find the general solution of the system of equations

$$\frac{dx}{dt} = x + 4y - 6 \qquad [1]$$

$$\frac{dy}{dt} = -2x - 3y + 7 \qquad [2]$$

b) Find the particular solution if $x = 0$ and $y = -2$ when $t = 0$.
c) Sketch the graphs of both x and y against t.
d) Plot the tangent field representing the relationship between x and y for the interval $x \in [-8, 8]$, $y \in [-4, 4]$.
e) Sketch the particular solution from part **c** on the tangent field.

SOLUTION

a) From [1] we have

$$\frac{d^2x}{dt^2} = \frac{dx}{dt} + 4\frac{dy}{dt}$$

Substituting from [2], we obtain

$$\frac{d^2x}{dt^2} = \frac{dx}{dt} - 8x - 12y + 28$$

Substituting from [1], we obtain

$$\frac{d^2x}{dt^2} = \frac{dx}{dt} - 8x - 3\left(\frac{dx}{dt} - x + 6\right) + 28$$

which gives

$$\frac{d^2x}{dt^2} + 2\frac{dx}{dt} + 5x = 10 \qquad [3]$$

The auxiliary equation is $m^2 + 2m + 5 = 0$, giving $m = -1 \pm 2i$.

The complementary function is $x = e^{-t}(A\cos 2t + B\sin 2t)$.

The trial solution is $x = P$, giving $\dfrac{d^2x}{dt^2} = \dfrac{dx}{dt} = 0$.

Substituting in [3], we obtain

$$5P = 10 \quad \Rightarrow \quad P = 2$$

Hence, the general solution for x is

$$x = e^{-t}(A\cos 2t + B\sin 2t) + 2 \qquad [4]$$

Differentiating [4], we have

$$\frac{dx}{dt} = e^{-t}[(-A + 2B)\cos 2t + (-2A - B)\sin 2t]$$

From [1], we have

$$y = \frac{1}{4}\left(\frac{dx}{dt} - x + 6\right)$$

$\Rightarrow \quad y = \frac{1}{4}\{e^{-t}[(-A+2B)\cos 2t + (-2A-B)\sin 2t] - x + 6\}$

and substituting from [4], we get the general solution for y:

$$y = -\tfrac{1}{2}e^{-t}[(A-B)\cos 2t + (A+B)\sin 2t] + 1 \qquad [5]$$

Hence, the pair of equations [4] and [5] is the general solution.

b) Applying the initial conditions, we have

$0 = A + 2 \quad \text{giving} \quad A = -2$

$-2 = -\tfrac{1}{2}(A - B) + 1 \quad \text{giving} \quad B = -8$

The particular solution is, therefore,

$x = e^{-t}(-2\cos 2t - 8\sin 2t) + 2 \qquad [6]$
$y = e^{-t}(-3\cos 2t + 5\sin 2t) + 1 \qquad [7]$

c) The graphs show the pair of equations [6] and [7].

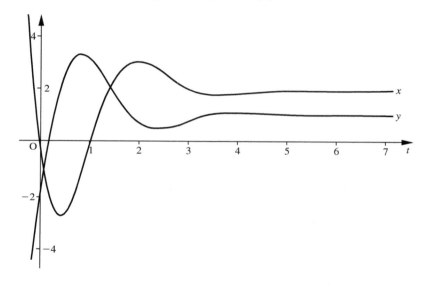

d), e) Graphing the tangent field is only practicable if you have access to suitable graphing software, although it is possible to graph the parametric particular solution on most graphic calculators.

From [1] and [2], we have

$$\frac{dy}{dx} = \frac{-2x - 3y + 7}{x + 4y - 6}$$

This gives the tangent field shown at the top of the next page.

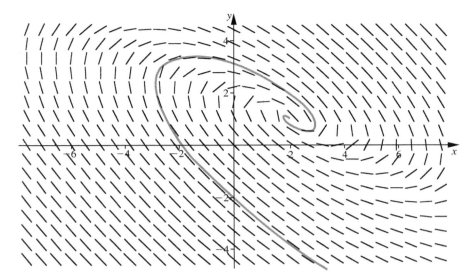

In this solution, as time increases the transient terms will diminish and both x and y will converge to the constant values $x = 2$, $y = 1$. The solution curves plotted of y against x form spirals converging to the point $(2, 1)$.

Exercise 3M

In Questions **1** to **6**:

a) Find the general solution of the system of equations.
b) Find the particular solution using the given boundary conditions.
c) Sketch the graphs of both x and y against t.

In addition, if you have suitable software, you should plot the tangent field showing the relationship between x and y (suggested intervals $x \in [-8, 8]$, $y \in [-4, 4]$) and/or sketch the particular solution from part **c**.

1 $\dfrac{dx}{dt} = x - 2y$

$\dfrac{dy}{dt} = -2x + y + 3$

$x = 0$ and $y = 1$ when $t = 0$

2 $\dfrac{dx}{dt} = x + 2y - 4$

$\dfrac{dy}{dt} = -2x + y + 3$

$x = 2$ and $y = 0$ when $t = 0$

3 $\dfrac{dx}{dt} = -x - y - 5$

$\dfrac{dy}{dt} = 5x - 3y - 1$

$x = 2$ and $y = -5$ when $t = 0$

4 $\dfrac{dx}{dt} = 6 - 3y$

$\dfrac{dy}{dt} = 3x - 12$

$x = 3$ and $y = 1$ when $t = 0$

CHAPTER 3 DIFFERENTIAL EQUATIONS

5 $\dfrac{dx}{dt} = 2x + 3y + 4$

$\dfrac{dy}{dt} = 3x + 2y - 9$

$x = 5$ and $y = 2$ when $t = 0$

6 $\dfrac{dx}{dt} = x - y + 1$

$\dfrac{dy}{dt} = -2x + 3y - 3$

$x = 2$ and $y = 0$ when $t = 0$

7 Given the system of equations

$$\dfrac{dx}{dt} = ax + by + c \qquad \dfrac{dy}{dt} = dx + ey + f$$

show that you obtain the same auxiliary equation whether you eliminate x or y in forming a second-order differential equation.

Numerical methods

The importance of differential equations and the fact that they are often very difficult, if not impossible, to solve by analytical methods, have led to the development of many numerical techniques for solving them. By solving numerically, we mean finding a series of points, starting from a known point, which correspond to the particular solution through that point. By their nature, numerical methods are approximate, and the further we get from the known point, the greater will be the error.

One of the simplest numerical techniques is **Euler's method**, which we examine here.

Euler's method

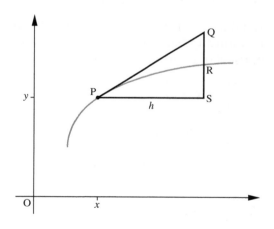

This can be adapted for differential equations of higher orders but we will look at the method applied to first-order equations of the form $\dfrac{dy}{dx} = f(x, y)$

If we know that the curve passes through the point $P(x, y)$, we need to find the coordinates of the 'next' point, R, on the curve, whose x-coordinate is $(x + h)$. We cannot find R exactly, but provided the step size, h, is small we can obtain a good approximation by finding the point Q on the tangent to the curve at P and with the same x-coordinate as R.

The gradient of PQ is $f(x, y)$ from the differential equation. From the diagram, we see that

$$\dfrac{QS}{h} = f(x, y) \quad \Rightarrow \quad QS = h f(x, y)$$

Hence, the coordinates of Q are $(x + h, y + h f(x, y))$.

Thus, we have a pair of recurrence relations

$$x_{n+1} = x_n + h$$
$$y_{n+1} = y_n + hf(x_n, y_n)$$

which we can use to find successive points (x_n, y_n) on the curve.

Of course, every time we apply the method, we introduce an error (the length QR in the diagram). Eventually, the total error will be unacceptable and so we must take care when applying this method. We can reduce the error by making h smaller. We can also apply a correction technique (described later) or use other, more accurate methods which are outside the scope of this book.

Numerical methods are well suited to spreadsheet work, and you are encouraged to solve the exercises by this approach.

Example 27 Use Euler's method with a step size of 0.1 to find $y(0.5)$ for the equation

$$\frac{dy}{dx} = 2x + y$$

where $y(0) = 1$.

SOLUTION

The recurrence relations are

$$x_{n+1} = x_n + 0.1$$
$$y_{n+1} = y_n + 0.1(2x_n + y_n)$$

The results are given in the spreadsheet table below.

x	y	$f(x, y)$
0	1.000 00	1.000 00
0.1	1.100 00	1.300 00
0.2	1.230 00	1.630 00
0.3	1.393 00	1.993 00
0.4	1.592 30	2.392 30
0.5	1.831 53	2.831 53

Thus Euler's method gives $y(0.5) = 1.83$ (to 3 sf).

Note The differential equation in Example 27 can be solved analytically. Its solution is

$$y = 3e^x - 2x - 2$$

To get an idea of the way the errors build up, we can compare the values obtained by Euler's method with the true values shown in the table on the right.

The accuracy can be improved by reducing the step size. For example, taking $h = 0.01$ results in $y(0.5) = 1.9339$, and $h = 0.001$ gives $y(0.5) = 1.9449$.

x	y
0	1.000 00
0.1	1.115 51
0.2	1.264 21
03	1.449 58
0.4	1.675 47
0.5	1.946 16

Exercise 3N

Use Euler's method to solve each of the following questions using the step length given.

1 $\dfrac{dy}{dx} = x + y$

Find $y(2)$ if $y(0) = 2$ and $h = 0.2$.

2 $\dfrac{dy}{dx} = x^2(y + 1)$

Find $y(1)$ if $y(0) = 2$ and $h = 0.2$.

3 $\dfrac{dy}{dx} = x^2 + y^2$

Find $y(1)$ if $y(0) = 1$ and $h = 0.1$.

4 $\dfrac{dy}{dx} = \dfrac{xy + 1}{xy - 1}$

Find $y(1)$ if $y(0) = 1$ and $h = 0.2$.

5 $\dfrac{dy}{dx} = \sin x + \cos y$

Find $y(1)$ if $y(0) = 1$ and $h = 0.2$.

6 $\dfrac{dy}{dx} = xy + 2x$

Find $y(2)$ if $y(0) = 1$ and $h = 0.2$.

Improving the accuracy

In Example 27, we used Euler's method to solve $\dfrac{dy}{dx} = 2x + y$, where $y(0) = 1$.

The correct solution, obtained analytically, is 1.946 16 (to 5 dp).

The results obtained by Euler's method for various step sizes are

h	$y(0.5)$	Error
0.1	1.8315	−0.1147
0.01	1.9339	−0.0123
0.001	1.9449	−0.0013

We can see that the result is gradually getting closer to the correct value, and that the error is approximately proportional to h.

If we plot a graph of the calculated $y(0.5)$ values against these, and other, h values, we can see how our result approaches the correct one. (See graph at the top of the next page.)

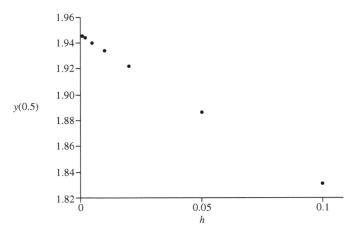

In general, the error in Euler's method is proportional to the step size, h, provided that h is small enough. We can use this fact to address the accuracy problem in two ways.

The first approach is to estimate $y(0.5)$ for two values of h, then find the intercept of the line defined by these two points on the vertical axis.

The equation of this line will be

$$y = Y + kh$$

where Y is the intercept, y is the calculated value and k is the gradient of the line.

Using the values of y for $h = 0.002$ and $h = 0.001$, we get

$$1.943\,6946 = Y + 0.002k$$
$$1.944\,9282 = Y + 0.001k$$

We solve these to give

$$Y = 1.946\,1618 \quad \text{and} \quad k = -1.2336$$

This result is now accurate to five decimal places, a good improvement over the two calculated values in the table.

The second approach is to calculate the step size required to obtain the result to a desired degree of accuracy. If we required a result with an error of less than 10^{-7}, we would need

$$|Y - y| < 10^{-7}$$
$$\Rightarrow \quad |kh| < 10^{-7}$$
$$\Rightarrow \quad h < \frac{10^{-7}}{1.2336}$$
$$\Rightarrow \quad h < 0.000\,000\,081\,0636$$

A sensible value would be $h = 0.000\,000\,08$. This would take $\dfrac{0.5}{h} = 6\,250\,000$ steps.

Note that we chose a value of h so that the number of steps is an integer.

It is clear that Euler's method is not practicable if we required this degree of accuracy. We would need to use a more efficient method.

Exercise 3O

1 Euler's method was used to solve the differential equation $\frac{dy}{dx} = 1 - xy$ to find the value of $y(2)$, given that $y(1) = 0$, using various values of step length h. The results are shown in the table.

h	$y(2)$
0.100	0.500 502 219
0.050	0.489 094 562
0.010	0.480 389 706
0.005	0.479 326 271
0.001	0.478 479 345

  a) Find an improved estimate of the solution.
  b) Choose a suitable value of h so that Euler's method would find the value of $y(2)$ with an error of less than 10^{-5}.

2 A reflecting dish is built so that a signal travelling in the direction of the negative y-axis is reflected by the surface of the dish and then passes through the origin where the receiver is placed. The position of points (x, y) on the surface of the dish satisfy the differential equation

$$\frac{dy}{dx} = \frac{y + \sqrt{x^2 + y^2}}{x}$$

Euler's method was used to solve the differential equation in order to design a dish with an edge radius of 10 m. $y(5)$ was chosen to be zero and the value of $y(0.2)$ was found using various values of the step length h. The results are shown in the table.

h	$y(2)$
-0.100	$-2.570\,800\,477$
-0.050	$-2.514\,672\,905$
-0.010	$-2.499\,733\,353$
-0.005	$-2.497\,866\,603$
-0.001	$-2.496\,373\,309$

  a) Show that the differential equation given does satisfy the conditions of the problem.
  b) Find an improved estimate of the solution, correct to four decimal places.
  c) Choose a suitable value of h so that Euler's method can be used without correcting the estimate to find the value of $y(0.2)$ with an error of less than 10^{-5}.

4 General motion in one dimension

I don't know who's ahead – it's either Oxford or Cambridge.
JOHN SNAGGE

Motion with variable mass

Newton's second law (the equation of motion) is the basic model for mechanical systems. In the situations we have encountered so far (see *Introducing Mechanics*, pages 76–8), the law is stated as

$$\mathbf{F} = m\mathbf{a}$$

In applying this, we have made the implicit assumption that the mass of the body involved is constant. This is usually a safe assumption. Even with a motor vehicle, which burns fuel and so gets lighter as it travels, the change of mass is so small and so slow that it can effectively be ignored. However, in some circumstances large and rapid changes of mass can occur, and for these we need to modify our approach.

Newton originally stated his second law as 'The rate of change of momentum of a body is proportional to the force applied, and is in the direction of the force' (except that he wrote it in Latin). With the usual choice of units, the constant of proportionality is 1, so we can say

Applied force = Rate of change of momentum

Suppose that a body of mass m, moving with velocity $\mathbf{v}$, gains a small amount of mass δm during a small time interval δt, after which it is travelling with velocity $\mathbf{v} + \delta\mathbf{v}$. Suppose further that, prior to its being incorporated into the body, the mass δm was travelling with velocity $\mathbf{u}$. Let $\mathbf{F}$ be the average resultant force acting on the system during δt. We then have

$$\mathbf{F} \approx \frac{\text{Change of momentum}}{\delta t}$$

The initial momentum of the system is $\quad m\mathbf{v} + \delta m \mathbf{u}$
The final momentum of the system is $\quad (m + \delta m)(\mathbf{v} + \delta\mathbf{v})$

Hence, we have

$$\mathbf{F} \approx \frac{(m + \delta m)(\mathbf{v} + \delta\mathbf{v}) - (m\mathbf{v} - \delta m\mathbf{u})}{\delta t}$$

$$\Rightarrow \quad \mathbf{F} \approx m\frac{\delta\mathbf{v}}{\delta t} + (\mathbf{v} - \mathbf{u})\frac{\delta m}{\delta t} + \frac{\delta m \delta\mathbf{v}}{\delta t}$$

When we let $t \to 0$, this becomes

$$\mathbf{F} = m\frac{d\mathbf{v}}{dt} + (\mathbf{v} - \mathbf{u})\frac{dm}{dt} \qquad [1]$$

Notice that $\mathbf{v} - \mathbf{u}$ is the velocity of the body relative to the mass increment.

CHAPTER 4 GENERAL MOTION IN ONE DIMENSION

In simple cases, such as a raindrop falling through a slow moving cloud, the velocity **u** is effectively zero. Equation [1] then becomes

$$\mathbf{F} = m\frac{d\mathbf{v}}{dt} + \mathbf{v}\frac{dm}{dt} \qquad [2]$$

Notice that equation [2] is equivalent to $\mathbf{F} = \dfrac{d(m\mathbf{v})}{dt}$.

Example 1 A raindrop is formed when a small droplet falls through a cloud of water vapour. Small particles of water, initially stationary, attach themselves to the droplet so that its mass increases at a constant rate of $0.0002 \, \text{kg s}^{-1}$. The droplet takes 10 seconds to fall through the cloud. Assuming that initially the droplet has zero mass and velocity, and ignoring air resistance, find the speed of the drop as it leaves the cloud.

SOLUTION

Take downwards to be the positive direction. Motion takes place in a straight line, and the velocity of the droplet at time t is v downwards.

The only force acting in the direction of motion is the weight of the droplet.

The rate of change of the mass is $\dfrac{dm}{dt} = 0.0002$.

At time t, the mass of the droplet is $m = 0.0002t$.

As the cloud is stationary, the equation of motion is

$$\mathbf{F} = m\frac{d\mathbf{v}}{dt} + \mathbf{v}\frac{dm}{dt} = \frac{d(m\mathbf{v})}{dt}$$

$$\Rightarrow \quad 0.0002tg = \frac{d(0.0002tv)}{dt}$$

$$\Rightarrow \quad 0.0002tv = \int 0.0002tg \, dt$$

$$\Rightarrow \quad tv = \tfrac{1}{2}t^2 g + C$$

When $t = 0$, $v = 0$, and so $C = 0$. Hence, we have

$$v = \tfrac{1}{2}tg$$

When the drop leaves the cloud, $t = 10$, and so $v = 5g$.

The drop therefore leaves the cloud with a speed of $5g \, \text{m s}^{-1}$.

Note The equation of motion is a vector equation. Therefore, in Example 1, strictly we should have written the velocity as **v** and the force as $0.0002tg\mathbf{i}$. However, as all the motion takes place in a straight line, we can safely write the solution as shown, and we will adopt this practice in the rest of the chapter.

Extending the raindrop model

Example 1 involved the simplest assumption: namely, that the rate of change of the mass of the droplet was constant. We will first generalise this situation and then examine other, more realistic, assumptions.

Mass increases at a constant rate

Suppose that $\frac{dm}{dt} = k$. Suppose further that the droplet starts with zero mass and velocity. We have, at time t, mass $m = kt$ and velocity $= v$ downwards.

Our equation of motion is

$$\mathbf{F} = m\frac{d\mathbf{v}}{dt} + \mathbf{v}\frac{dm}{dt} = \frac{d(m\mathbf{v})}{dt}$$

which gives

$$ktg = \frac{d(ktv)}{dt}$$

$$\Rightarrow \quad tv = \int tg\,dt$$

$$\Rightarrow \quad tv = \tfrac{1}{2}t^2 g + C$$

The initial conditions give $C = 0$. Therefore, we have

$$v = \tfrac{1}{2}tg$$

This model therefore implies that the droplet has a constant acceleration of $\tfrac{1}{2}g$, independent of the rate at which its mass is increasing.

Other models

It is more realistic to suppose that a large raindrop will acquire extra water more rapidly than a small raindrop. Two ways in which we may model this are to assume that the rate of increase of mass is proportional either to the radius or to the surface area of the raindrop. We will now explore these two models. In each, we will assume that air resistance can be ignored.

- **Rate of increase proportional to radius,** $\dfrac{dm}{dt} \propto r$

  This can be written as

  $$\frac{dm}{dt} = kr$$

  We will assume that the raindrop is a sphere of water of density $1000\,\text{kg}\,\text{m}^{-3}$. This gives

  $$m = \frac{4000\pi r^3}{3}$$

  $$\Rightarrow \quad \frac{dm}{dt} = 4000\pi r^2 \frac{dr}{dt}$$

As $\frac{dm}{dt} = kr$, we have

$$\frac{dr}{dt} = \frac{k}{4000\pi r}$$

$$\Rightarrow \int 4000\pi r \, dr = \int k \, dt$$

$$2000\pi r^2 = kt + C$$

If, as before, $r = 0$ when $t = 0$, we have $C = 0$, and hence

$$r = \sqrt{\frac{kt}{2000\pi}}$$

$$\Rightarrow m = \frac{2}{3}\sqrt{\frac{k^3 t^3}{2000\pi}}$$

The equation of motion is

$$\mathbf{F} = m\frac{d\mathbf{v}}{dt} + \mathbf{v}\frac{dm}{dt} = \frac{d(m\mathbf{v})}{dt}$$

$$\Rightarrow mg = \frac{d(mv)}{dt}$$

$$\Rightarrow mv = \int mg \, dt$$

$$\Rightarrow \frac{2v}{3}\sqrt{\frac{k^3 t^3}{2000\pi}} = \int \frac{2}{3}\sqrt{\frac{k^3 g^2 t^3}{2000\pi}} \, dt$$

which gives

$$v\sqrt{t^3} = \tfrac{2}{5}g\sqrt{t^5} + C_1$$

When $t = 0$, $v = 0$, and so $C_1 = 0$. Hence, we have

$$v = \tfrac{2}{5}gt$$

This model therefore implies that the droplet has a constant acceleration of $\tfrac{2}{5}g$, independent of the constant of proportionality linking the mass and the radius of the raindrop.

- **Rate of increase proportional to surface area,** $\dfrac{dm}{dt} \propto A$

This can be written as

$$\frac{dm}{dt} = 4\pi k r^2$$

We will assume that the raindrop is a sphere of water of density $1000 \, \text{kg m}^{-3}$. This gives

$$m = \frac{4000\pi r^3}{3}$$

$$\Rightarrow \frac{dm}{dt} = 4000\pi r^2 \frac{dr}{dt}$$

As $\frac{dm}{dt} = 4\pi k r^2$, we have

$$\frac{dr}{dt} = \frac{k}{1000}$$

$$\Rightarrow \quad r = \frac{kt}{1000} + C$$

If, as before, $r = 0$ when $t = 0$, we have $C = 0$, and hence

$$r = \frac{kt}{1000}$$

$$\Rightarrow \quad m = \frac{4\pi k^3 t^3}{3\,000\,000}$$

The equation of motion is

$$\mathbf{F} = m\frac{d\mathbf{v}}{dt} + \mathbf{v}\frac{dm}{dt} = \frac{d(m\mathbf{v})}{dt}$$

$$\Rightarrow \quad mg = \frac{d(mv)}{dt}$$

$$\Rightarrow \quad mv = \int mg\,dt$$

$$\Rightarrow \quad \frac{4\pi k^3 t^3}{3\,000\,000}v = \int \frac{4\pi k^3 g t^3}{3\,000\,000}\,dt$$

which gives

$$vt^3 = \tfrac{1}{4}gt^4 + C_1$$

When $t = 0$, $v = 0$, and so $C_1 = 0$. Hence, we have

$$v = \tfrac{1}{4}gt$$

This model therefore implies that the droplet has a constant acceleration of $\tfrac{1}{4}g$, independent of the constant of proportionality linking the mass and the surface area of the raindrop.

Modelling the motion of a rocket

As a rocket moves, burnt fuel is ejected. The mass of the rocket therefore decreases with time. The equation of motion is the same as for a raindrop gathering water, except that in the case of the rocket the rate of change of mass is negative.

Example 2 A rocket of mass 2500 kg, including its fuel, is in space and is at rest relative to a chosen frame of reference. The motor is then fired. The fuel burns at a rate of $20\,\text{kg s}^{-1}$, and the material is emitted at a speed of $3000\,\text{m s}^{-1}$. If the rocket initially contains 2000 kg of fuel, find its speed when all the fuel has been used.

SOLUTION

The equation of motion is

$$\mathbf{F} = m\frac{d\mathbf{v}}{dt} + (\mathbf{v} - \mathbf{u})\frac{dm}{dt}$$

where $\mathbf{v} - \mathbf{u}$ is the velocity of the rocket relative to the ejected material. In this case, $\mathbf{v} - \mathbf{u} = 3000\,\text{m s}^{-1}$.

There are no external forces acting on the rocket.

At time t, the mass of the rocket is $m = 2500 - 20t$.

The mass of the rocket is reducing by $20\,\text{kg s}^{-1}$, hence we have

$$\frac{dm}{dt} = -20$$

The equation of motion is, therefore,

$$0 = (2500 - 20t)\frac{dv}{dt} + 3000 \times (-20)$$

$$\Rightarrow \quad \frac{dv}{dt} = \frac{3000}{125 - t}$$

$$\Rightarrow \quad v = \int \frac{3000}{125 - t}\,dt = -3000\ln(125 - t) + C$$

The initial conditions are $v = 0$ when $t = 0$, which gives $C = 3000\ln 125$. Hence, we have

$$v = 3000\ln\left(\frac{125}{125 - t}\right)$$

The fuel will be exhausted when $t = 100$. This gives

$$v = 3000\ln 5 = 4828.3$$

Hence, when the fuel is all used, the speed of the rocket is $4830\,\text{m s}^{-1}$ (to 3 sf).

Exercise 4A

1. A hailstone falls through a stationary cloud. Initially, it has mass $0.02\,\text{kg}$, and enters the cloud at a speed of $5\,\text{m s}^{-1}$. As it passes through the cloud, it accumulates mass at a rate of $0.001\,\text{kg s}^{-1}$. If it takes $5\,\text{s}$ to fall through the cloud, at what speed will it be travelling when it emerges?

2. A rocket is launched vertically upwards from rest. It has an initial mass of $5000\,\text{kg}$, of which $4000\,\text{kg}$ is fuel. The fuel is used at a rate of $40\,\text{kg s}^{-1}$, and is emitted at a rate of $3000\,\text{m s}^{-1}$ relative to the rocket. Find the speed of the rocket when all the fuel has been used.

3. A droplet of water, initially of radius zero, starts from rest and falls through a stationary cloud of mist. As it does so, water vapour condenses on the droplet at a rate proportional to the radius of the droplet. Show that $v = \dfrac{4\pi g r^2}{5k}$, where k is the constant of proportionality.

EXERCISE 4A

4 A droplet of water, initially of radius r_0, starts from rest and falls through a stationary cloud of mist. As it does so, water vapour condenses on the droplet at a rate proportional to the surface area of the droplet.
  a) State any modelling assumptions necessary to simplify the problem.
  b) Find expressions in terms of time for the velocity and acceleration of the droplet as it falls through the cloud.
  c) Show that
    i) the initial acceleration of the droplet is g
    ii) as t increases, the acceleration of the droplet approaches $\frac{1}{4}g$.

5 A long sledge, of mass m_0 kg, is sliding on a horizontal surface, with no resistance to its motion, at a constant speed v_0. At time $t = 0$, the sledge starts to pass under a hopper, which releases sand so that it falls gently onto the sledge at a constant rate of k kg s^{-1}. Show that, at time t, the length, s, of the sand trail on the sledge is given by

$$s = \frac{m_0 v_0}{k} \ln\left(\frac{m_0 + kt}{m_0}\right)$$

6 If the sledge in Question **5** is subject to a frictional force with coefficient of friction μ, show that its velocity at time t is given by

$$v = \frac{m_0 v_0 - \mu g(m_0 t + \frac{1}{2}kt^2)}{m_0 + kt}$$

7 A hot air balloon of total mass 2000 kg is floating in equilibrium at an altitude of 200 m. Included in its mass is a bag of sand of mass 40 kg. This is released at a constant rate of 2 kg s^{-1} until the bag is empty.

Assuming that there is no air resistance and that the upthrust on the balloon remains constant, find the height to which the balloon has risen, and its velocity, when all the sand has been released.

8 A rocket is in space and moving with a speed of 100 m s^{-1} relative to some chosen frame of reference, when the engines are fired. They burn fuel at a rate of 10 kg s^{-1}, ejecting it backwards at a speed of 1000 m s^{-1} relative to the rocket. If the initial mass of the rocket is 400 kg and the engines are fired for 10 s, what is the speed of the rocket at the end of this time?

9 A sledge of mass 30 kg is sliding on a horizontal sheet of ice, with no resistance to its motion, at a speed of 10 m s^{-1}. At time $t = 0$, it enters a snowstorm, and snow accumulates on it at a rate of 0.2 kg s^{-1}. Find the speed of the sledge when $t = 20$.

10 A raindrop of mass M is initially at rest. It falls under gravity through a stationary cloud and accumulates mass so that, at time t, its mass is $Me^{\lambda t}$, where λ is a constant. After T seconds, the mass of the raindrop has doubled.
  a) Show that the speed of the raindrop at that time is $\dfrac{gT}{2\ln 2}$.
  b) How far has the raindrop fallen in that time?

CHAPTER 4 GENERAL MOTION IN ONE DIMENSION

Variable forces

We now consider situations in which the force applied varies with time, with velocity or with displacement.

If the mass is constant, we can use the familiar form of Newton's second law

$$\mathbf{F} = m\mathbf{a}$$

As we saw in *Introducing Mechanics* (pages 104–8), this can be written, for motion in one dimension, as a first- or second-order differential equation by expressing the acceleration as $\frac{dv}{dt}$, $\frac{d^2s}{dt^2}$ or $v\frac{dv}{ds}$, depending on the nature of the function for the force.

Example 3 A body of mass 2 kg is moving in a straight line under the influence of a force given by $F = 18 \sin 3t$. Initially, the body is stationary at the origin. Find an expression for its displacement at time t.

SOLUTION

The equation of motion is

$$18 \sin 3t = 2\frac{dv}{dt}$$

$$\Rightarrow \quad v = \int 9 \sin 3t \, dt = -3 \cos 3t + C$$

When $t = 0$, $v = 0$, which gives $C = 3$. Hence, we have

$$v = \frac{ds}{dt} = 3 - 3 \cos 3t$$

$$\Rightarrow \quad s = \int 3 - 3 \cos 3t \, dt = 3t - \sin 3t + C_1$$

When $t = 0$, $s = 0$, which gives $C_1 = 0$. Hence, we have

$$s = 3t - \sin 3t$$

Example 4 A particle of mass 3 kg is attached to an elastic string of stiffness 9 N m^{-1}, the other end of which is attached to a fixed point A. The particle is held at rest vertically below A with the string just taut, and is then released. It falls through a resisting medium which gives a resistance of magnitude $12v$, where v is the speed of the particle. Find an expression for the displacement of the particle below its starting point at time t seconds after release.

SOLUTION

Let the displacement of the particle below its starting point be x. The equation of motion can be written as

$$3g - 9x - 12\frac{dx}{dt} = 3\frac{d^2x}{dt^2}$$

$$\Rightarrow \frac{d^2x}{dt^2} + 4\frac{dx}{dt} + 3x = g$$

To solve this second-order differential equation, we first find the complementary function from the auxiliary equation

$$m^2 + 4m + 3 = 0$$
$$\Rightarrow m = -1 \quad \text{or} \quad -3$$

This gives the complementary function $x = Ae^{-t} + Be^{-3t}$.

It is easy to see that the particular integral is $x = \tfrac{1}{3}g$, so the general solution is

$$x = Ae^{-t} + Be^{-3t} + \tfrac{1}{3}g$$

from which we have

$$\frac{dx}{dt} = -Ae^{-t} - 3Be^{-3t}$$

The initial conditions are $t = 0$, $x = 0$ and $\dfrac{dx}{dt} = 0$. Substituting these, we get

$$A + B = -\tfrac{1}{3}g$$

and $\quad -A - 3B = 0$

Solving, we get $A = -\tfrac{1}{2}g$, $B = \tfrac{1}{6}g$.

Hence, at time t, the particle has fallen a distance

$$x = \tfrac{1}{6}g(2 + e^{-3t} - 3e^{-t})$$

As $t \to \infty$, the particle approaches a point $\tfrac{1}{3}g$ m below its starting point. (This is the equilibrium position.)

Example 5 A metal particle of mass m is initially stationary at a distance a from the centre of a fixed magnet. When the particle is released, it is attracted towards the magnet by a force $\dfrac{k}{x^2}$, where x is the distance between the particle and the centre of the magnet. No other force acts on the particle. Find the speed at which the particle strikes the surface of the magnet (which is $0.1a$ from its centre).

SOLUTION

The equation of motion is

$$-\frac{k}{x^2} = m\frac{dv}{dt}$$

CHAPTER 4 GENERAL MOTION IN ONE DIMENSION

However, in this case, it is more appropriate to express this as

$$-\frac{k}{x^2} = mv\frac{dv}{dx}$$

$$\Rightarrow \int v\,dv = \int \frac{-k}{mx^2}\,dx$$

$$\Rightarrow \tfrac{1}{2}v^2 = \frac{k}{mx} + C$$

When $x = a$, $v = 0$, which gives $C = -\dfrac{k}{ma}$. Hence, we have

$$v = \sqrt{\frac{2k(a-x)}{amx}}$$

When $x = 0.1a$, this gives $v = \sqrt{\dfrac{18k}{am}}$.

Except in the simplest cases, models involving both variable mass and variable force lead to equations which are difficult or impossible to solve by analytical methods, and you are unlikely to encounter them at this level. An example is given for the sake of completeness.

Example 6 A raindrop, having initial mass and velocity zero, falls through a stationary cloud. It accumulates mass at a constant rate k. It is subject to air resistance having magnitude $\lambda k v$, where v is the speed of the raindrop and λ is a constant. Find an expression for the velocity of the raindrop at time t.

SOLUTION

The mass of the raindrop at time t is $m = kt$.

The equation of motion is

$$\mathbf{F} = m\frac{d\mathbf{v}}{dt} + \mathbf{v}\frac{dm}{dt}$$

which in this case gives

$$kgt - \lambda k v = kt\frac{dv}{dt} + kv$$

$$\Rightarrow \frac{dv}{dt} + \frac{(1+\lambda)v}{t} = g$$

This first-order linear differential equation can be solved by using the integrating factor

$$e^{\int \frac{(1+\lambda)}{t}dt} = t^{(1+\lambda)}$$

Multiplying through by the integrating factor, we get

$$t^{(1+\lambda)} \frac{dv}{dt} + (1+\lambda)t^\lambda v = gt^{(1+\lambda)}$$

$$\Rightarrow \frac{d(t^{(1+\lambda)}v)}{dt} = gt^{(1+\lambda)}$$

$$\Rightarrow t^{(1+\lambda)}v = \int gt^{(1+\lambda)} dt = \frac{gt^{(2+\lambda)}}{(2+\lambda)} + C$$

The initial conditions $t = 0$, $v = 0$ give $C = 0$, so we have

$$v = \frac{gt}{(2+\lambda)}$$

Impulse of a variable force

Suppose that the force acting on a particle of constant mass m is a function of time. That is,

Force = $F(t)$

Suppose that the force acts for a period T, during which the velocity of the particle changes from u to v.

The equation of motion is

$$m\frac{dv}{dt} = F(t)$$

$$\Rightarrow \int_u^v m\, dv = \int_0^T F(t)\, dt$$

$$\Rightarrow \int_0^T F(t)\, dt = mv - mu$$

The right-hand side of this equation represents the change of momentum of the particle, which is the impulse of the force. Hence, for a variable force, we have

$$\text{Impulse} = \int_0^T F(t)\, dt$$

Example 7 A ball of mass 0.5 kg, travelling at 20 m s^{-1}, is struck head on by a bat. The contact between the bat and ball lasts for 0.2 seconds, during which time the force exerted on the ball is $F = 37500t^2(1 - 5t)$. Find the speed of the ball after impact.

SOLUTION

The magnitude of the impulse exerted by the bat on the ball is

$$37\,500 \int_0^{0.2} (t^2 - 5t^3)\, dt = 37\,500 \left[\frac{t^3}{3} - \frac{5t^4}{4}\right]_0^{0.2} = 25\,\text{N s}$$

If we take the direction of this impulse to be positive, we have

$$\text{Initial momentum of ball} = 0.5 \times (-20) = -10\,\text{N s}$$
$$\text{Final momentum of ball} = -10 + 25 = 15\,\text{N s}$$
$$\Rightarrow \quad \text{Velocity after impact} = \frac{15}{0.5} = 30\,\text{m s}^{-1}$$

So, the direction of the ball is reversed and its initial speed is $30\,\text{m s}^{-1}$.

Work done by a variable force

Suppose that the force acting on a particle of constant mass m is a function of displacement. That is,

$$\text{Force} = F(s)$$

Suppose that the force acts over a distance S, during which the velocity of the particle changes from u to v.

The equation of motion is

$$mv\frac{\mathrm{d}v}{\mathrm{d}s} = F(s)$$

$$\Rightarrow \quad \int_u^v mv\,\mathrm{d}v = \int_0^S F(s)\,\mathrm{d}s$$

$$\Rightarrow \quad \int_0^S F(s)\,\mathrm{d}s = \tfrac{1}{2}mv^2 - \tfrac{1}{2}mu^2$$

The right-hand side of this equation represents the change of kinetic energy of the particle, which is the work done by the force. Hence, for a variable force

$$\text{Work done} = \int_0^S F(s)\,\mathrm{d}s$$

Example 8 On a certain planet, of radius R, the gravitational force acting on an object of mass m situated a distance x from the planet is $F = \dfrac{km}{x^2}$.

a) Find the work done in sending a body from the planet surface to a height of $3R$.

b) Show that, provided a projectile's initial velocity exceeds a certain value, it can 'escape' completely.

SOLUTION

a) The work done is

$$W = \int_R^{4R} \frac{km}{x^2}\,\mathrm{d}x = \left[\frac{-km}{x}\right]_R^{4R} = -\frac{3km}{4R}$$

b) The work required to send a body to a distance S from the centre of the planet is

$$W = \int_R^S \frac{km}{x^2}\, dx = \left[\frac{-km}{x}\right]_R^S$$

$$\Rightarrow \quad W = km\left(\frac{1}{R} - \frac{1}{S}\right)$$

As S increases, $W \to \dfrac{km}{R}$. Hence, provided the initial kinetic energy of the projectile is greater than this, the projectile will escape the planet's gravitational influence. This happens if

$$\tfrac{1}{2}mv^2 > \frac{km}{R} \quad \Rightarrow \quad v > \sqrt{\frac{2k}{R}}$$

Power

If a body of mass m is raised to a speed v by the application of a force F, the work done, W, is the kinetic energy imparted. That is,

$$W = \tfrac{1}{2}mv^2$$

Differentiating with respect to t, we get

$$\frac{dW}{dt} = mv\frac{dv}{dt}$$

By Newton's second law, $m\dfrac{dv}{dt}$ is the applied force, F.

Power is the rate of doing work, $\dfrac{dW}{dt}$.

Hence, we have

$$\text{Power} = Fv$$

Example 9 A truck of mass m moves from rest along a level road. Its engine exerts a constant power P. The total resistance acting against the truck is kv. Find an expression for the time taken to reach a speed V.

SOLUTION

If the forward force exerted by the truck is F and v is its speed at time t, we have

$$P = Fv$$

$$\Rightarrow \quad F = \frac{P}{v}$$

CHAPTER 4 GENERAL MOTION IN ONE DIMENSION

The equation of motion is then

$$\frac{P}{v} - kv = m\frac{dv}{dt}$$

$$\Rightarrow \int dt = \int \frac{mv}{P - kv^2}\, dv$$

$$\Rightarrow t = -\frac{m}{2k}\ln(P - kv^2) + C$$

When $t = 0$, $v = 0$, which gives $C = \frac{m}{2k}\ln P$. Hence, we have

$$t = \frac{m}{2k}\ln\left(\frac{P}{P - kv^2}\right)$$

Therefore, the time needed to reach a speed of V is $\frac{m}{2k}\ln\left(\frac{P}{P - kV^2}\right)$.

Note that this is only possible if $V < \sqrt{\frac{P}{k}}$.

Exercise 4B

1 A body of mass 1 kg moves along a straight line under the influence of a force, $F(t)$, which varies with time. When $t = 0$, the velocity of the body is v_0. In each of the following cases, find the relationship between v and t.

a) $F(t) = 3 - 4t$, $v_0 = 0$. **b)** $F(t) = 1 - \frac{1}{2(t+1)}$, $v_0 = 0$. **c)** $F(t) = 18\sin 3t$, $v_0 = -6$.

2 A body of mass 2 kg moves along a straight line under the influence of a force, $F(v)$, which varies with velocity. When $t = 0$, the velocity of the body is v_0. In each of the following cases, find the relationship between v and t.

a) $F(v) = \frac{1}{2v}$, $v_0 = 4$.

b) $F(v) = 3 - v$, $v_0 = 6$.

c) $F(v) = 9 - 4v^2$, $v_0 = 3$.

d) $F(v) = 12v - 6v^2$, $v_0 = 10$.

3 A body of mass 1 kg moves along a straight line under the influence of a force, F. When $t = 0$, the velocity of the body is v_0 and its displacement is s_0. In each of the following cases, find the relationship between v and s.

a) $F = \frac{1}{v}$, $v_0 = 4$, $s_0 = 0$.

b) $F = 3 - v$, $v_0 = 6$, $s_0 = 0$.

c) $F = 5 - 3v^2$, $v_0 = 3$, $s_0 = 0$.

d) $F = \frac{4}{s}$, $v_0 = 0$, $s_0 = 12$.

e) $F = \frac{6}{s^2}$, $v_0 = 0$, $s_0 = 24$.

EXERCISE 4B

4 A body of mass 2 kg moves along a straight line under the influence of a force F, which is given by $F = 5 - 2v$. At time $t = 0$, the body has velocity $v = 10\,\mathrm{m\,s^{-1}}$ and displacement $s = 0\,\mathrm{m}$.

  a) Find the relationship between v and t. When happens as t increases?

  b) By writing the acceleration as $v\dfrac{\mathrm{d}v}{\mathrm{d}s}$, show that
  $$2(v + s - 10) = 5\ln\left(\frac{15}{2v - 5}\right)$$

  c) Repeat part **a** for the case where the initial velocity is $1\,\mathrm{m\,s^{-1}}$.

5 A small electrically charged sphere, A, of mass m, is moving in a straight line towards another sphere B, which is identical to A but is fixed. The repulsive force between them is given by $F = \dfrac{k}{x^2}$, where k is a constant and x is the distance separating the spheres at time t. Initially, A is at a distance d from B and is moving with speed u.

  a) Write down the equation of motion of sphere A.
  b) Find the smallest distance between the spheres in the subsequent motion.

6 An object of mass 3 kg slides along a horizontal, smooth surface at a constant speed of $6\,\mathrm{m\,s^{-1}}$ until it hits a vertical wall. It remains in contact with the wall for 0.1 s, during which time the reaction force exerted by the wall on the object is $F(t) = 150\pi \sin 10\pi t$.

  a) Find the total impulse exerted on the object during this contact.
  b) Find the speed with which the object moves away from the wall.

7 A body of mass 0.1 kg is moving along a straight line through an origin O. Initially, its displacement from O is $s = 1\,\mathrm{m}$ and its velocity is $v = 12\,\mathrm{m\,s^{-1}}$. It is subject to a retarding force $F = -2s$.

  a) Find the work done by the retarding force as the body's displacement increases to 2 m.
  b) Find the velocity of the body at this point.
  c) Find the maximum displacement achieved by the body.

8 In the book *The Planiverse*, by A.K. Dewdney, the hero, Yndrd, lives on the rim of a circular, two-dimensional planet called Arde. Instead of the inverse square law of gravitation, which applies in our three-dimensional world, Ardean gravity obeys an inverse linear law. That is, an object of mass m which is at a distance of x from the centre of Arde is pulled downwards by a force $\dfrac{km}{x}$, where k is a constant (you might try to decide why this should be so). The radius of Arde is R.

  a) Find the work needed to raise an Ardean spaceship of mass m to a height of $3R$ above the planet's rim.
  b) Show that, unlike in three dimensions (see Example 8), no spaceship can escape from Arde's gravitational influence.

CHAPTER 4 GENERAL MOTION IN ONE DIMENSION

9 A vehicle of mass m is being driven by an engine with a constant power output, P. The vehicle starts from rest and moves along a horizontal road against a resistance, Rv, where R is a constant.

a) Write down the equation of motion and hence find an expression for the velocity at time t.

b) Show that the maximum speed of the vehicle is $\sqrt{\dfrac{P}{R}}$.

10 A truck of mass m travels along a horizontal road against a resistance of magnitude kv, where v is its velocity and k is a constant. Its displacement from its starting point is x at time t. The engine works at a constant rate of P. The maximum velocity the truck can attain is V.

a) Find an expression for V in terms of P and k.

b) Write down the equation of motion of the truck. Show that it can be rewritten in the form
$$P(V^2 - v^2) = mV^2 v^2 \frac{\mathrm{d}v}{\mathrm{d}x}$$

c) Show, by solving the given differential equation, that the distance travelled by the truck in reaching half its maximum speed is $\dfrac{mV^3}{2P}(\ln 3 - 1)$.

Examination questions

Chapters 1 to 4

Chapters 1 and 2

1. The coordinates of the points A, B and C are $(2, 1, 5)$, $(1, 0, 8)$ and $(-1, 4, 2)$ respectively
   a) Find $\overrightarrow{CA} \times \overrightarrow{BA}$.
   b) Hence find
      i) the area of triangle ABC
      ii) a cartesian equation of the plane ABC.

   D is the point with coordinates $(8, 2, 3)$.

   c) Find a vector equation of the line from D perpendicular to the plane.
   d) Find the coordinates of the point where the line in part c meets the plane ABC.

   Hence find the distance from D to the plane ABC. (AEB 99)

2. Three forces, $3\mathbf{i} + 2\mathbf{j}$, $-2\mathbf{i} + \mathbf{j}$ and $4\mathbf{i} - 8\mathbf{j}$, act at the points whose coordinates are $(1, 2)$, $(2, 1)$ and $(-3, 0)$ respectively. The three forces are equivalent to a single force, $\mathbf{F}$. Find
   a) the magnitude of $\mathbf{F}$
   b) a vector equation of the line of action of $\mathbf{F}$. (AEB 99)

3. A force of magnitude $4\,\text{N}$ acts on a lamina lying in the plane $x + y + z = 7$. The force acts through the points $2\mathbf{i} - \mathbf{j} + 2\mathbf{k}$ and $3\mathbf{i} + \mathbf{j} + \mathbf{k}$.
   a) Find an equation of the line of action of the force.
   b) Find the position vector of the point where the force acts on the lamina.
   c) Calculate the moment of the force about the origin. (AEB 97)

4. Two forces $(\mathbf{i} + 2\mathbf{j} - \mathbf{k})\,\text{N}$ and $(3\mathbf{i} - \mathbf{k})\,\text{N}$ act through a point O of a rigid body, which is also acted upon by a couple of moment $(\mathbf{i} + \mathbf{j} + 3\mathbf{k})\,\text{N m}$.
   a) Show that the couple and forces are equivalent to a single resultant force $\mathbf{F}$.
   b) Find a vector equation for the line of action of $\mathbf{F}$ in the form $\mathbf{r} = \mathbf{a} + \lambda \mathbf{b}$, where $\mathbf{a}$ and $\mathbf{b}$ are constant vectors and λ is a parameter. (EDEXCEL)

5. Two forces $\mathbf{F}_1$ and $\mathbf{F}_2$ act on a rigid body. $\mathbf{F}_1 = (21\mathbf{i} - 12\mathbf{j} + 12\mathbf{k})\,\text{N}$ and $\mathbf{F}_2 = (p\mathbf{i} + q\mathbf{j} + r\mathbf{k})\,\text{N}$, where p, q and r are constants. $\mathbf{F}_1$ acts through the point A with position vector $(3\mathbf{i} - 2\mathbf{j} + \mathbf{k})\,\text{m}$, relative to a fixed origin O. $\mathbf{F}_2$ acts through the point B with position vector $(\mathbf{i} + \mathbf{j} + \mathbf{k})\,\text{m}$ relative to O.

   The two forces $\mathbf{F}_1$ and $\mathbf{F}_2$ are equivalent to a single force $(25\mathbf{i} - 14\mathbf{j} + 12\mathbf{k})\,\text{N}$, acting through O, together with a couple $\mathbf{G}$.

   a) Find the values of p, q and r.
   b) Find the magnitude of $\mathbf{G}$. (EDEXCEL)

EXAMINATION QUESTIONS CHAPTERS 1 TO 4

6 Three forces $(-\mathbf{i} + 2\mathbf{j})\,\text{N}$, $(5\mathbf{i} + \mathbf{j} - 3\mathbf{k})\,\text{N}$ and $(-4\mathbf{i} - 3\mathbf{j} + 3\mathbf{k})\,\text{N}$ act on a rigid body at the points A, B and O respectively. The position vectors of A and B, relative to O, are $(2\mathbf{i} + \mathbf{k})\,\text{m}$ and $(\mathbf{i} + \mathbf{j})\,\text{m}$ respectively.

Show that the forces are equivalent to a couple and find the magnitude of this couple.

(EDEXCEL)

7 The forces $\mathbf{F}_1 = \mathbf{i} + 2\mathbf{j} + 3\mathbf{k}$ and $\mathbf{F}_2 = -\mathbf{i} + 2\mathbf{k}$ act on a rigid body at the points whose position vectors relative to the origin O are $\mathbf{r}_1 = \mathbf{j} + \mathbf{k}$ and $\mathbf{r}_2 = 4\mathbf{i} + \mathbf{j} - \mathbf{k}$ respectively. A third force $\mathbf{F}_3 = P\mathbf{i} + Q\mathbf{j} + R\mathbf{k}$, acting at the point with position vector $\mathbf{p} = \alpha\mathbf{i} + \beta\mathbf{j} + 5\mathbf{k}$, is then added so that

$$\mathbf{F}_1 + \mathbf{F}_2 + \mathbf{F}_3 = \mathbf{0}$$

and the sum of the moments of the three forces about O is equal to $-7\mathbf{i} + 9\mathbf{j} - 6\mathbf{k}$.

i) Show that $P = 0$ and find the values of Q and R. Are the three forces in equilibrium? Give a reason for your answer.
ii) Find the moments of $\mathbf{F}_1$ and $\mathbf{F}_2$ about the origin.
iii) Hence show that $\alpha = 3$ and find the value of β.
iv) Find the sum of the moments of the three forces about the point $3\mathbf{i} - \mathbf{j}$.
v) Show that a fourth force $\mathbf{F}_4$, acting at the point $3\mathbf{i} + 3\mathbf{j} + \mathbf{k}$, can be added to the above three forces so that the sum of the moments of the four forces about O is zero. Are these four forces then in equilibrium? Give a reason for your answer. (MEI)

8 A rigid body rotates about an axis fixed in space which passes through the point O of the body. Relative to a set of cartesian axes fixed in space with origin at O, the point A of the body has position vector $\mathbf{a} = -2\mathbf{i} + \mathbf{j} - \mathbf{k}$ and velocity $\mathbf{v}_A = -5\mathbf{i} - 5\mathbf{j} + 5\mathbf{k}$.

i) Verify that $\mathbf{a} \cdot \mathbf{v}_A = 0$ and give a physical reason why this must be true

A second point B of the body has position vector $\mathbf{b} = 3\mathbf{i} + 2\mathbf{k}$ and velocity $\mathbf{v}_B = -4\mathbf{i} + 3\mathbf{j} + 6\mathbf{k}$.

ii) Verify that $(\mathbf{a} - \mathbf{b}) \cdot (\mathbf{v}_A - \mathbf{v}_B) = 0$ and give a physical reason why this must also be true.

The angular velocity $\boldsymbol{\omega}$ is given by $\boldsymbol{\omega} = \omega_1 \mathbf{i} + \omega_2 \mathbf{j} + \omega_3 \mathbf{k}$.

iii) By considering $\boldsymbol{\omega} \times \mathbf{a}$ and $\boldsymbol{\omega} \times \mathbf{b}$, find the values of ω_1, ω_2 and ω_3.

In a second motion about another axis fixed in the body through the same origin O,

$$\mathbf{a} = 2\mathbf{i} + 4\mathbf{j} + \mathbf{k} \qquad \mathbf{b} = 3\mathbf{i} + \mathbf{k}$$
$$\mathbf{v}_A = \lambda\mathbf{i} - \mathbf{j} + 2\mathbf{k} \qquad \mathbf{v}_B = -3\mathbf{i} + \mu\mathbf{j} + \nu\mathbf{k}$$

where λ, μ, ν are constants.

iv) Find the values of λ, μ, and ν and hence find the angular velocity in this case. (MEI)

Chapter 3

9 The differential equation $\dfrac{\mathrm{d}y}{\mathrm{d}x} = \dfrac{1}{\sqrt{x^2 + y^2}}$ is to be investigated, firstly by means of a tangent field and then numerically.

i) Describe the isocline $\dfrac{\mathrm{d}y}{\mathrm{d}x} = m$, where m is a positive constant.
ii) Sketch (using the same axes) the isoclines given by $\dfrac{\mathrm{d}y}{\mathrm{d}x} = 0.25, 0.5, 1$ and 2. Use these to draw a tangent field for the equation.

iii) On your tangent field, sketch the solution curves through the points $(0, 1)$, $(1, 0)$ and $(0.5, 0.5)$.

Points on the solution curve through $(0, 1)$ are to be found numerically using Euler's method. The algorithm is given by $x_{n+1} = x_n + h$, $y_{n+1} = y_n + hf(x_n, y_n)$, where h is the step length and $f(x, y) = \dfrac{dy}{dx}$.

iv) Use a step length of $h = 0.25$ to calculate an approximation for $y(1)$, the value of y when $x = 1$.

v) The table shows values calculated for $y(1)$ for a succession of smaller step lengths.

h	Approximation for $y(1)$
0.05	1.7066
0.02	1.7006
0.01	1.6986

Assuming that the error in the approximation for $y(1)$ is proportional to h, use the last two entries to find a better estimate for $y(1)$. (MEI)

10 The current in an electrical circuit, consisting of an inductor, resistor and capacitor in series with an alternating power source, is described by the equation

$$\frac{d^2I}{dt^2} + 25\frac{dI}{dt} + 100I = -170 \sin 20t$$

where I is the current in amperes and t is the time in seconds after the power source is switched on.

i) Find the general solution.

When $t = 0$, $\dfrac{dI}{dt} = I = 0$.

ii) Find the solution.

The exponentially decaying terms in the solution describe what is known as the transient current. The non-decaying terms describe the steady-state current.

iii) Write down an expression for the steady-state current for the solution in part **ii**. Why would this expression remain unchanged if the initial conditions were different?

iv) Express the steady-state current in the form $R \sin(20t + \alpha)$, where R and α are to be determined. Verify that, after only 1 second, the magnitude of the transient current is close to 1% of the steady-state amplitude, R. (WJEC)

11 In a chemical process, compound P reacts with an abundant supply of a gas to form compound Q. The process is governed by two reaction rates, one for the forward reaction, and a lesser one for the reverse reaction, where Q decomposes into P and the gas. P is introduced into a reaction chamber at a constant rate of 21 kg per hour and Q is extracted at a rate proportional to the quantity of Q in the chamber. The equations which describe the process are

$$\frac{dp}{dt} = -5p + q + 21 \qquad \frac{dq}{dt} = -8q + 10p$$

where p and q are the quantities (in kg) of P and Q respectively and t is the time in hours.

i) Calculate the values of p and q for which $\dfrac{\mathrm{d}p}{\mathrm{d}t}$ and $\dfrac{\mathrm{d}q}{\mathrm{d}t}$ are both zero.

ii) Eliminate q to show that $\dfrac{\mathrm{d}^2 p}{\mathrm{d}t^2} + 13\dfrac{\mathrm{d}p}{\mathrm{d}t} + 30p = 168$.

iii) Find the general solution of the differential equation in part ii. Hence find the particular solutions for p and q for which $p = q = 0$ when $t = 0$.

iv) Sketch the graphs of the solutions, showing the significance of your answers to part i.

(MEI)

12 At time t seconds the position vector of a particle P relative to a fixed origin O is $\mathbf{r}$ metres and its velocity is $\mathbf{v}\,\mathrm{m\,s^{-1}}$, where

$$\dfrac{\mathrm{d}\mathbf{v}}{\mathrm{d}t} = -2\mathbf{v}$$

At time $t = 0$, $\mathbf{r} = \mathbf{i} + \mathbf{j}$ and $\mathbf{v} = -2\mathbf{i} - 4\mathbf{j}$.

Find $\mathbf{r}$ in terms of t. (EDEXCEL)

13 The position vector $\mathbf{r}$ metres of a particle relative to a fixed origin O, at time t seconds, satisfies the vector differential equation

$$\dfrac{\mathrm{d}^2 \mathbf{r}}{\mathrm{d}t^2} + 4\mathbf{r} = \mathbf{0}$$

Given that, when $t = 0$, the particle is at the point with position vector $2\mathbf{i}$ m and is moving with velocity $6\mathbf{j}\,\mathrm{m\,s^{-1}}$, find a cartesian equation for the path of the particle. (EDEXCEL)

Chapter 4

14 A spherical raindrop falls under gravity through a stationary cloud. Initially the drop is at rest and its radius is a. As it falls, water from the cloud condenses on the drop in such a way that the radius of the drop increases at a constant rate k. At time t, the speed of the drop is v.

a) Show that

$$(a + kt)\dfrac{\mathrm{d}v}{\mathrm{d}t} + 3kv = g(a + kt)$$

b) Hence show that, when the drop has doubled its radius, its speed is $\dfrac{15ga}{32k}$.

15 A raindrop, initially at rest, falls through a stationary cloud of water vapour, absorbing water from the cloud. At time t, the raindrop has mass m and downward speed v, and it has fallen a distance x. The rate of increase of mass is modelled by $\dfrac{\mathrm{d}m}{\mathrm{d}t} = kmv$, where k is a positive constant, and air resistance is ignored.

i) Derive the equation of motion for the raindrop, and hence show that $\dfrac{\mathrm{d}v}{\mathrm{d}t} + kv^2 = g$.

ii) Give a reason why the terminal speed of the raindrop is $\sqrt{\left(\dfrac{g}{k}\right)}$.

iii) Write down a differential equation relating v and x. Hence find x in terms of g, k and v.

iv) Given that the terminal speed is $18\,\mathrm{m\,s^{-1}}$, find the distance fallen by the raindrop when its speed is $17\,\mathrm{m\,s^{-1}}$. (OCR)

16 A hailstone falls under gravity in still air and as it falls its mass increases. Its initial mass is m_0. The rate of increase of its mass is proportional to its speed v.

a) Show that, when the hailstone has fallen a distance x, its mass m is given by

$$m = m_0(1 + \lambda x)$$

where λ is a constant.

Assuming that there is no air resistance,

b) show that

$$\frac{d}{dx}(v^2) + \frac{2\lambda}{1 + \lambda x}(v^2) = 2g$$

Given that $v = 0$ when $x = 0$,

c) find an expression for v^2 in terms of x, λ and g. (EDEXCEL)

17 A rocket initially has total mass M. It propels itself by its motor ejecting burnt fuel. When all of its fuel has been burned its mass is kM, $k < 1$. It is moving with speed U when its motor is started. The burnt fuel is ejected with constant speed c, relative to the rocket, in a direction opposite to that of the rocket's motion. Assuming that the only force acting on the rocket is that due to the motor, find the speed of the rocket when all of its fuel has been burned.
(EDEXCEL)

18 A starship is travelling in outer space (where external forces are negligible). It is slowing down by firing its rocket motors so that burnt fuel is ejected directly forwards with constant speed c relative to the starship. Initially the mass of the starship is m_0 and its speed is u. After time t, the mass is m and the speed is v. Derive the equation of motion

$$m\frac{dv}{dt} - c\frac{dm}{dt} = 0$$

Verify that the solution of this equation is $v - c \ln m = K$, where K is a constant, and hence show that the mass of the starship when it comes to rest is $m_0 e^{-u/c}$.

The starship comes to rest when $t = T$. Assuming that $m = m_0 e^{-\lambda t^2}$, where λ is a constant, express λ in terms of u, c and T, and show that the distance travelled by the starship while it is slowing down is $\frac{2}{3}uT$. (OCR)

19 An object of mass 1 kg falls vertically from rest. A resisting force of magnitude $2v$ N acts on the object during its descent, where v m s^{-1} is the speed of the object at time t seconds after it begins to fall. The object is modelled as a particle. Show that

$$v = 4.905(1 - e^{-2t})$$

Calculate the increase in the speed of the object during the fifth second of its descent.

Sketch the (t, v) graph for the motion of the object, showing clearly the behaviour of v as t becomes large. (OCR)

20 A particle of mass 0.2 kg moves along a horizontal straight line under the action of a propulsive force of magnitude e^{3t} N and a resistive force of magnitude $0.4v$ N, where v m s^{-1} is the speed of the particle at time t s.

a) Show that $\dfrac{dv}{dt} + 2v = 5e^{3t}$.

b) Find an integrating factor for this differential equation and hence find its general solution.
c) Given that the particle is at rest when $t = 0$, determine, correct to two decimal places, the value of v when $t = 1$.   (WJEC)

21 A ball of mass m is thrown vertically upwards. When the ball is at a height x above its initial position, its speed is v. The only forces acting on the ball are its weight and a resisting force of magnitude mkv, where k is a positive constant.

a) Show that, while the ball is moving upwards,
$$v\frac{dv}{dx} = -(g + kv)$$

The initial speed of the ball is $\dfrac{3g}{k}$.

b) Show that the maximum height of the ball above its initial position is
$$\frac{g}{k^2}(3 - \ln 4) \qquad \text{(EDEXCEL)}$$

5 Impulses and impulsive tensions

The bigger they come, the harder they fall.
ROBERT FITZSIMMONS

In this chapter, we look again at the instantaneous changes caused by collisions and impulsive tensions. In *Introducing Mechanics* (pages 275–300), we dealt with one-dimensional problems involving impacts or jerks along the line of centres of the objects, and considered some simple two-dimensional problems involving oblique impact. For the sake of completeness and coherence, some of what follows is a repetition of work done in *Introducing Mechanics*.

During a collision or when an inelastic string becomes instantaneously taut:

- momentum is conserved
- energy is not conserved (unless the collision is perfectly elastic).

Oblique impact

We know from Newton's third law that the contact forces acting on each body are equal and opposite, and hence that the impulses acting on each body are equal and opposite. The impulse, of course, acts normal to the plane of contact. These ideas were used to solve the problem of straight-line collisions and derive the law of conservation of momentum.

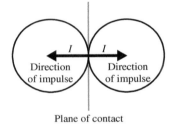

When the collision is not in a straight line, the problem is a two-dimensional one. Our choice of base directions (unit direction vectors) affects the simplicity of our equations. We usually choose the direction of the impact (the normal to the plane of contact) as one of the directions. The other must, of course, be perpendicular to this.

When we model these situations we will assume that:

- **Surfaces are smooth**.
  - There are no frictional forces acting along the plane of contact, i.e. perpendicular to the impulse. The important consequence of this is that there is no change in the linear momentum of a body in a direction perpendicular to an impulse.
  - There is no frictional force acting between the objects and the plane surface on which they are moving.

- **Objects are modelled as particles**.
  This avoids difficulties caused by rotation. Many of the shots made by professional snooker players depend upon rotational effects to achieve their objective.

- **Newton's law of restitution is valid in the direction of an impact**.

CHAPTER 5 IMPULSES AND IMPULSIVE TENSIONS

There are three basic problems to solve:
- A sphere hits a wall (the wall cannot move).
- A sphere strikes a second, stationary, sphere.
- Two moving spheres collide.

Example 1 A particle, of mass 1 kg, slides on the surface of a smooth table with a speed $2\,\text{m s}^{-1}$. It collides with a smooth vertical wall, striking the wall at an angle of $45°$ to its direction of motion. It then rebounds. The coefficient of restitution between the particle and the vertical surface is 0.6. Find

a) the velocity of the particle after impact
b) the kinetic energy lost during the collision.

SOLUTION

a)

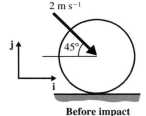
Before impact

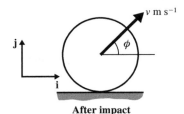
After impact

Let the velocity of the particle after impact be $v\,\text{m s}^{-1}$ in the direction making an angle ϕ with the wall, as shown in the diagram.

We choose unit direction vectors parallel and perpendicular to the direction of the impact, as shown.

Velocity before impact: $\mathbf{u} = 2\cos 45°\,\mathbf{i} - 2\sin 45°\,\mathbf{j} \quad \Rightarrow \quad \mathbf{u} = \sqrt{2}\,\mathbf{i} - \sqrt{2}\,\mathbf{j}$
Velocity after impact: $\mathbf{v} = v\cos\phi\,\mathbf{i} + v\sin\phi\,\mathbf{j}$

The wall is smooth and so the linear momentum parallel to it (the **i**-direction) will be conserved. So, we have

$$\sqrt{2} = v\cos\phi \qquad [1]$$

In the **j**-direction (perpendicular to the wall), Newton's law of restitution applies. Hence, we have

Separation speed $= e \times$ Approach speed
$$v\sin\phi = e \times \sqrt{2} = 0.6 \times \sqrt{2} \qquad [2]$$

Dividing [2] by [1], we get

$$\tan\phi = 0.6 \quad \Rightarrow \quad \phi = 30.96°$$

Squaring and adding [1] and [2], we get

$$v^2 = 2 + 2 \times 0.36 = 2.72$$
$$\Rightarrow \quad v = 1.6492$$

Hence, after impact, the particle has a speed of $1.65\,\text{m s}^{-1}$, making an angle of $31.0°$ with the wall (both results correct to 3 sf).

b) Before impact: $KE = \frac{1}{2} \times 1 \times 2^2 = 2\,J$

After impact: $KE = \frac{1}{2} \times 1 \times 2.72 = 1.36\,J$

So, the kinetic energy lost during impact is $(2 - 1.36) = 0.64\,J$.

Example 2 A small sphere, A, of mass 1 kg, moving with a speed of $4\,m\,s^{-1}$, collides with a second, identical, sphere, B, which is initially at rest. At the point of contact, the line joining their centres makes an angle of $60°$ with the direction of motion of sphere A. If the coefficient of restitution between the spheres is 0.5, what are the velocities of the two spheres after the collision?

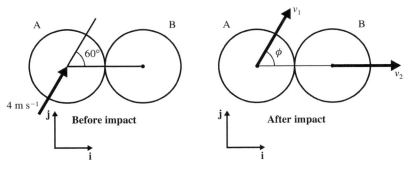

SOLUTION

The impact between the spheres is along the line of centres and this is chosen as one of the base directions. The second base direction is perpendicular to this, as shown.

The second sphere, B, must move in the direction of the impulse acting on it, i.e. along the line of centres.

Before impact, we have

Velocity of A: $\mathbf{u}_1 = 4\cos 60°\mathbf{i} + 4\sin 60°\mathbf{j} = 2\mathbf{i} + 2\sqrt{3}\mathbf{j}$

Velocity of B: $\mathbf{u}_2 = 0$

After impact, we have

Velocity of A: $\mathbf{v}_1 = v_1 \cos\phi\,\mathbf{i} + v_1 \sin\phi\,\mathbf{j}$

Velocity of B: $\mathbf{v}_2 = v_2\,\mathbf{i}$

Momentum is conserved in the **i**-direction. There is no impulse in the **j**-direction and the velocity components in this direction remain unchanged. Hence, we have

In the **i**-direction: $2 = v_1 \cos\phi + v_2$ [1]

In the **j**-direction: $2\sqrt{3} = v_1 \sin\phi$ [2]

From Newton's law of restitution, we get

$v_2 - v_1 \cos\phi = e \times 2 = 1$ [3]

We now have the three independent equations needed to find the unknowns v_1, v_2 and ϕ.

From [1] and [3], we have
$$2v_2 = 3 \quad \Rightarrow \quad v_2 = 1.5$$
Substituting into [1], we obtain
$$v_1 \cos \phi = 2 - v_2 = 0.5 \qquad [4]$$
Dividing [2] by [4], we get
$$\tan \phi = 4\sqrt{3} \quad \Rightarrow \quad \phi = 81.787°$$
Squaring and adding [2] and [4], we have
$$v_1^2 = (2\sqrt{3})^2 + (0.5)^2 = 12.25$$
$$\Rightarrow \quad v_1 = 3.5$$

Hence, after the collision, sphere A moves with a speed of $3.5 \,\text{m s}^{-1}$ in a direction making an angle of $81.8°$ with the line of centres, and sphere B moves along the line of centres with a speed of $1.5 \,\text{m s}^{-1}$.

Example 3 Two identical small spheres of mass 1 kg collide. The first, A, moves with a speed of $6 \,\text{m s}^{-1}$ and the second, B, with a speed of $8 \,\text{m s}^{-1}$. Their directions of motion make angles of $30°$ and $45°$ respectively with their line of centres at the instant of contact, as shown. If the coefficient of restitution between the spheres is 0.4, find

a) the velocities of the spheres after the collision
b) the kinetic energy lost during the collision.

SOLUTION

a

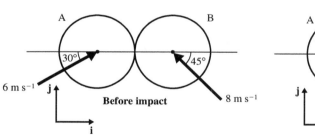
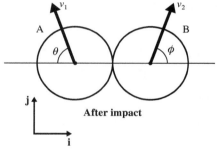

We will assume that the speeds and directions of the spheres after the collision are as shown in the figure. The base directions are as shown.

Before impact, we have

Velocity of A: $\mathbf{u}_1 = 6\cos 30° \mathbf{i} + 6\sin 30° \mathbf{j} = 3\sqrt{3}\mathbf{i} + 3\mathbf{j}$
Velocity of B: $\mathbf{u}_2 = -8\cos 45° \mathbf{i} + 8\sin 45° \mathbf{j} = -4\sqrt{2}\mathbf{i} + 4\sqrt{2}\mathbf{j}$

After impact, we have

Velocity of A: $\mathbf{v}_1 = -v_1 \cos \theta \mathbf{i} + v_1 \sin \theta \mathbf{j}$
Velocity of B: $\mathbf{v}_2 = v_2 \cos \phi \mathbf{i} + v_2 \sin \phi \mathbf{j}$

There is no impulse on either sphere in the **j**-direction, so the velocity components in this direction remain unchanged. Hence, we have

For A: $\quad 3 = v_1 \sin \theta \quad$ [1]

For B: $\quad 4\sqrt{2} = v_2 \sin \phi \quad$ [2]

In the **i**-direction, linear momentum is conserved, so we have

$$3\sqrt{3} + (-4\sqrt{2}) = -v_1 \cos \theta + v_2 \cos \phi$$
$$\Rightarrow \quad 3\sqrt{3} - 4\sqrt{2} = -v_1 \cos \theta + v_2 \cos \phi \quad [3]$$

Also in the **i**-direction, Newton's law of restitution applies, so we have

$$v_2 \cos \phi - (-v_1 \cos \theta) = 0.4\left(3\sqrt{3} - (-4\sqrt{2})\right)$$
$$\Rightarrow \quad v_2 \cos \phi + v_1 \cos \theta = 1.2\sqrt{3} + 1.6\sqrt{2} \quad [4]$$

We now have the four independent equations needed to find the unknowns v_1, v_2, θ and ϕ.

Adding [3] and [4], we obtain

$$2v_2 \cos \phi = 4.2\sqrt{3} - 2.4\sqrt{2}$$
$$\Rightarrow \quad v_2 \cos \phi = 2.1\sqrt{3} - 1.2\sqrt{2} \quad [5]$$

Subtracting [3] from [4], we obtain

$$2v_1 \cos \theta = -1.8\sqrt{3} + 5.6\sqrt{2}$$
$$\Rightarrow \quad v_1 \cos \theta = -0.9\sqrt{3} + 2.8\sqrt{2} \quad [6]$$

Dividing [1] by [6], we obtain

$$\tan \theta = \frac{3}{-0.9\sqrt{3} + 2.8\sqrt{2}} \quad \Rightarrow \quad \theta = 51.329°$$

Squaring and adding [1] and [6], we have

$$v_1^2 = 3^2 + (-0.9\sqrt{3} + 2.8\sqrt{2})^2$$
$$\Rightarrow \quad v_1 = 3.8425$$

Dividing [2] by [5], we obtain

$$\tan \phi = \frac{4\sqrt{2}}{2.1\sqrt{3} - 1.2\sqrt{2}} \quad \Rightarrow \quad \phi = 71.068°$$

Squaring and adding [2] and [5], we obtain

$$v_2^2 = (4\sqrt{2})^2 + (2.1\sqrt{3} - 1.2\sqrt{2})^2$$
$$\Rightarrow \quad v_2 = 5.9803$$

Hence, after the collision, A has a speed of $3.84 \, \text{m s}^{-1}$ and moves along a line making an angle of $51.3°$ with the line of centres. B has a speed of $5.98 \, \text{m s}^{-1}$ and moves along a line making an angle of $71.1°$ with the line of centres. The angles are measured as shown in the diagram.

CHAPTER 5 IMPULSES AND IMPULSIVE TENSIONS

b) Before impact, we have

$$\text{Total KE} = \tfrac{1}{2} \times 1 \times 6^2 + \tfrac{1}{2} \times 1 \times 8^2 = 50\,\text{J}$$

After impact, we have

$$\text{Total KE} = \tfrac{1}{2} \times 1 \times 14.76 + \tfrac{1}{2} \times 1 \times 35.76 = 25.26\,\text{J}$$

Thus the KE lost during the collision is $(50 - 25.26) = 24.7\,\text{J}$ (to 3 sf).

Exercise 5A

1 A snooker ball, moving with a speed of $12\,\text{m s}^{-1}$, collides with the cushion at an angle of $25°$. If the coefficient of restitution between the ball and cushion is 0.7, what is the velocity of the ball after rebounding?

2 A snooker ball, of mass $0.2\,\text{kg}$, hits the cushion with a speed of $10\,\text{m s}^{-1}$ so that its direction of motion makes an angle of $60°$ with the cushion. If the coefficient of restitution is 0.7, how much kinetic energy is lost during the collision?

3 A ball of mass $0.5\,\text{kg}$ falls from a height of $2\,\text{m}$ onto a surface inclined at an angle $30°$ to the horizontal. If the coefficient of restitution between the ball and the surface is 0.8, what is the velocity of the ball when it rebounds?

4 A ball, A, is projected along a smooth, horizontal surface towards a second, identical ball, B, which is at rest. The speed of A is $4\,\text{m s}^{-1}$ and it moves in a direction making an angle of $30°$ with the line of centres. After the collision, B has a speed of $2\,\text{m s}^{-1}$. Find

   a) value of the coefficient of restitution
   b) the velocity of A after the impact.

5 Two smooth spheres of equal radii collide. The mass of A is $3\,\text{kg}$ and that of B is $2\,\text{kg}$. Sphere A has a velocity of $2\mathbf{i} + 3\mathbf{j}$ whilst B has a velocity of $-3\mathbf{i} - 4\mathbf{j}$, where the **i**-direction is along the line of centres from A to B at the moment of impact. The coefficient of restitution between the two spheres is 0.8. Find

   a) the velocities of the spheres after the impact
   b) the kinetic energy lost during the collision.

6 An object slides along smooth, horizontal ground towards a wall. It hits the wall and rebounds so that it is now moving perpendicular to its original direction of motion. If it were originally moving at an angle of $60°$ to the wall, find the value of the coefficient of restitution between the object and the wall.

7 A small sphere, A, moving in the direction of $3\mathbf{i} + 4\mathbf{j}$, collides with a second, identical sphere, B, which is at rest. After the collision, B moves in the **i**-direction whilst A moves in the direction of $3\mathbf{i} + 16\mathbf{j}$. Show that this can happen whatever the initial speed of A and find the coefficient of restitution between the balls.

8 Two identical balls are moving towards each other, each with a speed of $4\,\text{m s}^{-1}$ but the lines along which they are moving are parallel and separated by the radius of the balls. If the coefficient of restitution between the balls is $\tfrac{3}{4}$, find the speed of each ball after the impact and the angle through which the path of each has been deflected.

9 A ball, A, of mass 1 kg, collides obliquely with a second ball, B, of mass m kg, which is at rest. The velocity of A before the collision is $(a\mathbf{i} + b\mathbf{j})$, where $a > 0$. After the collision, the velocity of A is $(c\mathbf{i} + d\mathbf{j})$ whilst that of B is $w\mathbf{i}$. The coefficient of restitution between the balls is e.

   a) Show that c will be negative if $m > \dfrac{1}{e}$.

   b) Find w if $m = \dfrac{1}{e}$.

10 A billiard ball hits the cushion at an angle of α to the table edge. After the collision, it rebounds from the edge at an angle β to the table edge. If the coefficient of restitution between the ball and the table cushion is e, show that $\tan \beta = e \tan \alpha$.

11 A ball, A, collides with a second identical ball, B, which is at rest. The direction of A makes an angle of $30°$ with the line of centres at the instant of collision. If the coefficient of restitution between the balls is e, show that A is deflected by the collision through an angle θ, where

$$\tan \theta = \frac{(1+e)\sqrt{3}}{5 - 3e}$$

12 A sphere collides obliquely with a second, identical sphere. The collision is perfectly elastic. Show that

   a) after the collision, the two spheres move in directions perpendicular to each other
   b) no energy is lost during the collision.

13 A billiard ball, of mass m kg, slides on the surface of a smooth table with a speed u m s^{-1}. It hits the cushion, its direction of motion making an angle θ with the edge of the table, and rebounds. If the coefficient of restitution between the ball and the cushion is e, find the velocity of the ball after impact.

14 A small sphere, A, moving with a speed of u, collides with a second, identical sphere, B. B is initially at rest and, at the point of contact, the line joining their centres makes an angle θ with the direction of motion of A. If the coefficient of restitution between the spheres is e, what are the velocities of the two spheres after the collision?

Impulsive tensions

When a string suddenly becomes taut or an impulse is applied to one end of a taut string, there is a tension induced within it as it resists stretching.

This tension creates an impulse which acts on the objects at each end of the string. We refer to these as **impulsive tensions**. Note that they are not tensions in the sense of forces, as they have the same dimensions as impulse and their unit is N s.

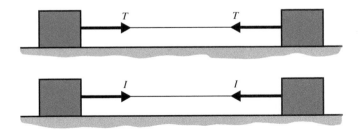

CHAPTER 5 IMPULSES AND IMPULSIVE TENSIONS

The other property that is used in the solution of such problems is that if a string is inextensible, both ends must move with the same speed in the direction of the string.

We have already explored impulsive tensions in one dimension in *Introducing Mechanics* (pages 284–5). We now examine some more complex situations.

Example 4 A block, A, of mass 3 kg, rests on the ground. It is connected to a light rope which passes over a smooth pulley and which has another block, B, of mass 2 kg, attached to its other end. The block B is supported so that the rope is slack. B is dropped and falls for 2 m before the rope becomes taut. Find the speed of the blocks immediately after the rope becomes taut and the impulsive tension in the rope.

SOLUTION

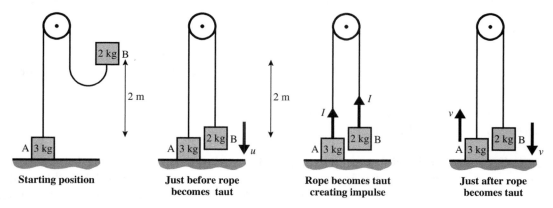

| Starting position | Just before rope becomes taut | Rope becomes taut creating impulse | Just after rope becomes taut |

When the rope becomes taut, block B has fallen a distance of 2 m. At this point, its speed will be given by

$$u^2 = 2g \times 2 = 4g$$

Also at this point, it will be subject to an impulse, I, from the rope.

From the impulse/momentum equation for block B, we have

$$-I = 2v - 2(2\sqrt{g}) \qquad [1] \quad \text{(positive downwards)}$$

There will be an equal but opposite impulse acting on block A.

For block A, we have

$$I = 3v \qquad [2] \quad \text{(positive upwards)}$$

Adding [1] and [2], we get

$$0 = 5v - 4\sqrt{g} \quad \Rightarrow \quad v = \frac{4\sqrt{g}}{5}$$

Substituting this into [2], we have

$$I = \frac{12\sqrt{g}}{5}$$

Hence, the speed of the blocks immediately after the rope becomes taut is $\frac{4\sqrt{g}}{5}$ m s^{-1}, and the impulsive tension in the rope is $\frac{12\sqrt{g}}{5}$ N s.

Example 5 Three particles, A, B and C, are connected by two inextensible strings, as shown. The strings are just taut so that the tension in each is zero. A has a mass of 2 kg, B has a mass of 3 kg and C has a mass of 5 kg. Particle C is given an impulse of 10 N s in the direction BC. What are the velocities of the three particles immediately after the impulse?

SOLUTION

The impulse given to C will induce an impulsive tension, I, in string BC and this, in turn, will induce an impulsive tension, J, in string AB.

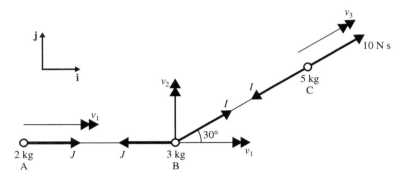

Let the speeds of the particles A and C be v_1 and v_3 respectively. The direction of v_1 must be along the string AB, since the impulse is in this direction. Similarly, the direction of v_3 must be in the direction BC.

B has velocity components parallel and perpendicular to AB. Its component in the direction AB must be v_1, since string AB cannot be extended. If we choose unit vectors **i** and **j** as in the diagram, the velocity of B after the impulse can be expressed as $\mathbf{v} = v_1\mathbf{i} + v_2\mathbf{j}$.

The impulse/momentum equation for the particles are:

$$\text{For A:} \qquad J = 2v_1 \qquad [1]$$

$$\text{For B, in the } \mathbf{i}\text{-direction:} \quad I\cos 30° - J = 3v_1$$
$$\Rightarrow \quad I\sqrt{3} - 2J = 6v_1 \qquad [2]$$

$$\text{in the } \mathbf{j}\text{-direction:} \quad I\sin 30° = 3v_2$$
$$\Rightarrow \quad I = 6v_2 \qquad [3]$$

$$\text{For C:} \qquad 10 - I = 5v_3 \qquad [4]$$

We now have four independent equations in five variables. Thus we need one more equation. Our final equation is obtained by realising that the component of the velocity of B in the direction of BC must be v_3. Hence, we have

$$v_1 \cos 30° + v_2 \sin 30° = v_3$$
$$\Rightarrow \quad v_1\sqrt{3} + v_2 = 2v_3 \qquad [5]$$

We now solve equations [1] to [5].

From [1] and [2], we have

$$I\sqrt{3} = 10v_1 \qquad [6]$$

From [3], [4], [5] and [6], we have

$$\frac{3I}{10} + \frac{I}{6} = 4 - \frac{2I}{5}$$

$$\Rightarrow \quad I = \frac{60}{13}$$

Hence, we obtain:

From [6]: $v_1 = \dfrac{6\sqrt{3}}{13}$

From [3]: $v_2 = \dfrac{10}{13}$

From [4]: $v_3 = \dfrac{14}{13}$

From [1]: $J = \dfrac{12\sqrt{3}}{13}$

Thus, the impulsive tension in AB is $\dfrac{12\sqrt{3}}{13}$ N s, the impulsive tension in BC is $\dfrac{60}{13}$ N s, the velocity of A is $\dfrac{6\sqrt{3}}{13}\mathbf{i}$ m s^{-1}, of B is $\left(\dfrac{6\sqrt{3}}{13}\mathbf{i} + \dfrac{10}{13}\mathbf{j}\right)$ m s^{-1} and of C is $\left(\dfrac{7\sqrt{3}}{13}\mathbf{i} + \dfrac{7}{13}\mathbf{j}\right)$ m s^{-1}.

Note These velocities are instantaneous only. Once the particles start to move, string AB no longer lies in the **i**-direction, so the force acting on A now has a different direction from the original impulse. Modelling the motion after the initial impulse is quite complicated, leading to a pair of simultaneous differential equations.

Exercise 5B

1 Two blocks, A, of mass 5 kg, and B, of mass 3 kg, are fastened to the ends of a rope which passes over a pulley. Block A rests on a table vertically below the pulley, whilst B is held above the table with the rope slack. B then falls from rest a distance of 1 m, at which point the rope becomes taut in a vertical position.

   a) What will be the speeds of the blocks immediately after this happens?
   b) What modelling assumptions are involved in your solution?

2 Two identical blocks, A and B, each of mass 2 kg, are fastened to the ends of a rope 2 m long. Block A rests on a smooth table, 1 m from the edge. The rope passes over a pulley at the edge of the table and block B is held at the height of the tabletop, as shown. B is then released.

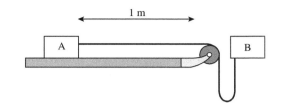

How long does it take for A to reach the edge of the table, assuming that B does not, in the meantime, reach the floor?

3 Three identical particles, A, B and C, each of mass 3 kg, are at rest on a smooth, horizontal surface. They are connected by light, inextensible strings AB and BC, and are positioned so that the strings are just taut and angle ABC is 135°. Particle C then receives an impulse of 20 N s in the direction BC. Find the initial velocities of the particles and the impulsive tensions in the strings.

4 An object of mass $3m$ is connected by means of a light, inextensible string to a scale pan of mass m. The object rests on the ground. The string passes over a smooth pulley and the scale pan hangs suspended. An object of mass m, falling from rest a distance h above the scale pan, lands on it and does not bounce. Show that the mass of $3m$ will rise to a height of $\dfrac{h}{5}$.

5 Particles A and B, each of mass m, are connected together by means of a light, inextensible string of length $2a$. They are at rest a distance a apart on a smooth horizontal plane. A is then projected with speed u along the plane at right angles to AB. Find the velocities of the particles immediately after the string becomes taut.

6 Particles A, B and C, each of mass 2 kg, are connected together by light, inextensible strings AB and BC, each of length 2 m. The particles lie at rest and in a straight line ABC on a smooth, horizontal plane, so that AB = 2 m and BC = 1 m. Particle C is then projected along the plane at 15 ms^{-1} in a direction perpendicular to the line ABC. Find the impulsive tensions in the strings at the moment when BC becomes taut, and find the velocity with which A starts to move.

7 Three identical particles, A, B and C, each of mass 2 kg, are at rest on a smooth, horizontal surface. They are connected by light, inextensible strings AB and BC, and are positioned so that the strings are just taut and angle ABC is 135°. Particle C then receives an impulse of 35 N s in the direction AB. Find the impulsive tensions in the strings and the speed with which A starts to move.

8 A particle A, of mass 3 kg, is connected by light, inextensible strings to particles B and C, of masses 5 kg and 1 kg respectively. They lie at rest on a smooth horizontal plane, with the strings just taut and angle BAC = θ. A then receives an impulse directed away from B and C along the bisector of the angle BAC. Show that A starts to move at an angle α to the direction of the impulse, where

$$\tan \alpha = \frac{2 \sin \theta}{6 - 3 \cos \theta}$$

9 ABCD is a square formed by four inextensible strings connecting particles A, B and C, each of mass m, and particle D of mass $2m$. The particles lie at rest on a smooth, horizontal surface with the strings just taut. Equal and opposite impulses of magnitude J are applied simultaneously to the two particles at opposite ends of a diagonal in the direction of the diagonal, so that the system is jerked into motion. Show that the initial motion of the particles is the same whichever pair of particles receive the impulses, and find the kinetic energy of the system after the impulses.

10 Particles A, B and C, each of mass m, lie at rest on a smooth, horizontal surface so that $AB = BC = a$ and angle $ABC = 90°$. A is connected to B, and B to C, by light, inextensible strings each of length $2a$. Particle C is set in motion with velocity u in the direction BC. Find the velocities of the particles immediately after AB has become taut.

6 Oscillating systems

*It don't mean a thing
If it ain't got that swing.*
IRVING MILLS

Damped oscillations

You have already studied the mathematics of simple oscillating systems, such as a mass on the end of a spring oscillating along a vertical line or a pendulum bob swinging from side to side. (You may also have studied simple cases of damped oscillations, and some of what follows also appears in *Introducing Mechanics*, pages 411–19.)

Simple spring or pendulum systems, with suitable modelling assumptions, lead to equations of motion of the form

$$\frac{d^2x}{dt^2} + \omega^2 x = 0$$

This is the familiar equation of simple harmonic motion, which has a general solution

$$x = P\cos\omega t + Q\sin\omega t$$

or $\quad x = A\cos(\omega t + \varepsilon)$

where P, Q, A and ε are arbitrary constants.

The most important modelling assumption made in cases such as those mentioned above is that no energy is lost during the motion. In practice, energy will be lost due to

- air resistance, and
- friction – both internal and external

If, for example, we clamp a metre rule to a table, allowing a large portion of its length to overhang the edge, as shown, then move the free end away from its equilibrium position and release it, the rule will perform oscillations. However, we will find that the oscillatory motion will quickly die down and cease.

There are situations in which we want the oscillations to die down, and steps are taken to ensure that this happens quickly. For example, when a car goes over a hump, it would clearly be undesirable for it to continue bouncing up and down for a significant time afterwards. The suspension system in a car therefore consists of two parts.

- For each wheel there is a spring designed to absorb the initial shock of the bump from the road. This will react in exactly the same way as a spring with a mass on its end.
- For each spring there is a shock absorber whose purpose is to prevent the spring from vibrating for too long. This is called **damping**, and so shock absorbers are also called suspension dampers.

We can represent a spring and damper system diagramatically, as shown on the right.

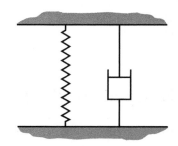

The shock absorber is an example of a **dashpot**. We can imagine it to be like a bicycle pump with holes in the piston allowing the fluid (gas or liquid) to pass through, but not too easily. (See diagram below right.)

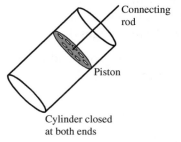

Connecting rod
Piston
Cylinder closed at both ends

Modelling a perfect dashpot

In order to model a dashpot, we assume the following properties of the perfect dashpot:

- The resistive force produced by the dashpot does not depend upon the position of the piston.
- The resistive force produced will at all times oppose the motion of the piston relative to the cylinder.
- The resistive force produced is proportional to the velocity of the piston relative to the cylinder.

Note The first two assumptions are drawn from practical observation whilst the third is the simplest we can make. This is **linear damping**.

From these assumptions, we can find an expression for the resistive force produced by the dashpot:

$$\mathbf{R} = -r\mathbf{v}$$

where r is the constant of proportionality – the **damping constant** – and $\mathbf{v}$ is the velocity of the piston relative to the cylinder.

Alternatively, we can write the resistive force as

$$\mathbf{R} = -r\frac{dx}{dt}$$

(We will meet this again in pages 145–52 in relation to the linear model for air resistance.)

Let us consider the example of a particle of mass m, suspended from a fixed point O by a perfect spring of stiffness k and natural length l_o. The particle is also attached to a perfect dashpot providing linear damping whose damping constant is r.

First, we will find the equilibrium position of the particle. Then we will derive the equation of motion of the particle when its displacement, x, is measured from that position.

We make the following three assumptions:

- The mass is a particle.
- The spring is light and provides no resistance.
- The dashpot is perfect.

The displacement, x, is measured from the equilibrium position, E. This provides the simplest algebra.

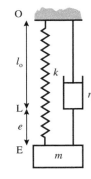

The forces on the particle are the tension, T, in the spring, the resistance, R, of the dashpot and the weight, mg, of the particle.

First, we consider the mass hanging at rest in the equilibrium position. Taking downwards as positive, and using Hooke's law, we have

$$T = -ke$$

As the particle is stationary, we have $R = 0$. So, resolving vertically, we obtain

$$mg - ke = 0 \qquad [1]$$

$$\Rightarrow \quad e = \frac{mg}{k}$$

Next, we consider the particle at P, with displacement x below E. We have

$$T = -k(e + x) \quad \text{and} \quad R = -r\frac{dx}{dt}$$

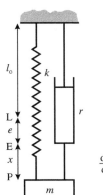

The equation of motion is

$$mg - k(e + x) - r\frac{dx}{dt} = m\frac{d^2x}{dt^2}$$

Substituting from [1] and rearranging, we obtain

$$m\frac{d^2x}{dt^2} + r\frac{dx}{dt} + kx = 0 \qquad [2]$$

This is a second-order homogeneous differential equation.

Note where the three terms of this equation come from:

- $m\dfrac{d^2x}{dt^2}$ is the mass–acceleration term from Newton's second law of motion.

- $r\dfrac{dx}{dt}$ is the term generated by the damping effects of the dashpot.

- kx is the tension in the spring above that necessary for equilibrium.

CHAPTER 6 OSCILLATING SYSTEMS

Solving the mathematical model

In order to solve equation [2],
$$m\frac{d^2x}{dt^2} + r\frac{dx}{dt} + kx = 0$$
we solve the auxiliary equation $m\lambda^2 + r\lambda + k = 0$ (see pages 56–60). The roots of this are
$$\lambda = \frac{-r \pm \sqrt{r^2 - 4mk}}{2m}$$
We will call the two roots λ_1 and λ_2. Where it helps to simplify the expressions, we will write
$$\rho = -\frac{r}{2m} \quad \text{and} \quad \Omega = \frac{\sqrt{4mk - r^2}}{2m}$$
We have three cases to consider:
- $r^2 - 4mk > 0$: The general solution has the form $x = Ae^{\lambda_1 t} + Be^{\lambda_2 t}$
- $r^2 - 4mk = 0$: The general solution has the form $x = e^{\rho t}(A + Bt)$
- $r^2 - 4mk < 0$: The general solution has the form
$$x = e^{\rho t}(A\cos\Omega t + B\sin\Omega t) \quad \text{or} \quad x = Re^{\rho t}\cos(\Omega t + \phi)$$

where A, B, R and ϕ are arbitrary constants.

Interpreting the solution

There is a spreadsheet DAMPEDHM available on the Oxford University Press website (http://www.oup.co.uk/mechanics) with which to investigate the solutions to this equation.

You may like to try the following suggestions.

- With $m = 2$, $k = 2$, $A = 1$ and $B = 0$, investigate the effects on the solution graph of changing the parameter r. In particular, what happens when $r = 4$?
- With $m = 2$, $k = 2$, $A = 4$ and $B = -2$, investigate the effects on the solution graph of changing the parameter r. In particular, what happens when $r = 4$?
- Try other combinations of the parameters.
- Try $r = 0$, which corresponds to undamped motion. The oscillations should continue for ever.

We can draw the following conclusions about the different forms of the general solution.

- $r^2 - 4mk > 0$

The general solution has the form $x = Ae^{\lambda_1 t} + Be^{\lambda_2 t}$. This is the condition for **strong damping**.

When $x = Ae^{\lambda_1 t} + Be^{\lambda_2 t}$, we have
$$\frac{dx}{dt} = \lambda_1 A e^{\lambda_1 t} + \lambda_2 B e^{\lambda_2 t}$$

λ_1 and λ_2 are both negative, so we have a combination of diminishing exponential terms. Hence, as $t \to \infty$, $x \to 0$ and $\dfrac{dx}{dt} \to 0$.

When $t = 0$, we have

$$x = A + B \quad \text{and} \quad \frac{dx}{dt} = \lambda_1 A + \lambda_2 B$$

If we have a given set of initial conditions, we can solve these equations for A and B to find the particular solution.

The graph of x against t crosses the t-axis when $x = 0$ and has a turning point when $\dfrac{dx}{dt} = 0$.

If $x = 0$, we have

$$A e^{\lambda_1 t} = -B e^{\lambda_2 t}$$

$$\Rightarrow \quad \frac{e^{\lambda_1 t}}{e^{\lambda_2 t}} = -\frac{B}{A}$$

$$\Rightarrow \quad t = \frac{1}{\lambda_1 - \lambda_2} \ln\left(\frac{-B}{A}\right) \qquad [3]$$

If $\dfrac{dx}{dt} = 0$, we have

$$\lambda_1 A e^{\lambda_1 t} = -\lambda_2 B e^{\lambda_2 t}$$

$$\Rightarrow \quad \frac{e^{\lambda_1 t}}{e^{\lambda_2 t}} = -\frac{\lambda_2 B}{\lambda_1 A}$$

$$\Rightarrow \quad t = \frac{1}{\lambda_1 - \lambda_2} \ln\left(\frac{-\lambda_2 B}{\lambda_1 A}\right) \qquad [4]$$

If A and B are of opposite sign and $A \neq 0$, each of equations [3] and [4] yields a unique value of t (which may be negative). Otherwise, neither equation has a real root. This means that in strong damping, the system does not perform a complete oscillation, but approaches the equilibrium position asymptotically, having passed through it at most once. Two examples of strong damping are shown in the figure below.

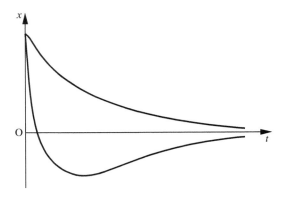

CHAPTER 6 OSCILLATING SYSTEMS

- $r^2 - 4mk = 0$

The general solution has the form $x = e^{\rho t}(A + Bt)$, giving

$$\frac{dx}{dt} = e^{\rho t}(\rho A + B + \rho B t)$$

This is the condition for **critical damping**. Since ρ is negative, we have a diminishing exponential term. Hence, as $t \to \infty$, $x \to 0$ and $\frac{dx}{dt} \to 0$.

When $t = 0$, we have

$$x = A \quad \text{and} \quad \frac{dx}{dt} = \rho A + B$$

If we have a given set of initial conditions, we can solve these equations for A and B to find the particular solution.

The graph of x against t crosses the t-axis when $x = 0$ and has a turning point when $\frac{dx}{dt} = 0$.

If $x = 0$, we have

$$A = -Bt$$

$$\Rightarrow \quad t = -\frac{A}{B} \qquad [5]$$

If $\frac{dx}{dt} = 0$, we have

$$\rho A + B = -\rho B t$$

$$\Rightarrow \quad t = -\frac{A}{B} - \frac{1}{\rho} \qquad [6]$$

If A and B are of opposite sign and $B \neq 0$, each of equations [5] and [6] yields a unique value of t. Otherwise, equation [5] has a single negative root and equation [6] has a single root which may be positive or negative. This means that in critical damping, as in strong damping, the system does not perform a complete oscillation, but approaches the equilibrium position asymptotically, having passed through it at most once. The importance of critical damping is that the system approaches the equilibrium position at the fastest possible rate. Three examples of critical damping are shown in the figure below.

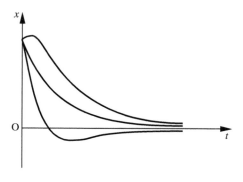

- $r^2 - 4mk < 0$

The general solution has the form $x = Re^{\rho t} \cos(\Omega t + \phi)$, giving

$$\frac{dx}{dt} = Re^{\rho t}[\rho \cos(\Omega t + \phi) - \Omega \sin(\Omega t + \phi)]$$

This is the condition for **weak damping**. The motion will therefore be oscillatory, but as $e^{\rho t}$ is a diminishing exponential, the amplitude of the oscillation will diminish with time. That is, as $t \to \infty$, $x \to 0$.

When $t = 0$, we have

$$x = R \cos\phi \quad \text{and} \quad \frac{dx}{dt} = R(\rho \cos\phi - \Omega \sin\phi)$$

If we have a given set of initial conditions, we can solve these equations for R and ϕ to find the particular solution.

The graph of x against t crosses the t-axis when $x = 0$ and has a turning point when $\frac{dx}{dt} = 0$. In this case, the equations $x = 0$ and $\frac{dx}{dt} = 0$ have an infinite number of solutions.

The turning points of the motion occur when

$$\rho \cos(\Omega t + \phi) = \Omega \sin(\Omega t + \phi)$$
$$\Rightarrow \quad \tan(\Omega t + \phi) = \frac{\rho}{\Omega}$$

When considering the period of oscillation, we note that successive maxima occur when $(\Omega t + \phi)$ has increased by 2π.

This corresponds to an increase in t of $\frac{2\pi}{\Omega}$. The motion therefore has a constant period of $\frac{2\pi}{\Omega}$.

An example of weak damping, with the bounding curves, is shown below.

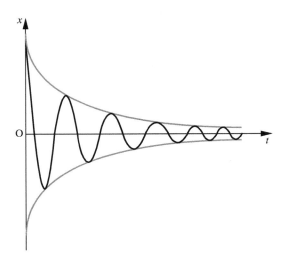

CHAPTER 6 OSCILLATING SYSTEMS

The arbitrary constants

The values of A and B, or R and ϕ, depend upon the particular situation we are dealing with. We need initial conditions to find the **particular solution** (see pages 64–6).

Example 1 A particle, P, of mass 1 kg, rests on a smooth, horizontal surface. It is attached to two points, A and B, 3 m apart, by two springs AP and BP, as shown. The natural length of each spring is 1 m and the stiffnesses are $2\,\text{N}\,\text{m}^{-1}$ and $3\,\text{N}\,\text{m}^{-1}$ respectively. The particle is fastened to a dashpot with a damping constant $r = 4\,\text{N}\,\text{s}\,\text{m}^{-1}$.

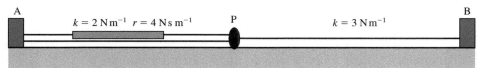

a) Find the position of equilibrium of the particle.

The particle is then displaced a distance 0.5 m from the equilibrium position and released.

b) Find the equation of motion of the particle in its subsequent motion.
c) Find the particular solution of the equation of motion of the particle.
d) Find the period of the oscillation.
e) Find the time taken for the amplitude of the oscillations to reduce to 0.2 m

SOLUTION

a)

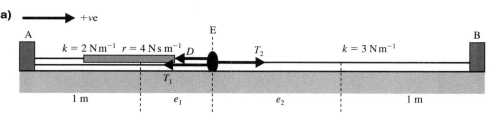

The forces acting on the particle are the tensions, T_1 and T_2, and the resistance, D, of the dashpot. When the particle is in the equilibrium position, E, we have

$$T_1 = 2e_1 \quad T_2 = 3e_2 \quad D = 0$$

where e_1 and e_2 are the extensions of the springs, as shown in the diagram.

Taking positive to the right, when the particle is in equilibrium, the equation of motion is:

$$T_2 - T_1 = 0$$
$$\Rightarrow \quad 3e_2 - 2e_1 = 0 \qquad [1]$$

Also we have $AB = 3\,\text{m}$, giving

$$e_1 + e_2 = 1 \qquad [2]$$

Solving [1] and [2], we get $e_1 = \frac{3}{5}, e_2 = \frac{2}{5}$. So, the equilibrium position is $AE = 1.6\,\text{m}$.

b)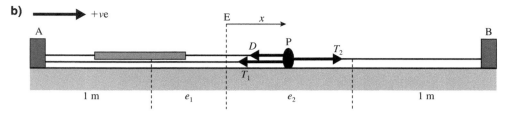

Suppose the particle is in a general position, P, x m to the right of the equilibrium position, E, as shown.

The forces acting on the particle are

$$T_1 = 2(e_1 + x) \qquad T_2 = 3(e_2 - x) \qquad D = 4\frac{dx}{dt}$$

The equation of motion is

$$T_2 - T_1 - D = \frac{d^2x}{dt^2}$$

$$\Rightarrow \quad 3(e_2 - x) - 2(e_1 + x) - 4\frac{dx}{dt} = \frac{d^2x}{dt^2}$$

Substituting the values of e_1 and e_2 obtained in part **a**, we have

$$\frac{d^2x}{dt^2} + 4\frac{dx}{dt} + 5x = 0$$

which is the equation of motion of the particle.

c) To solve the equation of motion, we use the auxiliary equation

$$\lambda^2 + 4\lambda + 5 = 0$$
$$\Rightarrow \quad \lambda = -2 \pm i$$

So, the general solution is

$$x = e^{-2t}(A\cos t + B\sin t)$$

Differentiating, we get

$$\frac{dx}{dt} = e^{-2t}[(B - 2A)\cos t - (A + 2B)\sin t]$$

The initial conditions are $t = 0$, $x = 0.5$, $\frac{dx}{dt} = 0$. Hence, we have

$$t = 0, x = 0.5 \quad \Rightarrow \quad 0.5 = A$$

$$t = 0, \frac{dx}{dt} = 0 \quad \Rightarrow \quad 0 = B - 2A \quad \Rightarrow \quad B = 1$$

So, the particular solution is

$$x = e^{-2t}(\tfrac{1}{2}\cos t + \sin t) \qquad [3]$$

d) The periodic time of the oscillations is $\dfrac{2\pi}{1} = 2\pi$ s.

e) We can write equation [3] as

$$x = \frac{\sqrt{5}}{2} e^{-2t} \cos(t - 1.107)$$

The amplitude is, therefore, $a = \dfrac{\sqrt{5}\, e^{-2t}}{2}$.

For the amplitude to reduce to 0.2 m, we have

$$0.2 = \frac{\sqrt{5}\, e^{-2t}}{2}$$

$$\Rightarrow \quad e^{2t} = \frac{5\sqrt{5}}{2}$$

$$\Rightarrow \quad t = 0.5 \ln\left(\frac{5\sqrt{5}}{2}\right) = 0.86\,\text{s}$$

Note Example 1, part **e**, concerning the amplitude of the motion, needs a little explanation. When the 'amplitude' is 0.2, the curve is not at one of its extremes. The value $t = 0.86$ corresponds to the point where the enveloping curve has an x-value of 0.2. This particular oscillation dampens down quite quickly, as can be seen from the graph (shown below). Had the damping been much weaker, there would have been a larger number of significant oscillations and an extreme point with an amplitude close to 0.2 m would have occurred.

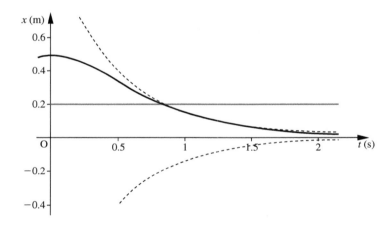

Exercise 6A

1 A particle moves along a straight line, Ox, such that its displacement, x, from the equilibrium point, O, is satisfied by the equation of motion

a) $\dfrac{d^2x}{dt^2} + 3\dfrac{dx}{dt} + 5x = 0$

b) $2\dfrac{d^2x}{dt^2} + 5\dfrac{dx}{dt} + 2x = 0$

c) $3\dfrac{d^2x}{dt^2} + 6\dfrac{dx}{dt} + 3x = 0$

d) $2\dfrac{d^2x}{dt^2} + 5\dfrac{dx}{dt} + 3x = 0$

e) $2\dfrac{d^2x}{dt^2} + 4\dfrac{dx}{dt} + 4x = 0$

f) $8\dfrac{d^2x}{dt^2} + 8\dfrac{dx}{dt} + 2x = 0$

In each case, determine whether the motion will be strongly damped, critically damped or oscillatory. If it is oscillatory, write down the period of the motion.

2 A block of mass 2 kg is attached to a spring of stiffness $10\,\text{N m}^{-1}$. The block is suspended in a resistive medium producing linear damping with a damping constant of $4\,\text{N s m}^{-1}$. The block is pulled 3 m down from the equilibrium position and released with an initial velocity of $3\,\text{m s}^{-1}$ upwards. Take $g = 10\,\text{m s}^{-2}$.

a) Determine the equation of motion of the block.
b) Show that the motion is weakly damped harmonic and find the period.
c) Find the particular solution of the equation of motion.
d) Show that successive maximum displacements are in geometric progression, and find the common ratio of this geometric progression.

3 The pointer on a measuring instrument oscillates on the scale, gradually settling down to its final value. If x is the displacement from the final position at time t, then the position of the pointer can be described by the second-order differential equation

$$\dfrac{d^2x}{dt^2} + 3\dfrac{dx}{dt} + 5x = 0$$

a) Find the general solution of the differential equation.
b) If at time $t = 0$, $x = 0.2$ and $\dfrac{dx}{dt} = 0$, find the particular solution.
c) How long will it take for the amplitude to fall to within 10% of its original value?

4 A light elastic string, of natural length 2 m and modulus of elasticity 60 N, has a particle of mass 3 kg attached to its mid-point. One end of the string is fixed to a point A and the other end to a point B, a distance 4 m vertically below A. Take $g = 10\,\text{m s}^{-2}$.

a) Find the equilibrium position of the particle.

The particle is held a distance of 3 m below A and released from rest. Its motion is subject to a resistance of $12v$ N, where v is the speed of the particle.

b) Find and solve the equation of motion of the particle.
c) Find the maximum height to which the particle rises above the equilibrium position.

5 A particle of mass 2 kg is attached to the end of a light elastic string of natural length 1 m and modulus of elasticity 40 N. The string is hung from a fixed point A and the particle is held at rest a distance of 1 m below. It is then released. Its motion is subject to a resistance force of magnitude kv, where v is the speed of the particle. Take $g = 10\,\text{m s}^{-2}$.

a) Find the equation of motion of the particle.
b) Find the value of k which corresponds to critical damping. Solve the equation of motion for this value of k.
c) Find the position of the particle after 2 seconds.

CHAPTER 6 OSCILLATING SYSTEMS

6 A particle of mass 3 kg hangs in equilibrium on the end of an elastic string of natural length 1 m and modulus of elasticity $3g$ N. The other end of the string is fastened to a fixed point O. The particle is then pulled down a further 1 m and released from rest. The subsequent motion is subject to linear damping with damping constant $3\sqrt{3g}$ N s m^{-1}. Find the distance below O at which the particle next comes to rest.

7 A particle of mass 1 kg is attached to the end of a light, elastic string of natural length 1 m and modulus of elasticity g N. The string is hung from a fixed point, A, and the particle hangs in equilibrium at a point E.

a) Find the distance AE.

The particle is now held at rest a distance of 1 m below A and is released. Its motion is subject to a resistance force of magnitude $v\sqrt{3g}$ N, where v is the speed of the particle. The displacement of the particle above E is x m.

b) Find the equation of motion of the particle.
c) Show that the particle's motion is weakly damped harmonic and find the period of oscillation.
d) Show that, if the motion continues indefinitely, the particle approaches a position of rest at E. Find the total distance travelled by the particle.

8 A particle of mass 1 kg is attached to one end of a light, elastic string of natural length 1 m and modulus of elasticity 5 N. The other end of the string is attached to a fixed point, O, on the surface of a smooth, horizontal table. The particle is held at rest at a point on the table 2 m from O, and is released. Its motion is resisted by a force of magnitude $2v$ N, where v is the speed of the particle.

a) Find the time which elapses before the string becomes slack.
b) Find the speed of the particle at the instant that the string goes slack.
c) Find the distance from O at which the particle comes to rest.

Forced oscillations

In many practical situations, the vibrations in an object occur because it is subject to a variable force. For example, the suspension of a car is subject to forces arising from undulations in the road. If we can express the applied force as a function of time, we can include it in our model.

In certain cases, this driving force is also oscillatory. For example, vehicle engines are attached to the vehicle by rubber mountings. The action of the engine causes vibrations which are 'felt' by the mounting as an oscillatory applied force. The mounting can be thought of as a combination of a spring and a dashpot, which damps the engine vibrations. Without such mountings, the vehicle body would vibrate with the engine and the experience would be most uncomfortable.

We will confine ourselves to simple oscillatory driving forces. These are easier to analyse than non-oscillatory ones. Moreover, when we have more complicated driving forces, we can use a technique called Fourier analysis to approximate them by a combination of sinusoidal oscillations.

We will use the same model for driven systems as we did for free vibrations with damping. All that is required is to add another term to the equation of motion.

FORCED OSCILLATIONS

Example 2 A particle of mass m is suspended from a fixed point, O, by a perfect spring of stiffness k and natural length l_o.

a) Find the equilibrium position of the particle.

The particle is acted upon by a downward force, $F\cos\omega t$.

b) Find the equation of motion of the particle when its displacement from the equilibrium position is x.

SOLUTION

a) We make the usual simplifying assumptions.

When the particle is in equilibrium, we have

$$mg - ke = 0 \qquad [1]$$

$$\Rightarrow \quad e = \frac{mg}{k}$$

So, the equilibrium position is a distance $l_o + \frac{mg}{k}$ below O.

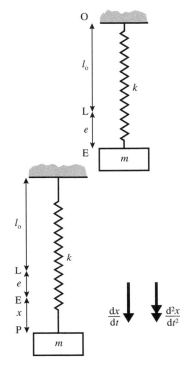

b) We will take x to be positive downwards from E.

At P, we have

$$T = -k(e + x)$$

The equation of motion is, therefore,

$$mg + F\cos\omega t - k(e + x) = m\frac{d^2x}{dt^2}$$

Substituting from [1] and rearranging, we obtain

$$m\frac{d^2x}{dt^2} + kx = F\cos\omega t \qquad [2]$$

This second-order non-homogeneous differential equation is similar to the one previously obtained without a driving force, but now we have the forcing term on the right-hand side of the equation.

In order to solve equation [2], we must first find the complementary function by solving the equivalent homogeneous equation

$$m\frac{d^2x}{dt^2} + kx = 0 \qquad [3]$$

This we should recognise as the equation for simple harmonic motion whose solution is

$$x = R\cos(\Omega t + \phi) \quad \text{where} \quad \Omega^2 = \frac{k}{m}$$

We then find a particular integral by using a trial solution.

The form of the right-hand side of [2] indicates a trial solution of the form

$$x = P\cos\omega t + Q\sin\omega t \quad (\text{provided } \omega \neq \Omega)$$

CHAPTER 6 OSCILLATING SYSTEMS

which gives

$$\frac{dx}{dt} = -P\omega \sin \omega t + Q\omega \cos \omega t$$

and

$$\frac{d^2x}{dt^2} = -P\omega^2 \cos \omega t - Q\omega^2 \sin \omega t$$

We substitute these into [2] to find the value of P and Q by comparing coefficients in the resulting identity:

$$m(-P\omega^2 \cos \omega t - Q\omega^2 \sin \omega t) + k(P\cos \omega t + Q\sin \omega t) = F\cos \omega t$$

For $\cos \omega t$: $\quad -Pm\omega^2 + Pk = F \quad$ [4]

For $\sin \omega t$: $\quad -Q\omega^2 + Qk = 0 \quad$ [5]

From [4] and [5], we find

$$P = \frac{F}{k - m\omega^2} \quad \text{and} \quad Q = 0$$

The general solution will therefore be

$$x = R\cos\left(\sqrt{\frac{k}{m}}t + \phi\right) + \frac{F\cos \omega t}{k - m\omega^2}$$

Interpreting the solution to Example 2

The first part of this general solution, $R\cos\left(\sqrt{\frac{k}{m}}t + \phi\right)$, is an undamped oscillation.

The particular integral, $\dfrac{F\cos \omega t}{k - m\omega^2}$, is a constant oscillation whose frequency is that of the driving force.

Example 3 A particle of mass m is suspended from a fixed point, O, by a perfect spring of stiffness k and natural length l_0. The particle is also attached to a perfect dashpot providing linear damping with damping constant r, where $r^2 < 4mk$.

a) Find the equilibrium position of the particle.

The particle is acted upon by a force $F\cos \omega t$.

b) Find the equation of motion of the particle when its displacement, x, is measured from the equilibrium position.

SOLUTION

We make the usual simplifying assumptions.

a) When the particle is in equilibrium, we have

$$mg - ke = 0 \quad [1]$$

$$\Rightarrow \quad e = \frac{mg}{k}$$

b) We will take x to be positive downwards from E.

At P, we have

$$T = -k(e+x) \quad \text{and} \quad R = -r\frac{dx}{dt}$$

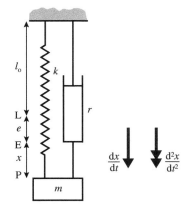

The equation of motion is

$$mg + F\cos\omega t - k(e+x) - r\frac{dx}{dt} = m\frac{d^2x}{dt^2}$$

Substituting [1] and rearranging, we obtain

$$m\frac{d^2x}{dt^2} + r\frac{dx}{dt} + kx = F\cos\omega t \qquad [2]$$

This is a second-order non-homogeneous differential equation. As before, this equation is similar to the one obtained from damped harmonic motion without a driving force, but here we have a forcing term, a function of t, on the right-hand side of the equation.

To solve equation [2], we first find the complementary function from the equivalent homogeneous equation:

$$m\frac{d^2x}{dt^2} + r\frac{dx}{dt} + kx = 0 \qquad [3]$$

We have already done this on page 118. As $r^2 - 4mk < 0$, the solution is oscillatory. The complementary function is, therefore,

$$x = R e^{pt} \cos(\Omega t + \phi)$$

We then find a particular integral by using a trial solution.

The form of the right-hand side of [2] indicates a trial solution of the form

$$x = P\cos\omega t + Q\sin\omega t \quad (\text{provided } \omega \neq \Omega)$$

which gives

$$\frac{dx}{dt} = -P\omega \sin\omega t + Q\omega \cos\omega t$$

and

$$\frac{d^2x}{dt^2} = -P\omega^2 \cos\omega t - Q\omega^2 \sin\omega t$$

We substitute these into [2] to find the values of P and Q by comparing coefficients in the resulting identity:

$$m(-P\omega^2 \cos\omega t - Q\omega^2 \sin\omega t) + r(-P\omega \sin\omega t + Q\omega \cos\omega t) +$$
$$+ k(P\cos\omega t + Q\sin\omega t) = F\cos\omega t$$

For $\cos\omega t$: $\quad -Pm\omega^2 + Qr\omega + Pk = F$

For $\sin\omega t$: $\quad -Qm\omega^2 - Pr\omega + Qk = 0$

CHAPTER 6 OSCILLATING SYSTEMS

Solving these, we obtain

$$P = \frac{(k - m\omega^2)F}{r^2\omega^2 + (k - m\omega^2)^2} \quad \text{and} \quad Q = \frac{r\omega F}{r^2\omega^2 + (k - m\omega^2)^2}$$

The general solution is, therefore,

$$x = Re^{\rho t}\cos(\Omega t + \phi) + F\left[\frac{(k - m\omega^2)\cos\omega t + r\omega\sin\omega t}{r^2\omega^2 + (k - m\omega^2)^2}\right]$$

Interpreting the solution to Example 3

We know that the first part of this general solution, $Re^{\rho t}\cos(\Omega t + \phi)$, is a damped oscillation.

The particular integral, $P\cos\omega t + Q\sin\omega t$, is a constant oscillation whose frequency is that of the driving force.

As t increases, the amplitude of the complementary function diminishes. This is the **transient** part of the solution. The solution becomes ever closer to the particular integral. This is the **steady-state** part of the solution.

If the relation between r, m and k had been different, the complementary function would have corresponded to strong damping or to critical damping, rather than the weak damping shown in Example 3. However, for all of these situations, the particular integral becomes dominant as the value of t increases. Hence, in all circumstances, the particle will eventually vibrate with the frequency of the driving force.

Example 4 A particle of mass m is suspended on a perfect spring of stiffness k and natural length l_0. The other end of the spring is fastened to a free beam. Initially, the beam is stationary at a point O and the particle hangs in equilibrium at E. The beam then starts to oscillates about O so that at time t its displacement below O is $y = a\sin\omega t$.

a) Find the equation of motion of the particle if x is its displacement below E.
b) Solve the equation to find an expression for the position of the particle at time t.

SOLUTION

We make the usual simplifying assumptions.

a) When the particle is in equilibrium, we have

$$mg - ke = 0 \qquad [1]$$

$$\Rightarrow \quad e = \frac{mg}{k}$$

We will take x to be positive downwards from E.

There are only two forces acting on the particle – its weight and the tension in the spring.

To find the tension, we need the extension of the spring. This is $(e + x - y)$.

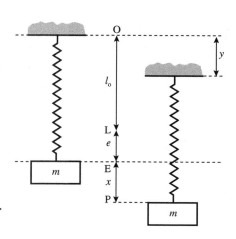

At P, we have
$$T = -k(e + x - y)$$

The equation of motion is
$$mg - k(e + x - y) = m\frac{d^2x}{dt^2}$$

Substituting [1] and rearranging, we obtain
$$m\frac{d^2x}{dt^2} + kx = ky$$

But $y = a \sin \omega t$, so we have
$$m\frac{d^2x}{dt^2} + kx = ka \sin \omega t \qquad [2]$$

b) Equation 2 is a second-order non-homogeneous differential equation. The homogeneous auxiliary equation is the familiar SHM equation and the forcing term, $ka \sin \omega t$, appears on the right-hand side, as before.

The complementary function is, therefore,
$$x = R \cos(\Omega t + \phi) \quad \text{where} \quad \Omega^2 = \frac{k}{m}$$

We then find a particular integral by using a trial solution.

The form of the right-hand side of [2] indicates a trial solution of the form
$$x = P \cos \omega t + Q \sin \omega t \quad (\text{provided } \omega \neq \Omega)$$

which gives
$$\frac{dx}{dt} = -P\omega \sin \omega t + Q\omega \cos \omega t$$

and
$$\frac{d^2x}{dt^2} = -P\omega^2 \cos \omega t - Q\omega^2 \sin \omega t$$

We substitute these into [2] to find the values of P and Q by comparing coefficients in the resulting identity:

$$m(-P\omega^2 \cos \omega t - Q\omega^2 \sin \omega t) + k(P \cos \omega t + Q \sin \omega t) = ka \sin \omega t$$

For $\cos \omega t$: $\quad -Pm\omega^2 + Pk = 0$

For $\sin \omega t$: $\quad -Qm\omega^2 + Qk = ka$

Solving these, we obtain
$$P = 0 \quad \text{and} \quad Q = \frac{ka}{k - m\omega^2}$$

The general solution is, therefore,
$$x = R\cos\left(\sqrt{\frac{k}{m}}\,t + \phi\right) + \left(\frac{ka}{k - m\omega^2}\right) \sin \omega t$$

CHAPTER 6 OSCILLATING SYSTEMS

Investigation

If you have access to suitable equipment, you can see the effects of forced vibrations by carrying out the following investigation.

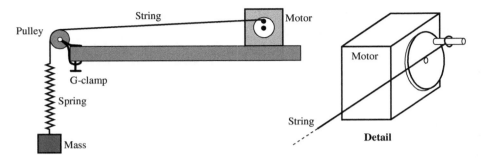

Attach a mass to the end of a spring, and attach the other end of the spring to a string which passes over a pulley to an eccentric spindle mounted on an electric motor, as shown.

Gradually increase the speed of the motor and notice what happens to the mass. Its motion for any given motor speed will be cyclic but the cycles will be erratic in many cases, as the motion combines the natural frequency of the spring and the frequency of the driving force. For one particular motor speed, you should get a catastrophic event, which we discuss in the next section.

You can also investigate the effects of damping (literally!) by submerging the mass in a container of water.

If you have access to a computer-graphing package, it is worth measuring the stiffness, k, of your spring and investigating graphically the predicted motion of the system for various values of ω. A suitable model for the system is that described in Example 4.

Resonance

In all our solutions to problems of forced harmonic motion, we had to state that $\omega \neq \Omega$. The question arises as to what happens if this is not so. If you carried out the investigation just described, you will have found a catastrophic event for a particular motor speed.

This phenomenon is called **resonance**. It occurs when the driving frequency is the **same** as the natural frequency of the system being driven.

Mechanical resonance can be extremely dangerous. On 7 November 1940, the suspension bridge at Tacoma Narrows (Washington, USA) collapsed in wind gusts of no more than $70 \, \text{km} \, \text{h}^{-1}$. Because its stiffening system was unable to damp out the buffetings, the 850 m main span developed torsional oscillations which steadily built up until the hangers were torn from the deck. This was a landmark event in suspension bridge engineering, leading to a reassessment of span-stiffening systems.

It is also the reason why marching soldiers break step as they cross a bridge. Marching in time may cause the bridge to collapse (as happened once). Other examples are the breaking of a glass by singing at the right frequency, and the tendency of an unbalanced wheel on a car to cause steering wobble at particular speeds.

So, when designing structures, engineers need to be aware of the destructive possibilities of resonance and take steps to damp it out.

In other ways, we use resonance, particularly electrical resonance. For example, when we tune a radio or television set to a particular station or channel, we adjust the resonant frequency of the tuned circuit in the set to match the frequency of that station or channel. In this way, the tuned circuit, which is connected directly to the antenna, selects the required signal from the many signals present in the antenna and passes it on to the next stage in the set.

Mathematically, the problem stems from our differential equation.

The equation from Example 2 is

$$m\frac{d^2x}{dt^2} + kx = F\cos\omega t$$

The complementary function is

$$x = R\cos(\Omega t + \phi) \quad \text{where} \quad \Omega^2 = \frac{k}{m}$$

If $\omega = \Omega$, the complementary function contains an element from the right-hand side of the differential equation. Hence, we must adjust our trial solution. Our new trial solution is

$$x = t(P\cos\omega t + Q\sin\omega t)$$

which gives

$$\frac{dx}{dt} = t(-P\omega\sin\omega t + Q\omega\cos\omega t) + (P\cos\omega t + Q\sin\omega t)$$

and

$$\frac{d^2x}{dt^2} = t(-P\omega^2\cos\omega t - Q\omega^2\sin\omega t) + 2(-P\omega\sin\omega t + Q\omega\cos\omega t)$$

Substituting these into the original differential equation and comparing coefficients, we have

For $\cos\omega t$: $\quad 2mQ\omega = F$ $\qquad$ [1]
For $\sin\omega t$: $\quad -2mP\omega = 0$ $\qquad$ [2]
For $t\cos\omega t$: $\quad -Pm\omega^2 + kP = 0$ $\qquad$ [3]
For $t\sin\omega t$: $\quad -Qm\omega^2 + kQ = 0$ $\qquad$ [4]

Equations [3] and [4] are satisfied by the conditions $\Omega^2 = \frac{k}{m}$ and $\omega = \Omega$ for any values of P and Q.

Equation [1] gives $Q = \frac{F}{2m\omega}$ and [2] gives $P = 0$.

Thus, our general solution is

$$x = R\cos(\omega t + \phi) + \frac{F}{2m\omega}t\sin\omega t$$

CHAPTER 6 OSCILLATING SYSTEMS

We now have a term in $t \sin \omega t$. As t increases, this also increases.
A typical solution curve is shown below.

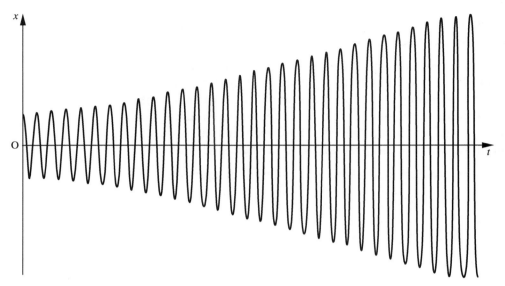

Exercise 6B

1. A block of mass 2 kg is placed on a smooth table. It is fastened to the end of a spring of length 0.5 m and stiffness $50 \, \text{N m}^{-1}$. The other end of the spring is attached to a piston. Initially, the system is at rest. The piston is then set in motion so that, at time t, it is a distance $y = 0.1 \sin 2t$ from its initial position, O.

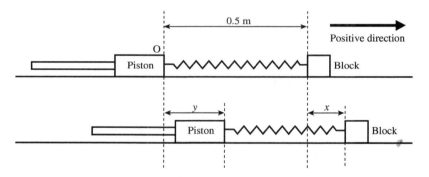

   a) If x is the distance of the block from its initial position, find the equation of motion of the block.
   b) Solve the equation to find an expression for x at time t.
   c) Sketch the graph of the solution. (A computer or graphics calculator would be useful for this.)

2. The motion of the piston in Question **1** is altered to $y = 0.2 \sin \omega t$.

   a) State the value of ω for which the system will resonate.
   b) Find and solve the equation of motion for this value of ω.

EXERCISE 6B

3 The block in Question **1** is now fastened to a dashpot with $r = 12\,\text{N}\,\text{s}\,\text{m}^{-1}$. As before, the piston moves with $y = 0.1\sin 2t$.

   a) Find the equation of motion of the block.
   b) Solve this to find an expression for x in terms of t.

4 A particle of mass m is hanging in equilibrium on the end of an elastic string of stiffness m. At time $t = 0$, the upper end of the string starts to move so that its displacement, y, below its initial position is given by $y = \sin t$. Let x be the displacement of the particle below its initial position.

   a) Find the equation of motion of the particle.
   b) Show that $x = \frac{1}{2}(\sin t - t\cos t)$.
   c) Show that, after slightly more than 21.6 seconds, the string goes slack for the first time.

5 The situation described in Question **4** is now modified so that the motion takes place in a resisting medium. The resistance force is $2mkv$, where v is the speed of the particle and k is a constant.

   a) Find the equation of motion of the particle.
   b) Solve the equation for $k = 1$.
   c) What is the long-term, steady-state, behaviour of the particle?

6 A body of mass m hangs from a light spring. When the body hangs in equilibrium, the extension of the spring is a. The top of the spring is then made to move so that its displacement from its initial position is given by $y = b\sin pt$. The motion of the body is subject to linear damping with damping constant mr.

   Show that, when the transient part of the motion of the body has died away, it moves in an oscillation with amplitude

   $$\frac{bg}{\sqrt{(ap^2 - g)^2 + a^2p^2r^2}}$$

7 A light spring, AB, has natural length 1 m and modulus of elasticity $4mg$ N. The spring is vertical, with its lower end, A, fixed and a particle of mass m kg rests on a light platform at B.

   a) Find the length AB when the system is in equilibrium.

   The end A is now made to move so that at time t its upward displacement from its original position is $y = b\sin(4\sqrt{g}t)$. The upward displacement of B at time t from its original position is x.

   b) Find x in terms of b and t.
   c) Show that the particle loses contact with the platform if b exceeds about 0.284.

8 AB is a light, elastic string of natural length 1 m and modulus of elasticity 4 N. A particle of mass 1 kg is attached to B and rests on a smooth, horizontal table. End A is held on the table so that the string is just taut, and is then made to move away from B at a constant velocity $V\,\text{ms}^{-1}$.

   a) Find the time which elapses before the string goes slack.
   b) Find the distance travelled by the particle before it catches up with end A.

CHAPTER 6 OSCILLATING SYSTEMS

9 A light spring, AB, of natural length 2 m and modulus of elasticity 8 N, lies on a smooth horizontal table with AB = 2 m. A particle of mass 1 kg is attached to the mid-point, C, of the spring and the whole system is at rest.

The ends are simultaneously set in motion. At time t s, the displacement of A from its initial position is given by $y = 0.3 \sin t$, and the displacement of B from its initial position is given by $z = 0.3 \sin 2t$. The resulting displacement of the particle from its initial position is denoted by x. All displacements are positive in the direction AB.

a) Write down the equation of motion of the particle.
b) Solve the equation to find an expression for x at time t.
c) The design of the spring is such that its coils touch if it is compressed to a length of 1.2 m. Use a graphics calculator or a computer-graphing package to decide if the coils will ever touch in the motion described.

10 A light spring, AB, of natural length $2a$ and modulus of elasticity $2mg$, has a particle of mass m attached to its mid-point. The end B is fixed to the ground and end A is held at a distance of $8a$ vertically above B. The particle hangs in equilibrium at a point E.

a) Find the distance EB.

The end A is now set in motion so that at time t its downward displacement from its initial position is $y = \sin\left(t\sqrt{\dfrac{g}{a}}\right)$. The downward displacement of the particle from its initial position is denoted by x.

b) Write down the equation of motion of the particle.
c) Solve the equation to find an expression for x at time t.
d) Find the minimum distance between the particle and the ground.

7 Projectiles

I shot an arrow into the air,
It fell to earth, I knew not where.
HENRY WADSWORTH LONGFELLOW

Projectiles on an inclined plane

You have already met the problem of modelling the motion of a projectile, which is covered in *Introducing Mechanics* (pages 123–34). One of the questions addressed there was that of finding the range of a projectile fired on a horizontal plane. In this chapter, we extend this to include a projectile fired on an inclined plane.

One approach would be to use the equation of the path of the projectile and the equation of the plane in the form $y = mx$ to find where the two meet. However, there is an alternative approach which is the subject of this section.

The problem

A projectile, of mass m, is fired up a plane, inclined at an angle α, with a speed of U at an angle θ to the plane.

Modelling assumptions

- The projectile is a particle.
- There is no air resistance.
- The projectile starts from the point O when $t = 0$.
- The value of g is constant.

The model

We choose unit vectors parallel and perpendicular to the plane.

The initial velocity of the projectile is

$$\mathbf{U} = U\cos\theta\,\mathbf{i} + U\sin\theta\,\mathbf{j} \qquad [1]$$

The acceleration of the projectile is

$$\mathbf{a} = -g\sin\alpha\,\mathbf{i} - g\cos\alpha\,\mathbf{j} \qquad [2]$$

Solving the model

Since the acceleration is constant, we can use the constant-acceleration formulae. Hence, if the velocity at time t is **v** and the position vector is **r**, we have

$$\mathbf{v} = (U\cos\theta - gt\sin\alpha)\mathbf{i} + (U\sin\theta - gt\cos\alpha)\mathbf{j} \qquad [3]$$

and

$$\mathbf{r} = (Ut\cos\theta - \tfrac{1}{2}gt^2\sin\alpha)\mathbf{i} + (Ut\sin\theta - \tfrac{1}{2}gt^2\cos\alpha)\mathbf{j} \qquad [4]$$

To find the range, we need to find the time of flight.

The projectile will hit the plane again when the **j**-component of position is zero. That is, when

$$Ut\sin\theta - \tfrac{1}{2}gt^2\cos\alpha = 0$$
$$\Rightarrow \quad t(U\sin\theta - \tfrac{1}{2}gt\cos\alpha) = 0$$

The solutions to this are $t = 0$ (the initial point) and

$$t = \frac{2U\sin\theta}{g\cos\alpha} \qquad [5]$$

The range is given by the **i**-component of the position vector when $t = \dfrac{2U\sin\theta}{g\cos\alpha}$. Hence, we have

$$R = U\left(\frac{2U\sin\theta}{g\cos\alpha}\right)\cos\theta - \frac{1}{2}g\left(\frac{2U\sin\theta}{g\cos\alpha}\right)^2\sin\alpha$$

$$\Rightarrow \quad R = \frac{2U^2\sin\theta(\cos\alpha\cos\theta - \sin\theta\sin\alpha)}{g\cos^2\alpha}$$

$$\Rightarrow \quad R = \frac{2U^2\sin\theta\cos(\alpha + \theta)}{g\cos^2\alpha} \qquad [6]$$

Two other quantities which are of interest are the maximum perpendicular distance from the plane attained by the projectile and the maximum vertical distance it reaches above the plane.

Maximum perpendicular distance from the plane

The projectile is at its maximum distance from the plane when the **j**-component of its velocity is zero.

Applying this condition, we have

$$U\sin\theta - gt\cos\alpha = 0$$
$$\Rightarrow \quad t = \frac{U\sin\theta}{g\cos\alpha} \qquad [7]$$

Note that, as in the case of the maximum height above a horizontal plane, this time is half the time of flight.

The position of the projectile at this time is given by

$$\mathbf{r} = \left[U\left(\frac{U\sin\theta}{g\cos\alpha}\right)\cos\theta - \frac{1}{2}g\left(\frac{U\sin\theta}{g\cos\alpha}\right)^2\sin\alpha \right]\mathbf{i} +$$

$$+ \left[U\left(\frac{U\sin\theta}{g\cos\alpha}\right)\sin\theta - \frac{1}{2}g\left(\frac{U\sin\theta}{g\cos\alpha}\right)^2\cos\alpha \right]\mathbf{j}$$

$$\Rightarrow \quad \mathbf{r} = \frac{U^2\sin\theta\,(2\cos\theta\cos\alpha - \sin\theta\sin\alpha)}{2g\cos^2\alpha}\mathbf{i} + \frac{U^2\sin^2\theta}{2g\cos\alpha}\mathbf{j}$$

The perpendicular distance of the projectile from the plane is the **j**-component of the position vector. Thus, the maximum perpendicular distance of the projectile from the plane is

$$\frac{U^2\sin^2\theta}{2g\cos\alpha} \qquad [8]$$

Note that, unless $\alpha = 0$, the **i**-component (the distance along the plane) at which this takes place is **not** half the range.

Maximum vertical distance above the plane

We can see from the diagram on the right that if the vertical distance of the projectile above the plane is h and the perpendicular distance is k then

$$h = \frac{k}{\cos\alpha}$$

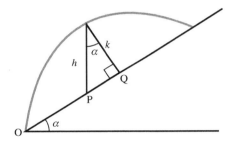

The maximum value for h occurs when k attains its maximum value. We found in [8] that the maximum value of k is $\dfrac{U^2\sin^2\theta}{2g\cos\alpha}$. Thus, the maximum vertical height above the plane is

$$h = \frac{U^2\sin^2\theta}{2g\cos^2\alpha} \qquad [9]$$

In the diagram, OQ is the **i**-component of the position vector. That is,

$$OQ = \frac{U^2\sin\theta\,(2\cos\theta\cos\alpha - \sin\theta\sin\alpha)}{2g\cos^2\alpha}$$

Also we have

$$PQ = k\tan\alpha = \frac{U^2\sin^2\theta\sin\alpha}{2g\cos^2\alpha}$$

Hence, the distance, OP, along the plane, when h is a maximum, is

$$OP = \frac{U^2\sin\theta\,(2\cos\theta\cos\alpha - \sin\theta\sin\alpha)}{2g\cos^2\alpha} - \frac{U^2\sin^2\theta\sin\alpha}{2g\cos^2\alpha}$$

$$= \frac{U^2\sin\theta\,(2\cos\theta\cos\alpha - 2\sin\theta\sin\alpha)}{2g\cos^2\alpha}$$

CHAPTER 7 PROJECTILES

which gives

$$\text{OP} = \frac{U^2 \sin\theta \cos(\alpha+\theta)}{g\cos^2\alpha} \qquad [10]$$

We note that this **is** half the range.

Maximum range up the plane

From [6], we have

$$R = \frac{2U^2 \sin\theta \cos(\alpha+\theta)}{g\cos^2\alpha}$$

$$\Rightarrow \quad \frac{dR}{d\theta} = \frac{2U^2[\cos\theta\cos(\alpha+\theta) - \sin\theta\sin(\alpha+\theta)]}{g\cos^2\alpha}$$

$$= \frac{2U^2\cos(2\theta+\alpha)}{g\cos^2\alpha}$$

For maximum range, $\dfrac{dR}{d\theta} = 0$. Thus, we have

$$\cos(2\theta+\alpha) = 0$$
$$\Rightarrow \quad 2\theta + \alpha = 90°$$
$$\Rightarrow \quad \theta = \tfrac{1}{2}(90° - \alpha)$$

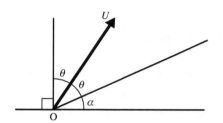

So, to attain the greatest range on an inclined plane, the projectile must be fired at an angle bisecting the plane and the vertical.

Note If $\alpha = 0$, $\theta = 45°$, so this result is a generalisation of the fact that the maximum range on a horizontal plane occurs when the angle of projection is 45°.

The maximum range up the plane is, therefore,

$$R = \frac{2U^2\cos\tfrac{1}{2}(90°+\alpha)\sin\tfrac{1}{2}(90°-\alpha)}{g\cos^2\alpha}$$

Using $\sin A - \sin B = 2\cos\tfrac{1}{2}(A+B)\sin\tfrac{1}{2}(A-B)$, we can write this as

$$R = \frac{U^2(\sin 90° - \sin\alpha)}{g\cos^2\alpha}$$

$$\Rightarrow \quad R = \frac{U^2(1-\sin\alpha)}{g(1-\sin^2\alpha)}$$

$$\Rightarrow \quad R = \frac{U^2}{g(1+\sin\alpha)} \qquad [11]$$

Landing perpendicular to the plane

A projectile fired on a horizontal plane can never be moving perpendicular to the plane (unless it is fired vertically upwards). However, if the plane is inclined, this becomes possible and, in particular, the projectile can be made to strike the plane perpendicularly.

For this to happen, two conditions must be satisfied:
- The **j**-component of the position vector must be zero (the projectile has reached the plane).
- The **i**-component of the velocity vector must be zero (the projectile is travelling in the **j**-direction).

The first of these is satisfied by [5]:
$$t = \frac{2U \sin \theta}{g \cos \alpha}$$

From [3], the second condition is satisfied when
$$U \cos \theta - gt \sin \alpha = 0$$

Combining these, we have
$$U \cos \theta - \frac{g(2U \sin \theta) \sin \alpha}{g \cos \alpha} = 0$$
$$\Rightarrow \quad \cos \theta \cos \alpha = 2 \sin \theta \sin \alpha$$
$$\Rightarrow \quad \tan \theta = \tfrac{1}{2} \cot \alpha \qquad [12]$$

Note that if $\alpha \to 0$, $\cot \alpha \to \infty$ and hence $\tan \theta \to \infty$, $\theta \to 90°$, agreeing with our earlier statement.

Projection down an inclined plane

We can tailor all the results just derived to fit the case in which the projectile is fired down the slope, by replacing α with $-\alpha$.

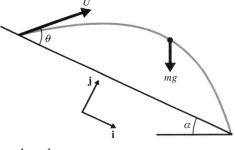

For the time of flight, we have
$$t = \frac{2U \sin \theta}{g \cos(-\alpha)} = \frac{2U \sin \theta}{g \cos \alpha}$$

So, the time of flight is unchanged.

We also find that the maximum perpendicular distance from the plane, $\dfrac{U^2 \sin^2 \theta}{2g \cos \alpha}$, and the maximum vertical height above the plane, $\dfrac{U^2 \sin^2 \theta}{2g \cos^2 \alpha}$, are unchanged.

For the range, we have
$$R = \frac{2U^2 \sin \theta \cos(\theta - \alpha)}{g \cos^2 \alpha}$$

which, not unexpectedly, is greater than the range if the projectile were fired up the plane.

We also find that the maximum range still occurs when the initial direction bisects the angle between the plane and the vertical, i.e. $\theta = \tfrac{1}{2}(90 + \alpha)$, and is given by
$$R = \frac{U^2}{g(1 - \sin \alpha)}$$

CHAPTER 7 PROJECTILES

It is not possible to make the projectile strike the plane at 90°, since the condition $\tan\theta = \frac{1}{2}\cot\alpha$ implies a negative angle of projection if α is negative.

Example 1 A projectile is fired with a speed of $50\,\mathrm{m\,s^{-1}}$, at an angle of elevation of 60° to the horizontal, up a plane which is inclined at an angle of 30° to the horizontal. Take $g = 10\,\mathrm{m\,s^{-2}}$. Find

a) the time of flight
b) the range up the plane
c) the direction, relative to the plane, upon landing.

SOLUTION

We take unit vectors parallel and perpendicular to the plane.

The initial velocity is

$$\mathbf{U} = 50\cos 30°\,\mathbf{i} + 50\sin 30°\,\mathbf{j}$$
$$= 25\sqrt{3}\,\mathbf{i} + 25\,\mathbf{j}$$

The acceleration is

$$\mathbf{a} = -10\sin 30°\,\mathbf{i} - 10\cos 30°\,\mathbf{j}$$
$$= -5\,\mathbf{i} - 5\sqrt{3}\,\mathbf{j}$$

Hence, using the constant-acceleration formulae, we have

$$\mathbf{v} = (25\sqrt{3}\,\mathbf{i} + 25\,\mathbf{j}) + (-5\,\mathbf{i} - 5\sqrt{3}\,\mathbf{j})\,t$$
$$\Rightarrow\quad \mathbf{v} = (25\sqrt{3} - 5t)\,\mathbf{i} + (25 - 5\sqrt{3}t)\,\mathbf{j}$$

and
$$\mathbf{r} = (25\sqrt{3}\,\mathbf{i} + 25\,\mathbf{j})\,t + \tfrac{1}{2}(-5\,\mathbf{i} - 5\sqrt{3}\,\mathbf{j})\,t^2$$
$$\Rightarrow\quad \mathbf{r} = \left(25\sqrt{3}\,t - \frac{5}{2}t^2\right)\mathbf{i} + \left(25t - \frac{5\sqrt{3}}{2}t^2\right)\mathbf{j}$$

a) The projectile will land when the **j**-component of position is zero. So, we have

$$25t - \frac{5\sqrt{3}}{2}t^2 = 0$$

$$\Rightarrow\quad t = 0\ \ (\text{projection time})\quad \text{or}\quad t = \frac{10\sqrt{3}}{3}$$

b) The range is given by the **i**-component of position. So, we have

$$R = 25\sqrt{3} \times \frac{10\sqrt{3}}{3} - \frac{5}{2} \times \frac{(10\sqrt{3})^2}{9}$$

$$\Rightarrow\quad R = \frac{500}{3}\,\mathrm{m}$$

c) The velocity on landing is

$$\mathbf{v} = \left(25\sqrt{3} - 5 \times \frac{10\sqrt{3}}{3}\right)\mathbf{i} + \left(25 - 5\sqrt{3} \times \frac{10\sqrt{3}}{3}\right)\mathbf{j}$$

$$\Rightarrow \quad \mathbf{v} = \frac{25\sqrt{3}}{3}\mathbf{i} - 25\mathbf{j}$$

Thus, the direction of motion relative to the plane is given by

$$\tan \lambda = \frac{(-25) \times 3}{25\sqrt{3}} = -\sqrt{3} \quad \Rightarrow \quad \lambda = -60°$$

The projectile hits the plane at an angle of $60°$ to the plane.

Example 2 The collision between the projectile in Example 1 and the ground is perfectly elastic. Will the second bounce be higher up the plane than the first?

SOLUTION

The velocity on landing is

$$\mathbf{v} = \frac{25\sqrt{3}}{3}\mathbf{i} - 25\mathbf{j}$$

The impact between the projectile and the plane is in the **j**-direction, hence the **i**-component of velocity will be unchanged. Because the collision is perfectly elastic, the **j**-component of velocity will be reversed. The velocity after impact will, therefore, be

$$\mathbf{v}_1 = \frac{25\sqrt{3}}{3}\mathbf{i} + 25\mathbf{j}$$

The angle between the new direction of motion of the projectile and the plane is given by

$$\tan \phi = \sqrt{3} \quad \Rightarrow \quad \phi = 60°$$

This means that, since the angle between the plane and the horizontal is $30°$, the projectile is now travelling vertically upwards. The second bounce will, therefore, be in exactly the same position as the first.

Exercise 7A

Throughout this exercise, take $g = 10\,\text{m s}^{-2}$.

1 A projectile is fired with a speed of $60\,\text{m s}^{-1}$ from a point, O, on a plane inclined at an angle of $20°$ to the horizontal. The direction of the projectile is up the plane, along its line of greatest slope, and making an angle of $40°$ with the horizontal. Find

   a) the time of flight
   b) the range on the plane
   c) the greatest perpendicular distance between the projectile and the plane.

CHAPTER 7 PROJECTILES

2 Another projectile is fired as in Question **1**, but this time it is fired down the plane with the same speed and still making an angle of 40° with the horizontal. Find

   a) the time of flight
   b) the range on the plane
   c) the greatest perpendicular distance between the projectile and the plane.

3 A projectile is fired with a speed of $60\,\text{m s}^{-1}$ from a point, O, on a plane inclined at an angle of 20° to the horizontal. The angle of elevation at which the projectile is fired is adjusted so that the range is a maximum. Find the angle of elevation (measured from the horizontal) if the projectile is fired

   a) up the plane   **b)** down the plane.

4 A projectile is fired with a speed of $30\,\text{m s}^{-1}$ up a plane inclined at an angle of 20° to the horizontal. It lands a distance of 25 m away from its point of projection. Find the two possible values for the angle of projection.

5 The projectile in Question **4** is now fired down the plane with the same result. Find the two possible values for the angle of projection in this situation.

6 A projectile is fired up a plane inclined at an angle α to the horizontal so that its initial direction makes an angle θ with the plane. Show that, when it lands, it makes an angle λ with the plane where

$$\tan \lambda = \frac{\tan \alpha}{1 - 2\tan\alpha \tan\theta}$$

7 A projectile is fired with a speed of $U\,\text{m s}^{-1}$ up a plane inclined at an angle α to the horizontal. The initial direction makes an angle θ with the plane. When it hits the plane, the projectile is moving horizontally.

   a) Show that $\tan(\theta + \alpha) = 2\tan\alpha$.
   b) For the case where $\alpha = 20°$, find **i)** the time of flight and **ii)** the range up the plane.

8 A projectile is fired down a plane, inclined at an angle α to the horizontal, with a speed of $U\,\text{m s}^{-1}$. The angle between the line of projection and the plane is θ. Find

   a) the time of flight
   b) the range down the plane
   c) the greatest vertical distance from the plane
   d) the maximum range down the plane.

9 A ball is fired at $20\,\text{m s}^{-1}$ from a point A on a plane inclined at 30° to the horizontal. The path of the ball is up the line of greatest slope and its initial direction makes an angle of θ with the plane. The ball strikes the plane at 90° and rebounds. The coefficient of restitution between the ball and the ground is 0.8. The ball strikes the plane a second time at a point B. Find the distance AB.

10 A particle, projected with speed u up a plane inclined at an angle θ to the horizontal, has a maximum range R. When fired with the same speed down the plane, its maximum range is $2R$. Find R and θ.

11 A particle is projected from origin O with initial speed V and elevation α. The ground under the flight path is horizontal from O to a point A, where $OA = a$, and then slopes up at an angle α. If the particle hits this slope at right angles, find V in terms of a, g and α.

12 A projectile is fired from a point A on a plane inclined at $30°$ to the horizontal. The projectile has initial speed V, and an initial elevation of $70°$ to the horizontal. The projectile lands at point B. The projectile is then fired from B down the plane. The initial speed is the same and the projectile lands at A. Show that there are two possible angles of projection at B and find them.

13 A particle P is projected from a point O with speed V and at an angle of θ to the horizontal. Line OQ corresponds to its original direction, and QP is the vertical line through the particle at time t. The lengths OQ and QP are called the **Littlewood coordinates** of P, represented as (ξ, η).

  a) Show that ξ and η are independent of θ.
  b) Use Littlewood coordinates, together with the sine rule, to find the time of flight of a particle projected up a plane inclined at α to the horizontal, if the initial direction of the particle is at an angle β to the plane.

Air resistance

In many modelling situations, one of the first assumptions that we make is that air resistance can be neglected. This leads to simpler equations, which are easier to solve. However, the solutions may not be good enough when we have a real-world problem to solve. We then need to revise our assumptions and air resistance is something we might wish to include in our model.

Air resistance (or any resistance caused by a fluid) is the result of the interaction between a body and the air through which it passes. It arises from several causes, the main ones being:

- There is friction between the surface of the body and the air next to that surface.
- The air in front of the body has to be pushed out of the way before the body can pass. The impact causes loss of momentum of the body.
- When the body has passed, there is a partial vacuum behind it.

We simplify these causes and assume that the air resistance is a single force acting on the body.

As we would expect, the magnitude of the air-resistance force depends upon the size of the body. The larger the body, the more air has to be pushed out of the way. It also depends on the shape of the body, the nature of its surface and, crucially, on the speed of the body.

However, the problem is not simple. If you have read any books about chaos theory, you will have met the problem of the dripping tap. As the water flow is increased, the droplets of water join to form a constant, smoothly flowing stream. As well as the speed of the flow increasing, the **nature** of the flow undergoes significant and quite sudden change. Something similar happens with air resistance, which means that a model which gives good predictions for low speeds is unlikely to be suitable at higher speeds.

CHAPTER 7 PROJECTILES

A fairly successful model for air resistance is given by

$$R = k_1 v + k_2 v^2$$

where v is the speed of the body and the constants k_1 and k_2 depend on the size, shape and texture of the body. The direction of R is, of course, directly opposite to that of the velocity.

The term $k_1 v$ dominates at low speeds and the term $k_2 v^2$ dominates at higher speeds (although the model itself is less successful at speeds approaching the speed of sound). Hence, when we are constructing a model, we normally use the dominant term and ignore the other one. This simplifies the solution of the resulting equations.

Example 3 A ball of mass m is dropped from a bridge. Assuming a linear model for air resistance:
a) What is its equation of motion?
b) What is the terminal velocity of the ball?
c) Find an expression for the velocity of the ball at time t.
d) Find an expression for the position of the ball at time t.

SOLUTION

a) The only forces acting on the ball are its weight and the air resistance.

Take downwards to be the positive direction, as shown.

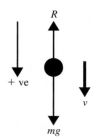

The equation of motion of the ball is

$$mg - R = m\ddot{x}$$
$$\Rightarrow \quad mg - kv = m\ddot{x} \qquad [1]$$

b) Initially, kv is small and so $mg - kv > 0$.

As v increases, $\ddot{x}$ will decrease, so that as $v \to \dfrac{mg}{k}$, $\ddot{x} \to 0$.

Zero acceleration means that the body must then move with a constant velocity, the **terminal velocity**, $\mathbf{v}_T$. Hence, we have

$$|\mathbf{v}_T| = \frac{mg}{k}$$

c) If we write $\ddot{x} = \dot{v}$, then [1] becomes

$$mg - kv = m\dot{v}$$

which gives

$$\int \frac{m \, dv}{mg - kv} = \int dt$$
$$\Rightarrow \quad -\frac{m}{k} \ln(mg - kv) = t + C_1$$

Using the initial condition $t = 0$, $v = 0$, we have $C_1 = -\dfrac{m}{k} \ln mg$.

146

Hence, we obtain

$$-\frac{m}{k}\ln(mg-kv) = t - \frac{m}{k}\ln mg$$

$$\Rightarrow \quad \ln\left(\frac{mg-kv}{mg}\right) = -\frac{kt}{m}$$

$$\Rightarrow \quad 1 - \frac{kv}{mg} = e^{-kt/m}$$

$$\Rightarrow \quad v = \frac{mg}{k}(1 - e^{-kt/m}) \qquad [2]$$

We note that, as t increases, the term $e^{-kt/m}$ diminishes and the speed approaches $\frac{mg}{k}$, as we deduced from consideration of the equation of motion. What this formula confirms is that the speed can **never exceed** $\frac{mg}{k}$.

d) From [2], we have

$$\frac{dx}{dt} = \frac{mg}{k}(1 - e^{-kt/m})$$

$$\Rightarrow \quad x = \int \frac{mg}{k}(1 - e^{-kt/m})\,dt$$

$$\Rightarrow \quad x = \frac{mg}{k}\left(t + \frac{m}{k}e^{-kt/m}\right) + C_2$$

Using the initial condition $x = 0$ when $t = 0$, we have $C_2 = -\frac{m^2 g}{k^2}$.
Hence, we obtain

$$x = \frac{mg}{k}t + \frac{m^2 g}{k^2}(e^{-kt/m} - 1) \qquad [3]$$

Note Equation [1] could be written as

$$\frac{d^2 x}{dt^2} + \frac{k}{m}\frac{dx}{dt} = g$$

which could be solved as a second-order linear differential equation using the complementary function and particular integral (see pages 56–66) to give equation [3].

Example 4 A free-fall parachutist, of mass m, has jumped from an aircraft. She is falling with a velocity of $2V$ when she opens her parachute. Assume air resistance is proportional to the square of velocity ($R = kv^2$) and that the parachute ultimately halves her speed.

a) What is her equation of motion after the parachute opens?
b) Find the constant k in terms of m, g and V.
c) Find an expression for her velocity at time t after the parachute opens.
d) Find an expression for her velocity at distance x below where the parachute opens.

SOLUTION

a) The only forces acting on the parachutist are the parachutist's weight and the air resistance.

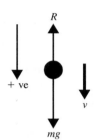

The equation of motion of the parachutist is
$$mg - R = m\ddot{x}$$
$$\Rightarrow \quad mg - kv^2 = m\ddot{x} \qquad [1]$$

b) Terminal velocity, v_T, corresponds to zero acceleration. So, we have
$$kv_T^2 = mg$$
Hence, the terminal velocity is
$$v_T = \sqrt{\frac{mg}{k}}$$
But the terminal velocity is V. Hence, we have
$$V = \sqrt{\frac{mg}{k}}$$
$$\Rightarrow \quad k = \frac{mg}{V^2} \qquad [2]$$

c) If we write $\ddot{x} = \dfrac{dv}{dt}$, then we have
$$mg - kv^2 = m\frac{dv}{dt}$$
$$\Rightarrow \quad V^2 - v^2 = \frac{m}{k}\frac{dv}{dt}$$
which gives
$$\int \frac{dv}{V^2 - v^2} = \frac{k}{m}\int dt$$
Since, for this problem, $v > V$, we can avoid the possibility of a negative logarithm by writing this as
$$\int \frac{dv}{v^2 - V^2} = -\frac{k}{m}\int dt$$

$$\Rightarrow \quad \frac{1}{2V}\int\left(\frac{1}{v-V}-\frac{1}{v+V}\right)dv = -\frac{k}{m}\int dt$$

$$\Rightarrow \quad \frac{1}{2V}[\ln(v-V) - \ln(v+V)] = -\frac{kt}{m} + C_1$$

To find C_1, we use the initial condition $t = 0$, when $v = 2V$, giving

$$C_1 = \frac{1}{2V}(\ln V - \ln 3V) = -\frac{1}{2V}\ln 3$$

Hence, we have

$$\frac{1}{2V}[\ln(v-V) - \ln(v+V)] = -\frac{kt}{m} - \frac{1}{2V}\ln 3$$

$$\Rightarrow \quad \ln\frac{3(v-V)}{(v+V)} = -\frac{2Vkt}{m}$$

Substituting from [2], we obtain

$$\ln\frac{3(v-V)}{(v+V)} = -\frac{2gt}{V}$$

$$\Rightarrow \quad \frac{3(v-V)}{(v+V)} = e^{-2gt/V}$$

which we rearrange to give

$$v = \frac{V(3 + e^{-2gt/V})}{3 - e^{-2gt/V}} \qquad [3]$$

We note that, as expected, $v \to V$ as t increases.

d) If we write $\dfrac{d^2x}{dt^2} = v\dfrac{dv}{dx}$, then from [1] we have

$$mg - kv^2 = mv\frac{dv}{dx}$$

$$\Rightarrow \quad V^2 - v^2 = \frac{mv}{k}\frac{dv}{dx}$$

which gives

$$\int\frac{v\,dv}{V^2 - v^2} = \frac{k}{m}\int dx$$

$$\Rightarrow \quad \int\frac{2v\,dv}{v^2 - V^2} = -\frac{2k}{m}\int dx$$

$$\Rightarrow \quad \ln(v^2 - V^2) = -\frac{2kx}{m} + C_2$$

To find C_2, we use the initial condition $x = 0$ when $v = 2V$, giving

$$C_2 = \ln 3V^2$$

Hence, we have
$$\ln(v^2 - V^2) = -\frac{2kx}{m} + \ln 3V^2$$
$$\Rightarrow \ln\left(\frac{v^2 - V^2}{3V^2}\right) = -\frac{2gx}{V^2} \quad \text{(from [2])}$$
$$\Rightarrow \frac{v^2 - V^2}{3V^2} = e^{-2gx/V^2}$$

which gives
$$v = V\sqrt{1 + 3e^{-2gx/V^2}}$$

We notice once again that, as x increases, $v \to V$.

Example 5 A ball of mass 1 kg is thrown directly upwards with an initial speed of $50\,\text{m s}^{-1}$. When the ball is travelling with speed v, air resistance is $0.08v$. Find the maximum height to which the ball rises and the time take to do so. Take $g = 10\,\text{m s}^{-2}$.

SOLUTION

Let the height of the ball be s m at time t s. From Newton's second law, we have
$$\frac{d^2s}{dt^2} = \frac{dv}{dt} = -g - 0.08v \qquad [1]$$
$$\Rightarrow \int \frac{0.08\,dv}{0.08v + 10} = -0.08 \int dt$$
$$\Rightarrow \ln(0.08v + 10) = -0.08t + C_1$$

Using the initial conditions $v = 50$ when $t = 0$, we have $C_1 = \ln 14$, which gives
$$\ln\left(\frac{0.08v + 10}{14}\right) = -0.08t \qquad [2]$$

When the ball reaches maximum height, $v = 0$, which gives $t = 4.21$.

We can rewrite equation [2] as
$$v = \frac{ds}{dt} = 175\,e^{-0.08t} - 125$$

Integrating this, we obtain
$$s = -2187.5\,e^{-0.08t} - 125t + C_2$$

Using the initial conditions $s = 0$ when $t = 0$, we have $C_2 = 2187.5$, which gives
$$s = 2187.5(1 - e^{-0.08t}) - 125t \qquad [3]$$

Putting $t = 4.21$, we get $s = 99.3$.

So, the ball rises to a height of 99.3 m and takes 4.21 s to do so.

AIR RESISTANCE

Some observations on Example 5

For comparison, a model assuming no air resistance would predict that the ball would rise to a maximum height of 125 m in a time of 5 s.

If our model were to assume no air resistance, the ball would return to ground level from its highest point in the same time as its upwards journey, and with a final speed equal to its initial speed. When the model allows for air resistance, this is no longer the case. In Example 5, equations [2] and [3] are satisfied by $s = 0$, $t = 0$ and $v = 50$, which are the initial values.

You should verify that these two equations are also satisfied by $s = 0$, $t = 8.94$ and $v = -39.4$, showing that the ball takes 4.73 s to return to the ground from its highest point, and arrives with a speed of $39.4 \, \text{m s}^{-1}$.

Equation [1] is valid throughout the motion because on the way down v is negative, which means that the air-resistance term will correspond to an upward force, as it should. Had the resistance been proportional to v^2, we would have had to adjust the equation of motion for the downward part of the journey.

Analysing the general motion of a projectile subject to air resistance can be quite complex. Here we look at a simple example to give a flavour of the situation.

Example 6 A projectile of mass 1 kg is fired from ground level at $50 \, \text{m s}^{-1}$ and with an elevation of $30°$. Air resistance is given by $\mathbf{R} = -0.05 \mathbf{v}$. Find the equation of the path of the projectile. Take $g = 10 \, \text{m s}^{-2}$.

SOLUTION

Taking **i** and **j** unit vectors in the horizontal and vertical directions respectively, we have the equation of motion of the projectile

$$\ddot{\mathbf{r}} = -10\mathbf{j} - 0.05\dot{\mathbf{r}} \qquad [1]$$

The **i**-component of [1] gives

$$\ddot{x} = -0.05\dot{x}$$

which we can write as the second-order linear differential equation

$$\frac{d^2x}{dt^2} + 0.05\frac{dx}{dt} = 0$$

Solving this by the usual method (see pages 56–60), we find

$$x = A + Be^{-0.05t}$$

Differentiating, we get

$$\dot{x} = -0.05 \, Be^{-0.05t}$$

Using the initial conditions $t = 0$, $x = 0$, $\dot{x} = 50\cos 30° = 25\sqrt{3}$, we have

$$A = 500\sqrt{3} \quad \text{and} \quad B = -500\sqrt{3}$$

So, we obtain

$$x = 500\sqrt{3}\,(1 - e^{-0.05t}) \qquad [2]$$

The **j**-component of [1] gives
$$\ddot{y} = -10 - 0.05\dot{y}$$
which we can write as the second-order linear differential equation
$$\frac{d^2 y}{dt^2} + 0.05 \frac{dy}{dt} = -10$$
Solving this by the usual method, we find
$$y = C + De^{-0.05t} - 200t$$
Differentiating, we get
$$\dot{y} = -0.05 D e^{-0.05t} - 200$$
Using the initial conditions $t = 0$, $y = 0$, $\dot{y} = 50 \sin 30° = 25$, we have
$$C = 4500 \quad \text{and} \quad D = -4500$$
So, we obtain
$$y = 4500(1 - e^{-0.05t}) - 200\,t \qquad [3]$$
From [2], we have
$$(1 - e^{-0.05t}) = \frac{x}{500\sqrt{3}} \quad \Rightarrow \quad t = 20 \ln\left(1 - \frac{x}{500\sqrt{3}}\right)$$
Substituting into [3], we find
$$y = 3x\sqrt{3} + 4000 \ln\left(1 - \frac{x}{500\sqrt{3}}\right) \qquad [4]$$
which is the equation of the path of the projectile.

Some observations on Example 6

$\dot{x} = 25\sqrt{3}\,e^{-0.05t}$ and $x = 500\sqrt{3}\,(1 - e^{-0.05t})$. As t increases, $\dot{x} \to 0$ and $x \to 500\sqrt{3}$. If the projectile were fired from the top of a cliff, its path would be asymptotic to the line $x = 500\sqrt{3}$.

$\dot{y} = -225 e^{-0.05t} - 200$. As t increases, $\dot{y} \to -200$, which is the projectile's terminal velocity as its path becomes effectively vertical.

You should compare equation [4] in Example 6 with the path obtained from using a model with no air resistance. This is
$$y = \frac{x\sqrt{3}}{3} - \frac{x^2}{375}$$
Use a graphics calculator or graph-drawing software to compare the two trajectories. Try experimenting with the air resistance, which in general is $\mathbf{R} = -k\mathbf{v}$, to ascertain the effect of altering the value of k.

Exercise 7B

Throughout this exercise, take $g = 10 \, \text{m s}^{-2}$.

1. A ballbearing of mass 0.001 kg is dropped into a vat of liquid. Assuming a linear model for the resistance to motion of the liquid in which $\mathbf{R} = -0.08\mathbf{v}$, find
   a) the equation of motion of the ballbearing
   b) the terminal velocity of the ballbearing
   c) an expression for the velocity at time t after being dropped
   d) an expression for the distance fallen at time t.

2. A ball of mass m is dropped from a bridge. Assuming air resistance is given by $R = kv^2$:
   a) What is the equation of motion of the ball?
   b) What is the terminal velocity of the ball?
   c) Find an expression for the velocity of the ball at time t.
   d) Find an equation relating the velocity, v, of the ball and the distance, x, which it has fallen

3. A parachutist of mass m opens his parachute when he is falling with a velocity $3V$, where V is his terminal velocity with the parachute open. Assuming a linear model $\mathbf{R} = -k\mathbf{v}$ for air resistance:
   a) What is his equation of motion?
   b) Find k in terms of m, g and V.
   c) Find an expression for the velocity of the parachutist at time t.
   d) Find an equation relating the velocity, v, of the parachutist and the distance, x, which he has fallen since opening his parachute.

4. A projectile of mass 1 kg is fired vertically upwards with a speed of $30 \, \text{m s}^{-1}$ from the ground. Assuming a linear model for air resistance of the form $\mathbf{R} = -k\mathbf{v}$, where the constant k takes the value $2 \times 10^{-2} \, \text{kg s}^{-1}$:
   a) What is the projectile's equation of motion?
   b) Find an expression for its velocity at time t.
   c) How long does it take to reach its maximum height?
   d) What is the greatest height reached by the projectile?
   e) Confirm that the projectile returns to the ground when $t = 5.8846 \, \text{s}$, and hence find the work done against air resistance during the flight.

5. A projectile of mass 1 kg is fired vertically upwards with a speed of $30 \, \text{m s}^{-1}$ from the ground. Assuming a quadratic model for air resistance of the form $R = kv^2$, where the constant k takes the value $4 \times 10^{-3} \, \text{kg m}^{-1}$:
   a) What is its equation of motion as the projectile rises?
   b) Find an expression for its velocity, as it rises, at time t.
   c) How long does it take to reach its maximum height?
   d) What is the greatest height reached by the projectile?
   e) What is its equation of motion as it falls?
   f) How long does it take to reach the ground?

6 A projectile of mass m is fired with initial speed V at an angle α to the horizontal. Air resistance is given by $\mathbf{R} = -mk\mathbf{v}$.

a) Show that the equation of its trajectory is

$$y = \left(\tan \alpha + \frac{g}{kV} \sec \alpha\right) x + \frac{g}{k^2} \ln\left(1 - \frac{k}{V} x \sec \alpha\right)$$

b) Show that the Littlewood coordinates of the projectile at time t (see Exercise 7A, Question **13**) are independent of α.

8 Motion of a particle in a plane

They sailed away for a year and a day,
To the land where the Bong-tree grows.
EDWARD LEAR

Cartesian coordinates

If a particle moves in the xy-plane, we can specify its position in several ways.

If we can establish the forces acting on the particle in two fixed, mutually perpendicular directions, it is convenient to use cartesian coordinates when formulating our equations of motion. We will find later that polar coordinates are preferable in other situations.

We have already met the relationship between the position, velocity and acceleration vectors for a particle, but we will rehearse them here for the sake of completeness.

Suppose a particle moves so that at time t it is at the point P(x, y), where

$$x = f(t) \quad \text{and} \quad y = g(t)$$

These give the parametric equations of the curve along which the particle is moving. Its position vector at time t is

$$\mathbf{r} = x\mathbf{i} + y\mathbf{i}$$

If we differentiate with respect to t, we get

$$\frac{d\mathbf{r}}{dt} = \frac{dx}{dt}\mathbf{i} + \frac{dy}{dt}\mathbf{j} \quad \text{or} \quad \dot{\mathbf{r}} = \dot{x}\mathbf{i} + \dot{y}\mathbf{j}$$

$\dot{x}$ and $\dot{y}$ are the velocity components in the x and y directions, so we have $\dot{\mathbf{r}} = \mathbf{v}$. In particular, the velocity direction makes an angle α with the x-direction, where

$$\tan \alpha = \frac{\dot{y}}{\dot{x}} = \frac{dy}{dx}$$

That is, the direction of motion is tangential to the curve.

Differentiating again, we have

$$\frac{d^2\mathbf{r}}{dt^2} = \frac{d\mathbf{v}}{dt} = \frac{d^2x}{dt^2}\mathbf{i} + \frac{d^2y}{dt^2}\mathbf{j} \quad \text{or} \quad \ddot{\mathbf{r}} = \dot{\mathbf{v}} = \ddot{x}\mathbf{i} + \ddot{y}\mathbf{j}$$

which gives the acceleration vector, $\mathbf{a}$.

Example 1 A particle of mass 1 kg is initially at rest at the origin. It is subject to a constant force of 5 N in the x-direction, and a variable force of magnitude $2v$ N acting at right angles to its path, as shown, where v is the speed of the particle. Find the equation for the path of the particle.

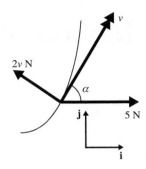

SOLUTION

If at time t the direction of motion makes an angle α with the x-direction, as shown, the resultant force acting on the particle is
$$\mathbf{F} = (5 - 2v\sin\alpha)\mathbf{i} + 2v\cos\alpha\,\mathbf{j}$$
By Newton's second law, we have
$$\mathbf{F} = m\mathbf{a} = m(\ddot{x}\mathbf{i} + \ddot{y}\mathbf{j})$$
$$\Rightarrow \quad \ddot{x} = 5 - 2v\sin\alpha \quad \text{and} \quad \ddot{y} = 2v\cos\alpha$$
But $v\cos\alpha = \dot{x}$ and $v\sin\alpha = \dot{y}$, so we have
$$\ddot{x} = 5 - 2\dot{y} \qquad [1]$$
$$\ddot{y} = 2\dot{x} \qquad [2]$$
Integrating equations [1] and [2], we obtain
$$\dot{x} = 5t - 2y + C_1$$
$$\dot{y} = 2x + C_2$$
As the particle is initially at rest at the origin, we have $C_1 = C_2 = 0$. Hence,
$$\dot{x} = 5t - 2y \qquad [3]$$
$$\dot{y} = 2x \qquad [4]$$
From equations [2] and [3], we have
$$\ddot{y} + 4y = 10t$$
We solve this second-order differential equation (see pages 56–66) to get
$$y = A\cos 2t + B\sin 2t + 2.5t$$
$$\Rightarrow \quad \dot{y} = -2A\sin 2t + 2B\cos 2t + 2.5$$
Initially, $y = \dot{y} = 0$, which gives $A = 0$, $B = -1.25$. Hence, we have
$$y = 1.25(2t - \sin 2t)$$
and from equation [4]
$$x = 1.25(1 - \cos 2t)$$
So, the particle moves along the curve with parametric equations
$$x = 1.25(1 - \cos 2t) \quad \text{and} \quad y = 1.25(2t - \sin 2t)$$

Distance travelled

To find the distance travelled by a particle moving along a curve, we need to be able to find the arc length of the curve. You may have met this in your study of pure mathematics, but we include it here.

If δs is a small element of arc, involving displacements δx and δy, as shown, we have

$$(\delta s)^2 \approx (\delta x)^2 + (\delta y)^2 \qquad [1]$$

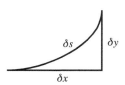

If the equation of the curve is given in the cartesian form $y = f(x)$, we write equation [1] as

$$\frac{\delta s}{\delta x} \approx \sqrt{1 + \left(\frac{\delta y}{\delta x}\right)^2}$$

If the equation of the curve is given in the parametric form $x = f(t)$, $y = g(t)$, we write equation [1] as

$$\frac{\delta s}{\delta t} \approx \sqrt{\left(\frac{\delta x}{\delta t}\right)^2 + \left(\frac{\delta y}{\delta t}\right)^2}$$

When we proceed to the limit, these become

$$\frac{ds}{dx} = \sqrt{1 + \left(\frac{dy}{dx}\right)^2}$$

and

$$\frac{ds}{dt} = \sqrt{\left(\frac{dx}{dt}\right)^2 + \left(\frac{dy}{dt}\right)^2}$$

Hence, we have

$$s = \int \sqrt{1 + \left(\frac{dy}{dx}\right)^2}\, dx \quad \text{or} \quad s = \int \sqrt{\left(\frac{dx}{dt}\right)^2 + \left(\frac{dy}{dt}\right)^2}\, dt$$

Example 2 A particle moves with constant acceleration $(\mathbf{i} + \mathbf{j})\,\text{m s}^{-2}$. Initially, it is at the origin and travelling at $2\,\text{m s}^{-1}$ in the x-direction. Find the distance travelled by the particle in the first 2 seconds.

SOLUTION

At time t, the velocity of the particle is $\mathbf{v} = (t+2)\mathbf{i} + t\mathbf{j}$, from which we have

$$\frac{dx}{dt} = t + 2 \quad \text{and} \quad \frac{dy}{dt} = t$$

Hence, we obtain

$$s = \int_0^2 \sqrt{(t+2)^2 + t^2}\, dt$$

$$\Rightarrow \quad s = \sqrt{2} \int_0^2 \sqrt{(t+1)^2 + 1}\, dt$$

We substitute $t+1 = \sinh u$, which gives

$$dt = \cosh u \, du \quad \text{and} \quad \sqrt{(t+1)^2 + 1} = \cosh u$$

When $t = 0$, $\sinh u = 1$, $\cosh u = \sqrt{2}$ and $u = \text{arsinh } 1$.

When $t = 2$, $\sinh u = 3$, $\cosh u = \sqrt{10}$ and $u = \text{arsinh } 3$.

Hence, we have

$$s = \sqrt{2} \int_{\text{arsinh } 1}^{\text{arsinh } 3} \cosh^2 u \, du$$

$$\Rightarrow \quad s = \frac{1}{\sqrt{2}} \int_{\text{arsinh } 1}^{\text{arsinh } 3} (1 + \cosh 2u) \, du$$

$$\Rightarrow \quad s = \frac{1}{\sqrt{2}} \left[u + \frac{1}{2} \sinh 2u \right]_{\text{arsinh } 1}^{\text{arsinh } 3} = 6.37$$

So, the particle travels 6.37 m in the first 2 seconds.

Exercise 8A

1. A particle of mass 2 kg moves under the action of a constant force $\mathbf{F} = (4\mathbf{i} + 2\mathbf{j})$ N. Initially, the particle is at the origin and travelling with velocity $(\mathbf{i} + \mathbf{j})$ m s^{-1}. Find the cartesian equation of the path of the particle.

2. A particle moves in the xy-plane so that at time t its acceleration is given by $\mathbf{a} = 4\mathbf{i} - 2y\mathbf{j}$. The particle starts from rest at $(0, 8)$. Find the cartesian equation of the path of the particle.

3. A particle moves so that at time t its acceleration is given by $\mathbf{a} = -\sin t \, \mathbf{i} + \cos t \, \mathbf{j}$. Initially, the particle is at the origin and moving with velocity $2\mathbf{i}$.

   a) Find the parametric equations of the path of the particle.
   b) Find the distance travelled by the particle in the first 2 seconds.

4. A particle moves in the xy-plane so that when it is at point P its acceleration is directed towards the origin O and has magnitude $\omega^2 \times$ OP. Initially, the particle is at the point $(4, 0)$ and is moving with velocity $3\omega \mathbf{j}$. Find the cartesian equation of the path of the particle.

5. A particle of mass 1 kg moves in the xy-plane so that when it is at point P it is subject to a force directed towards the origin O and of magnitude 2OP, and also to a resistive force of magnitude $3v$, where v is the speed of the particle. The particle is initially at $(4, 0)$ and moving with velocity $4\mathbf{j}$.

   a) Find the parametric equations of the path of the particle, and hence show that the cartesian equation of its path is
   $$x^2 - 2xy + y^2 - 4x + 8y = 0$$

   b) Deduce that the particle approaches O but will not reach it in a finite time.

6 A particle moves in three dimensions so that at time t it has acceleration given by
$$\ddot{\mathbf{r}} = 2\dot{\mathbf{r}} - 2\mathbf{r} + 2(t-1)\mathbf{k}$$
Initially, the particle is at the point $(1, 0, 0)$ and moving with velocity $\mathbf{i} + \mathbf{j} + \mathbf{k}$. Find the parametric equations of its path and describe the curve along which it is travelling.

7 A particle of mass m moves in the xy-plane under the action of a force $\mathbf{F} = 2m\mathbf{i} - my\mathbf{j}$. The particle starts from rest at the point $(0, 1)$. Find the cartesian equation of the path of the particle.

8 A particle of mass m is moving in the xy-plane subject to a constant force $\mathbf{F} = 3\mathbf{i} + 2\mathbf{j}$, and to a variable force of magnitude mv, where v is the speed of the particle, in the direction which is a 90° anticlockwise rotation of the particle's direction of motion. Initially, the particle is at rest at the origin. Find the parametric equations of the path of the particle.

Polar resolutes of acceleration

As a point, P, moves along a curve, its position can be specified in polar coordinates (r, θ) relative to an initial line Ox, as shown. In many examples involving the motion of a particle, it is convenient to use polar coordinates.

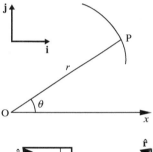

The position vector, $\mathbf{r}$, of P can be expressed in cartesian components as
$$\mathbf{r} = r(\cos\theta \, \mathbf{i} + \sin\theta \, \mathbf{j})$$

We now define $\hat{\mathbf{r}}$ as the unit vector in the direction OP (the **radial** direction) and $\hat{\mathbf{s}}$ as the unit vector in the direction of a 90° positive rotation of $\hat{\mathbf{r}}$ (the **transverse** direction). As can be seen from the diagram, we have

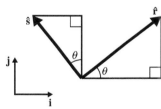

$$\hat{\mathbf{r}} = \cos\theta \, \mathbf{i} + \sin\theta \, \mathbf{j}$$
$$\hat{\mathbf{s}} = -\sin\theta \, \mathbf{i} + \cos\theta \, \mathbf{j}$$

As point P moves along the curve, θ changes and so $\hat{\mathbf{r}}$ and $\hat{\mathbf{s}}$ change. Their rates of change with respect to θ are

$$\frac{d\hat{\mathbf{r}}}{d\theta} = -\sin\theta \, \mathbf{i} + \cos\theta \, \mathbf{j} = \hat{\mathbf{s}} \quad \text{and} \quad \frac{d\hat{\mathbf{s}}}{d\theta} = -\cos\theta \, \mathbf{i} - \sin\theta \, \mathbf{j} = -\hat{\mathbf{r}}$$

and with respect to t are

$$\frac{d\hat{\mathbf{r}}}{dt} = \frac{d\hat{\mathbf{r}}}{d\theta} \times \frac{d\theta}{dt} = \dot{\theta}\hat{\mathbf{s}} \quad \text{and} \quad \frac{d\hat{\mathbf{s}}}{dt} = \frac{d\hat{\mathbf{s}}}{d\theta} \times \frac{d\theta}{dt} = -\dot{\theta}\hat{\mathbf{r}}$$

The position vector of P is
$$\mathbf{r} = r\hat{\mathbf{r}}$$

CHAPTER 8 MOTION OF A PARTICLE IN A PLANE

The velocity, **v** or $\dot{\mathbf{r}}$, of P is

$$\dot{\mathbf{r}} = \frac{\mathrm{d}(r\hat{\mathbf{r}})}{\mathrm{d}t}$$

$$\Rightarrow \quad \dot{\mathbf{r}} = \dot{r}\hat{\mathbf{r}} + r\frac{\mathrm{d}\hat{\mathbf{r}}}{\mathrm{d}t}$$

$$\Rightarrow \quad \dot{\mathbf{r}} = \dot{r}\hat{\mathbf{r}} + r\dot{\theta}\hat{\mathbf{s}}$$

The acceleration, **a**, $\dot{\mathbf{v}}$ or $\ddot{\mathbf{r}}$, of P is

$$\ddot{\mathbf{r}} = \frac{\mathrm{d}}{\mathrm{d}t}(\dot{r}\hat{\mathbf{r}} + r\dot{\theta}\hat{\mathbf{s}})$$

$$\Rightarrow \quad \ddot{\mathbf{r}} = \ddot{r}\hat{\mathbf{r}} + \dot{r}\frac{\mathrm{d}\hat{\mathbf{r}}}{\mathrm{d}t} + \dot{r}\dot{\theta}\hat{\mathbf{s}} + r\ddot{\theta}\hat{\mathbf{s}} + r\dot{\theta}\frac{\mathrm{d}\hat{\mathbf{s}}}{\mathrm{d}t}$$

$$\Rightarrow \quad \ddot{\mathbf{r}} = \ddot{r}\hat{\mathbf{r}} + \dot{r}\dot{\theta}\hat{\mathbf{s}} + \dot{r}\dot{\theta}\hat{\mathbf{s}} - r\dot{\theta}^2\hat{\mathbf{r}}$$

$$\Rightarrow \quad \ddot{\mathbf{r}} = (\ddot{r} - r\dot{\theta}^2)\hat{\mathbf{r}} + (r\ddot{\theta} + 2\dot{r}\dot{\theta})\hat{\mathbf{s}}$$

The components of velocity and acceleration in the $\hat{\mathbf{r}}$ direction are the **radial components**. The components in the $\hat{\mathbf{s}}$ direction are the **transverse components**.

Summary

For a point P(r, θ) moving in a plane, we have

$\mathbf{r} = r\hat{\mathbf{r}}$

$\mathbf{v} = \dot{\mathbf{r}} = \dot{r}\hat{\mathbf{r}} + r\dot{\theta}\hat{\mathbf{s}}$

$\mathbf{a} = \dot{\mathbf{v}} = \ddot{\mathbf{r}} = (\ddot{r} - r\dot{\theta}^2)\hat{\mathbf{r}} + (r\ddot{\theta} + 2\dot{r}\dot{\theta})\hat{\mathbf{s}}$

where $\hat{\mathbf{r}}$ and $\hat{\mathbf{s}}$ are unit vectors in the radial and transverse directions respectively.

Hence, we have

Radial component of velocity $= \dot{r}$
Transverse component of velocity $= r\dot{\theta}$
Radial component of acceleration $= \ddot{r} - r\dot{\theta}^2$
Transverse component of acceleration $= r\ddot{\theta} + 2\dot{r}\dot{\theta}$

Note The expression $r\ddot{\theta} + 2\dot{r}\dot{\theta}$ can be written as $\dfrac{1}{r}\dfrac{\mathrm{d}}{\mathrm{d}t}(r^2\dot{\theta})$.

Example 3 A particle moves on the curve $r = a(1 - \sin\theta)$ with constant angular velocity ω.

a) Find the radial and transverse components of velocity and acceleration at time t.

b) Find the values of θ for which the particle is instantaneously stationary.

SOLUTION

For all t, we have $\dot{\theta} = \omega$ and $\ddot{\theta} = 0$.

We also have
$$r = a(1 - \sin\theta)$$
which gives
$$\dot{r} = -a\cos\theta\,\dot{\theta} = -a\omega\cos\theta$$
and
$$\ddot{r} = a\omega\sin\theta\,\dot{\theta} = a\omega^2\sin\theta$$

a) Radial component of velocity: $\dot{r} = -a\omega\cos\theta$

Transverse component of velocity: $r\dot{\theta} = a\omega(1 - \sin\theta)$

Radial component of acceleration: $\ddot{r} - r\dot{\theta}^2 = a\omega^2\sin\theta - a\omega^2(1 - \sin\theta)$
$$= a\omega^2(2\sin\theta - 1)$$

Transverse component of acceleration: $r\ddot{\theta} + 2\dot{r}\dot{\theta} = -2a\omega^2\cos\theta$

b) The speed of the particle is given by
$$\sqrt{a^2\omega^2\cos^2\theta + a^2\omega^2(1 - \sin\theta)^2}$$

The particle is stationary when
$$\cos^2\theta + (1 - \sin\theta)^2 = 0$$
$$\Rightarrow \quad 2 - 2\sin\theta = 0$$
$$\Rightarrow \quad \theta = \frac{\pi}{2}$$

Example 4 A horizontal turntable, of radius 2 m and with centre O, is rotating at a constant angular velocity of 0.5 rad s^{-1}. A is a point on the circumference of the turntable, and there is a straight groove OA on the surface of the turntable. There is a small object of mass 0.1 kg which is free to move in the groove. Initially, the object is stationary at B, where OB = 1 m.

Explaining the modelling assumptions involved, find

a) the equation of the path followed by the object
b) the time taken for the object to reach A
c) the horizontal force exerted on the object by the side of the groove when the object is 1.5 m from O.

SOLUTION

Assumptions

- Friction is negligible.
- The object is a particle.
- The mass of the turntable is large compared with that of the moving object, so that the moment of inertia of the system is not significantly affected by the changing position of the object. If this is not the case, the angular velocity of the turntable could be affected, depending on the mechanism which is driving the turntable.

CHAPTER 8 MOTION OF A PARTICLE IN A PLANE

The model

We will take the position of OA at $t = 0$ to be the initial line for the polar coordinates. The initial conditions are, therefore, $t = 0, \theta = 0, r = 1, \dot{r} = 0$.

We also have $\dot{\theta} = 0.5, \ddot{\theta} = 0$ for all t.

a) As friction is negligible, there is no force acting in the direction OA. The radial acceleration is therefore zero.

$$\text{Radial acceleration} = \ddot{r} - r\dot{\theta}^2 \quad \Rightarrow \quad \ddot{r} - 0.25r = 0$$

Solving this differential equation (see pages 56–60), we obtain

$$r = A e^{0.5t} + B e^{-0.5t} \qquad [1]$$

$$\Rightarrow \quad \dot{r} = 0.5 A e^{0.5t} - 0.5 B e^{-0.5t} \qquad [2]$$

Substituting our initial conditions into [1] and [2], we get

$$A + B = 1$$

and $\quad A - B = 0$

Solving these, we have $A = B = 0.5$. Equation [1] then becomes

$$r = 0.5(e^{0.5t} + e^{-0.5t}) \quad \text{or} \quad r = \cosh 0.5t$$

As the angular velocity is $0.5\,\text{rad}\,\text{s}^{-1}$, we have $\theta = 0.5t$. The polar equation of the path of the object is, therefore,

$$r = \cosh \theta$$

b) When the object reaches A, we have $r = 2$, which gives

$$\cosh 0.5t = 2 \quad \Rightarrow \quad t = 2.634$$

So, the object takes 2.63 s to reach A.

c) The transverse acceleration is $r\ddot{\theta} + 2\dot{r}\dot{\theta}$. As $\ddot{\theta} = 0$ and $\dot{\theta} = 0.5$, we have

$$\text{Transverse acceleration} = \dot{r}$$

As $r = \cosh 0.5t$, we have, therefore,

$$\dot{r} = 0.5 \sinh 0.5t$$

$$\Rightarrow \quad \dot{r} = 0.5\sqrt{\cosh^2 0.5t - 1}$$

$$\Rightarrow \quad \dot{r} = 0.5\sqrt{r^2 - 1}$$

When $r = 1.5$, the transverse acceleration is

$$\dot{r} = 0.5\sqrt{1.5^2 - 1} = 0.559\,\text{m}\,\text{s}^{-2}.$$

Hence, from Newton's second law, we have

$$\text{Transverse force acting on object} = 0.1 \times 0.559 = 0.0559\,\text{N}$$

Modifying the model

In Example 4, the assumption most likely to be unjustified is that of negligible friction. If we assume that the groove in which the object is moving is rectangular in cross-section, there are two sources of friction.

- The contact between the object and the bottom of the groove will give a friction force of $\mu \times$ weight of object. This we could reasonably assume to be constant.
- The contact between the object and the side of the groove will give a friction force of $\mu \times$ transverse force. This will be variable, because the transverse force varies with the position of the object. However, the transverse force which we found in the example (0.0559 N) was much smaller than the weight (0.98 N) of the object, so it would seem safe to ignore this second source of friction.

We might reasonably modify the model by making the following assumption:

Friction is constant and equal to 0.98μ N

Try reworking Example 4 using this new model, taking, say, $\mu = 0.025$. (If μ is much larger than this, the object will not move at all.) You should find that the model now predicts that the time taken to reach A is 4.625 s and that when $r = 1.5$ m the transverse force is 0.026 N.

Example 5 A particle moves along a curve with constant angular velocity ω. At the point (r, θ), its radial component of acceleration is $5\omega^2(1-r)$. Initially, $\theta = 0$, $r = 5$ and the radial component of the velocity is 2. Find the polar equation of the curve.

SOLUTION

As $\dot{\theta} = \omega$, the radial component of acceleration is $\ddot{r} - \omega^2 r$. Hence, we have

$$\ddot{r} - \omega^2 r = 5\omega^2(1-r)$$
$$\Rightarrow \quad \ddot{r} + 4\omega^2 r = 5\omega^2$$

Solving this differential equation (see pages 56–66), we get

$$r = A \cos 2\omega t + B \sin 2\omega t + 1.25$$
$$\Rightarrow \quad \dot{r} = -2\omega A \sin 2\omega t + 2\omega B \cos 2\omega t$$

Substituting the initial conditions, we have

$$A + 1.25 = 5 \quad \Rightarrow \quad A = 3.75$$

and $\quad 2\omega B = 2 \quad \Rightarrow \quad B = \dfrac{1}{\omega}$

Thus, at time t, we have

$$r = 3.75 \cos 2\omega t + \frac{1}{\omega} \sin 2\omega t + 1.25$$

As the angular velocity is ω, we have $\theta = \omega t$, and hence the polar equation of the path is

$$4\omega r = 15\omega \cos 2\theta + 4 \sin 2\theta + 5\omega$$

CHAPTER 8 MOTION OF A PARTICLE IN A PLANE

Practical investigation

The nature of this topic makes it difficult to set up simple experiments. It is, however, relatively easy to explore the situation described in Example 4 on pages 161–2.

The main requirement is the turntable. By attaching a 4 m batten to an old record player, a track can be produced which rotates at 33 revolutions per minute.

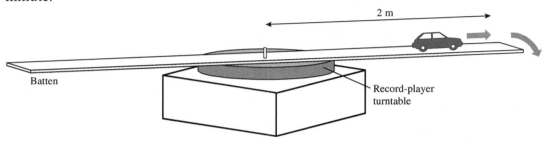

A small toy car with free running wheels can then be timed from rest until it falls off the end of the track (a distance of about a metre), and the time compared with that derived from the model.

The main difficulty with this setup is the relatively high speed of rotation, leading to travel times for the car in the order of 1 second, with consequent high relative error in the timing. If you have access to other equipment (such as the Leeds Mechanics Kit), you may be able to devise a more satisfactory apparatus.

Exercise 8B

1 A particle travels with constant angular velocity, ω, along the curve whose polar equation is $r = f(\theta)$. Find in terms of ω and θ the radial and transverse components of its velocity and acceleration at the point (r, θ) in the case where $f(\theta)$ is

  a) $a\theta$ **b)** $a\cos 2\theta$ **c)** $2\sin\theta + \cos\theta$ **d)** $\ln(1+\theta)$

2 A particle of mass m travels with constant angular velocity, ω, along the curve whose polar equation is $r = a(1 + \cos\theta)$. Show that, at the point (r, θ), the force on the particle is of magnitude $ma\omega^2\sqrt{5 + 4\cos\theta}$.

3 A particle P of mass m moves along the spiral $r = ae^\theta$ with constant angular velocity, ω, about the origin O. Show that the force acting on the particle is always perpendicular to OP, and find the magnitude of the force.

4 A particle travels with constant angular velocity, ω, along the curve whose polar equation is $r = a\sin\theta + b\cos\theta$.

  a) Find the radial and transverse components of its velocity and acceleration at the point $P(r, \theta)$.
  b) Show that the particle moves at a constant speed.

5 The radial and transverse components of velocity of a certain particle at the point (r, θ) are $2r^2$ and $3\theta^2$ respectively. Initially, $r = 1$, $\theta = 1$. Show that the polar equation of the path of the particle is

$$r^2 = \frac{3\theta}{4 - \theta}$$

Find the radial and transverse components of acceleration in terms of r and θ.

6 A particle moves on the curve $r(1 + \cos\theta) = a$ with constant angular velocity, ω. Show that the ratio of the transverse and radial components of velocity is $1 : \tan\frac{1}{2}\theta$.

7 A particle moves on the curve $r = a\sin 2\theta$ so that it has uniform angular velocity, ω, about the origin. Find the component of acceleration perpendicular to the initial line.

8 A particle starts from the origin with speed u and moves with constant angular velocity, ω, about the origin. The radial component of acceleration has constant magnitude ωu and is directed towards the origin. Taking the particle's initial direction as $\theta = 0$, show that the equation of the path of the particle is

$$r = \frac{u(1 - e^{-\theta})}{\omega}$$

9 A particle P of mass m moves with uniform angular speed, ω, about the origin O under the action of a variable force which is at all times in the direction of $\hat{\mathbf{s}}$, the transverse unit vector. Find the equation of the path of the particle (in terms of two arbitrary constants A and B), and show that the magnitude of the force can be written in the form

$$F = 2m\omega^2\sqrt{r^2 - k^2}$$

where k is a constant to be found in terms of A and B.

Motion under a central force

There is a class of problems in which a particle P moves relative to an origin O under the action of a force which is always directed along PO. Such a force is called a **central force**. Examples include planetary motion and the motion of particles in a magnetic or electric field or attached to an elastic string or spring. The force is usually directed towards O, although in some cases, such as magnetic repulsion, it could be directed away from O.

Area swept by the position vector

We will first examine the consequences of the fact that the force is a central force. If there is no transverse component to the force, the transverse component of acceleration is zero.

As we saw on page 160, the transverse component of acceleration can be written as $\frac{1}{r}\frac{d}{dt}(r^2\dot\theta)$. Hence, for a central force, we have

$$\frac{1}{r}\frac{d}{dt}(r^2\dot\theta) = 0$$
$$\Rightarrow \quad r^2\dot\theta = h \qquad [2]$$

where h is a constant.

Now consider the area swept out by the position vector as θ increases by a small amount $\delta\theta$.

The area of this elemental sector is given by

$$\delta A \approx \tfrac{1}{2} r^2 \delta\theta$$
$$\Rightarrow \quad \frac{\delta A}{\delta t} \approx \tfrac{1}{2} r^2 \frac{\delta\theta}{\delta t}$$

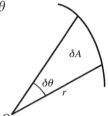

When we proceed to the limit, this becomes

$$\Rightarrow \quad \frac{dA}{dt} = \tfrac{1}{2} r^2 \dot\theta$$

So, from equation [2], we have

$$\frac{dA}{dt} = \tfrac{1}{2} h \qquad [3]$$

It follows that, for any central force, the area swept out by the position vector increases at a constant rate.

There is a second interpretation of h which is sometimes useful. From [2], we have

$$h = r^2\dot\theta = r \times r\dot\theta$$

But $r\dot\theta$ is the transverse component of velocity. So, it follows that h is the particle's **moment of velocity about the origin**.

Equation of the path

Suppose a particle P of mass m is moving under the action of a central force, F, directed towards an origin O. In general, F could be a function of both r and θ, but in most practical situations it is a function of r only, and it is this case which we will consider here.

The radial component of acceleration is $\ddot r - r\dot\theta^2$. The equation of motion of the particle is, therefore,

$$\ddot r - r\dot\theta^2 = -\frac{F}{m} \qquad [4]$$

We have to solve this equation to find the equation of the path of the particle.

This can best be effected by making the substitution $r = \dfrac{1}{u}$. Equation [2] then becomes

$$\dot{\theta} = hu^2 \qquad [5]$$

Also, we have

$$\dot{r} = -\frac{1}{u^2}\frac{du}{dt} = -\frac{1}{u^2}\frac{du}{d\theta}\frac{d\theta}{dt}$$

from which, using equation [5], we get

$$\dot{r} = -h\frac{du}{d\theta} \qquad [6]$$

Differentiating [6], we have

$$\ddot{r} = -h\frac{d^2u}{d\theta^2}\frac{d\theta}{dt}$$

from which, using equation [5], we get

$$\ddot{r} = -h^2u^2\frac{d^2u}{d\theta^2} \qquad [7]$$

Substituting from equations [5] and [7] into equation [4], we have

$$-h^2u^2\frac{d^2u}{d\theta^2} - h^2u^3 = -\frac{F}{m}$$

$$\Rightarrow \quad h^2u^2\left(\frac{d^2u}{d\theta^2} + u\right) = \frac{F}{m} \qquad [8]$$

If F is a function of r and therefore of u, equation [8] is a second-order differential equation in u and θ. If we can solve this, we have a relationship between u and θ, and hence between r and θ. In other words, we have the equation of the path of the particle.

Note It is inadvisable to try to memorise equation [8]. In an examination, you should instead derive it as needed.

Example 6 Two objects, each of mass m, are attached to the ends of a string AB of length 3 m. Initially, the objects lie at rest 3 m apart on a horizontal table, with the string against a fixed peg O, where OA is 2 m. The object at A is then given a velocity V perpendicular to the string, so that the string stays in contact with the peg in the subsequent motion. Find

a) the velocity of the object B when it reaches the peg
b) the equation of the path followed by A.

CHAPTER 8 MOTION OF A PARTICLE IN A PLANE

SOLUTION

Assumptions

- The string is light.
- The string is inextensible.
- The table is smooth.
- The peg is smooth.
- Air resistance is negligible.

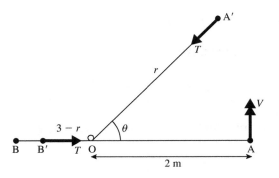

a) A′ and B′ show the positions of the objects at time t.

The radial component of acceleration at A′ is $\ddot{r} - r\dot{\theta}^2$. Hence, we have

$$\ddot{r} - r\dot{\theta}^2 = -\frac{T}{m} \qquad [1]$$

The acceleration at B′ is $\ddot{r}$ towards O. Hence, we also have

$$\ddot{r} = \frac{T}{m} \qquad [2]$$

We further know that $r^2\dot{\theta} = h$, where h is a constant given by the moment of velocity about O. As initially $r = 2$ and the transverse component of velocity is V, we have

$$r^2\dot{\theta} = 2V \qquad [3]$$

Using [2] and [3] to eliminate $\dfrac{T}{m}$ and $\dot{\theta}$ from [1], we get

$$\ddot{r} = \frac{2V^2}{r^3} \qquad [4]$$

The familiar formula $\dfrac{d^2s}{dt^2} = v\dfrac{dv}{ds}$ becomes $\ddot{r} = \dot{r}\dfrac{d\dot{r}}{dr}$ in the current notation. Using this, we can rewrite [4] as

$$\dot{r}\frac{d\dot{r}}{dr} = \frac{2V^2}{r^3}$$

$$\Rightarrow \quad \int \dot{r}\,d\dot{r} = \int 2V^2 r^{-3}\,dr$$

$$\Rightarrow \quad \dot{r}^2 = -2V^2 r^{-2} + c$$

Initially, $r = 2$ and $\dot{r} = 0$, giving $c = \tfrac{1}{2}V^2$. Hence, we have

$$\dot{r} = \frac{V}{r}\sqrt{\frac{r^2 - 4}{2}}$$

which is the speed of B′. When the object reaches the peg, the value of r is 3. So, we obtain

$$\text{Speed of object at peg} = V\sqrt{\frac{5}{18}}$$

b) From equations [1] and [2], we have
$$2\ddot{r} - r\dot{\theta}^2 = 0 \qquad [5]$$

Equation [4] relates to the transverse component of the motion, and equation [5] to the radial component. To find the equation of the path, we need to eliminate t between these equations. We use the substitution $r = \dfrac{1}{u}$. Equation [3] then becomes
$$\dot{\theta} = 2Vu^2 \qquad [6]$$

Also we have
$$\dot{r} = -\frac{1}{u^2}\frac{du}{dt} = -\frac{1}{u^2}\frac{du}{d\theta}\frac{d\theta}{dt}$$

from which, using equation [6], we obtain
$$\dot{r} = -2V\frac{du}{d\theta} \qquad [7]$$

Differentiating [7], we have
$$\ddot{r} = -2V\frac{d^2u}{d\theta^2}\frac{d\theta}{dt}$$

from which, using equation [6], we obtain
$$\ddot{r} = -4V^2u^2\frac{d^2u}{d\theta^2} \qquad [8]$$

Substituting from equations [6] and [8] into equation [5], we get
$$-8V^2u^2\frac{d^2u}{d\theta^2} - 4V^2u^3 = 0$$
$$\Rightarrow \quad \frac{d^2u}{d\theta^2} + \frac{u}{2} = 0$$

Solving this differential equation (see pages 56–60), we have
$$u = A\cos\left(\frac{\theta}{\sqrt{2}}\right) + B\sin\left(\frac{\theta}{\sqrt{2}}\right)$$
$$\Rightarrow \quad \frac{du}{d\theta} = -\frac{A}{\sqrt{2}}\sin\left(\frac{\theta}{\sqrt{2}}\right) + \frac{B}{\sqrt{2}}\cos\left(\frac{\theta}{\sqrt{2}}\right)$$

When $t = 0$, $\theta = 0$, $u = \frac{1}{2}$ and $\dfrac{du}{d\theta} = -\dfrac{\dot{r}}{2V} = 0$ (using equation [7]).

These give $A = \frac{1}{2}$, $B = 0$, and hence we have
$$u = \frac{1}{2}\cos\left(\frac{\theta}{\sqrt{2}}\right)$$

which gives the polar equation of the path as
$$r = 2\sec\left(\frac{\theta}{\sqrt{2}}\right)$$

CHAPTER 8 MOTION OF A PARTICLE IN A PLANE

Modifying the assumptions

Of the assumptions made in Example 6, the one most likely to be unjustified is that of negligible friction. Unfortunately, dropping this assumption makes the equations much more difficult to deal with, as the effect of friction for the object at A' depends on the angle ϕ between its instantaneous direction and the line OA'. This angle is given by

$$\tan \phi = \frac{r\dot{\theta}}{\dot{r}}$$

and so it is readily apparent that the resulting equations of motion will be complex and only soluble by numerical methods.

Planetary motion

Consider a small satellite of mass m in orbit around a planet of mass M. Newton's formula for gravitation tells us that each body is attracted towards the other by a force of magnitude $\frac{GMm}{r^2}$, where r is the distance between the centres of mass and G is the gravitational constant. Provided the planet is far more massive than the satellite, we can effectively model this as a particle of mass m moving under the action of a central force of magnitude $\frac{km}{r^2}$ directed towards a fixed point O (the centre of mass of the planet), because the planet will undergo negligible motion resulting from its being attracted by the satellite.

Making use of the usual substitution $r = \frac{1}{u}$, equation [8] on page 167 gave us the general differential equation for motion under a central force F:

$$h^2 u^2 \left(\frac{d^2 u}{d\theta^2} + u \right) = \frac{F}{m}$$

In our model of planetary motion, $F = \frac{km}{r^2} = kmu^2$, which gives us a differential equation:

$$\frac{d^2 u}{d\theta^2} + u = \frac{k}{h^2}$$

Solving this second-order differential equation (see pages 56–66), we have

$$u = A \cos \theta + B \sin \theta + \frac{k}{h^2}$$

$$\Rightarrow \quad \frac{du}{d\theta} = -A \sin \theta + B \cos \theta$$

The values of A and B depend on the initial conditions. Unless the path of the particle passes through O, there is always at least one point on the path at which it is travelling at right angles to its position vector. If we take the initial line as passing through this point, then we have $t = 0$, $\theta = 0$ and $\dot{r} = 0$. Also, differentiating $u = \dfrac{1}{r}$, we obtain

$$\frac{du}{d\theta} = -\frac{1}{r^2}\frac{dr}{d\theta} = -\frac{1}{r^2}\frac{dr}{dt}\frac{dt}{d\theta} = -\frac{1}{h}\dot{r}$$

and so, when $t = 0$, $\dfrac{du}{d\theta} = 0$. Using these initial conditions gives $B = 0$. Hence, we have

$$u = A\cos\theta + \frac{k}{h^2}$$

$$\Rightarrow \quad \frac{h^2 u}{k} = 1 + \frac{Ah^2}{k}\cos\theta$$

$$\Rightarrow \quad \frac{h^2}{kr} = 1 + \frac{Ah^2}{k}\cos\theta \qquad [9]$$

The polar equation of a conic is

$$\frac{l}{r} = 1 + e\cos\theta.$$

It follows that the path of the particle is a conic, with semi-latus rectum $l = \dfrac{h^2}{k}$ and eccentricity $e = \dfrac{Ah^2}{k}$. The path will be an ellipse, a parabola or a hyperbola depending on the value of e. Notice that the path is independent of the mass of the particle provided that it is small enough for our model of attraction towards a fixed point to be valid.

Example 7 A body of mass 2000 kg is in an elliptical orbit around a planet of mass 2.5×10^{26} kg. The body's furthest distance from the centre of the planet is 200 000 km, and at this point it is travelling at 27 000 km h^{-1}. Find

a) the closest distance of the body to the planet
b) its speed at this point
c) the time to complete one orbit.

Take $G = 6.67 \times 10^{-11}$ N m^2 kg^{-2}.

SOLUTION

a) In the foregoing section, we established that the equation of the path of a small body about a large one is

$$\frac{h^2}{kr} = 1 + \frac{Ah^2}{k}\cos\theta$$

where h is the moment of velocity about the centre and

$$k = G \times \text{Mass of large body}$$

In this example, we have
$$k = 6.67 \times 10^{-11} \times 2.5 \times 10^{26} = 1.6675 \times 10^{16}$$

Also, at the furthest point, the body is 2×10^8 m from the centre and travelling perpendicular to the radius vector at a speed of $27\,000\text{ km h}^{-1} = 7500\text{ m s}^{-1}$. This gives
$$h = 7500 \times 2 \times 10^8 = 1.5 \times 10^{12}$$

At the furthest point, $\theta = 180°$ and $r = 2 \times 10^8$ m. Substituting into the equation of the path, we get $A = 2.41 \times 10^{-9}$.

The orbit is therefore an ellipse with an eccentricity of $\dfrac{Ah^2}{k} = 0.325$.

Its equation is
$$\frac{1.349 \times 10^8}{r} = 1 + 0.325 \cos\theta$$

The body is at its point of closest approach when $\theta = 0°$. This gives
$$\frac{1.349 \times 10^8}{r} = 1.325$$
$$\Rightarrow \quad r = 1.018 \times 10^8$$

and so the closest approach is at a distance of $102\,000$ km (to 3 sf).

b) When the body is at its point of closest approach, it is travelling at right angles to the position vector. If its speed is V, the moment of its velocity is $1.018 \times 10^8\,V$. As this is the value of h, we have
$$V = \frac{1.5 \times 10^{12}}{1.018 \times 10^8} = 14\,730$$

The body is therefore travelling at $14\,730\text{ m s}^{-1} = 53\,000\text{ km h}^{-1}$ (to 3 sf).

c) Suppose the ellipse has semi-major axis a and semi-minor axis b. The area of the ellipse is then πab.

The major axis of the ellipse is the sum of the furthest and closest distances. This gives the semi-major axis as
$$a = 0.5 \times (2 \times 10^8 + 1.018 \times 10^8) = 1.509 \times 10^8$$

Using a standard result for an ellipse, we have
$$b = a\sqrt{1 - e^2} = 1.427 \times 10^8$$

The area is therefore $6.77 \times 10^{16}\text{ m}^2$.

We know from equation [3] on page 166 that the area is swept out at a constant rate given by $\dfrac{dA}{dt} = \tfrac{1}{2}h$. In this case, we have
$$\frac{dA}{dt} = 0.5 \times 1.5 \times 10^{12} = 7.5 \times 10^{11}$$

Hence, if the period of an orbit is T, we have

$$\frac{6.77 \times 10^{16}}{T} = 7.5 \times 10^{11}$$

$$\Rightarrow \quad T = 90\,204$$

So, one complete orbit takes $90\,200\,\text{s} = 25.1\,\text{h}$.

Period of orbit

In Example 7, we found the period of a particular orbit. Here, we tackle the general case.

For an ellipse with equation

$$\frac{l}{r} = 1 + e\cos\theta,$$

we have the standard results

$$\text{Semi-major axis } a = \frac{l}{1 - e^2}$$

$$\text{Semi-minor axis } b = a\sqrt{1 - e^2}$$

from which we have

$$l = a(1 - e^2) = \frac{b^2}{a}$$

For the orbit given by equation [9] on page 171, we have $l = \dfrac{h^2}{k}$, which gives

$$h = b\sqrt{\frac{k}{a}} \qquad [10]$$

The area of the ellipse is πab. If the period is T, the (constant) rate at which the area is swept is, therefore, $\dfrac{\pi ab}{T}$.

But we know from equation [3] on page 166 that $\dfrac{dA}{dt} = \tfrac{1}{2}h$. We therefore have

$$T = \frac{2\pi ab}{h}$$

and so, from equation [10],

$$T = 2\pi\sqrt{\frac{a^3}{k}}.$$

That is, T^2 is proportional to a^3.

Exercise 8C

1 A light, inextensible string, AB, of length 2 m, has a particle of mass 1 kg attached to each end. The particles lie at rest on a smooth, horizontal table so that AB = 2 m. The string passes through a small, fixed, smooth ring, O, on the table and initially O is at the mid-point of AB. A horizontal impulse of 8 N s is applied to A at an angle of $60°$ to the string so that both particles start to move.

  a) Find the initial radial and transverse components of A's velocity.

  b) Show that B reaches O with a speed of $\sqrt{22}$ m s^{-1}.

2 A particle A, of mass m, rests on a smooth, horizontal table. It is attached to a fixed point, O, on the table by means of a light, elastic string of natural length 1 m and modulus of elasticity mg. The particle is initially at rest with OA = 1 m. It is then projected along the table with initial velocity u perpendicular to the string.

  a) If the length of the string at time t is r, show that
$$\frac{d^2 r}{dt^2} = \frac{u^2}{r^3} - g(r-1)$$

  b) If the maximum length of the string during the subsequent motion is 2 m, show that
$$u = \sqrt{\frac{4g}{3}}$$

$\left[\text{\textbf{Hint} \quad Write } \dfrac{d^2 r}{dt^2} = \dot{r}\dfrac{d\dot{r}}{dr}\right]$

3 A particle P, of mass m, is travelling about a fixed point, O, under the influence of a force of magnitude $\dfrac{km}{r^2}$ directed towards O, where $r = $ OP. The initial conditions of the motion were such that the particle follows a circular orbit of radius a.

  a) Show that the speed of the particle is $\sqrt{\dfrac{k}{a}}$.

The particle now collides and coalesces with a second particle, also of mass m, which is at rest.

  b) Show that the combined particle moves in an elliptical orbit with eccentricity 0.75.

  c) Find the ratio between the periods of the two orbits.

4 A particle P, of unit mass, is moving in a plane under the influence of a force of magnitude $\dfrac{2}{r^3}$ directed towards a fixed point, O, where $r = $ OP. At time $t = 0$, $r = 1$, and the radial and transverse components of velocity are 1 and 2 respectively. Find an expression for r in terms of t.

5 A particle of unit mass is moving in a plane under the influence of a force of magnitude $\dfrac{k}{r^3}$ directed towards a fixed point, O. The particle is initially projected from a point P with velocity $2\sqrt{\dfrac{k}{3}}$ perpendicular to OP, where OP = 1. Show that the particle follows the curve $r\cos\tfrac{1}{2}\theta = 1$.

6 A satellite is in a circular orbit 500 km above the surface of the Earth.

a) Show that the speed of the satellite is approximately 7620 m s^{-1}.

The speed of the satellite is now suddenly boosted by 2000 m s^{-1}.

b) Show that the new orbit is elliptical, and find its eccentricity.

Take $G = 6.67 \times 10^{-11} \text{ N m}^2 \text{ kg}^{-2}$, the radius of the Earth as $6.37 \times 10^6 \text{ m}$ and the mass of the Earth as $5.98 \times 10^{24} \text{ kg}$.

Intrinsic coordinates

For some purposes, it is useful to define a point, P, on a curve in terms of its displacement, s, along the curve from some fixed reference point, P_o, and the angle, ψ, which the tangent at the point makes with some fixed reference direction. s and ψ are the **intrinsic coordinates** of P. We often choose these referents so that $\psi = 0$ when $s = 0$.

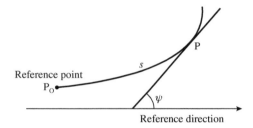

This enables us to define what we mean by **curvature**. Intuitively, we think of the degree of curvature as being the rate with which the direction changes as we move along a curve (for example, the amount by which a car steering wheel must be turned to negotiate a bend). We therefore define curvature as

$$\kappa = \frac{d\psi}{ds}$$

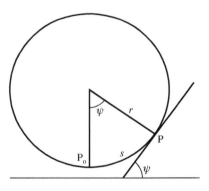

For a circle of radius r, ψ is equal to the angle swept by the radius vector in moving from P_o to P, as indicated in the diagram on the right.

It follows that $\psi = \dfrac{s}{r}$ and hence $\dfrac{d\psi}{ds} = \dfrac{1}{r}$.

For a general curve, we define the **circle of curvature** at a point P as being the circle which touches the curve at P and has the same curvature. The **centre of curvature** is the centre of the circle of curvature, and the **radius of curvature**, denoted by ρ, is the radius of the circle of curvature. Clearly, we have the relationship

$$\rho = \frac{1}{\kappa}$$

CHAPTER 8 MOTION OF A PARTICLE IN A PLANE

Tangential and normal components

When considering motion along a curve defined in terms of intrinsic coordinates, it is natural to take unit vectors, $\hat{\mathbf{t}}$ and $\hat{\mathbf{n}}$, which at a general point P are tangential and normal to the curve at P. $\hat{\mathbf{n}}$ is conventionally taken as being an anticlockwise rotation of $\hat{\mathbf{t}}$ through 90°.

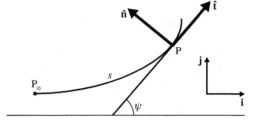

If the cartesian **i**- and **j**-directions are taken parallel and perpendicular to the reference direction, as shown, we have

$$\hat{\mathbf{t}} = \cos\psi\, \mathbf{i} + \sin\psi\, \mathbf{j}$$
$$\hat{\mathbf{n}} = -\sin\psi\, \mathbf{i} + \cos\psi\, \mathbf{j}$$

from which we can see that

$$\frac{d\hat{\mathbf{t}}}{d\psi} = \hat{\mathbf{n}} \quad \text{and} \quad \frac{d\hat{\mathbf{n}}}{d\psi} = -\hat{\mathbf{t}}$$

We can now establish the tangential and normal components of velocity and acceleration for a particle moving along a curve. If at a point P the particle has speed v, we have

$$\mathbf{v} = v\hat{\mathbf{t}}$$

Differentiating, we get

$$\dot{\mathbf{v}} = \dot{v}\hat{\mathbf{t}} + v\frac{d\hat{\mathbf{t}}}{dt}$$

$$\Rightarrow \quad \dot{\mathbf{v}} = \dot{v}\hat{\mathbf{t}} + v\frac{d\hat{\mathbf{t}}}{d\psi}\frac{d\psi}{ds}\frac{ds}{dt}$$

But $\dfrac{d\hat{\mathbf{t}}}{d\psi} = \hat{\mathbf{n}}$, $\dfrac{ds}{dt} = v$, and $\dfrac{d\psi}{ds} = \kappa = \dfrac{1}{\rho}$, where ρ is the radius of curvature at P. Hence, we have

$$\dot{\mathbf{v}} = \dot{v}\hat{\mathbf{t}} + \frac{v^2}{\rho}\hat{\mathbf{n}}$$

Hence, the tangential and normal components of acceleration are $\dot{v}$ and $\dfrac{v^2}{\rho}$ respectively. These are, of course, the same as if the particle were travelling along the instantaneous circle of curvature at P.

Example 8 A curve is defined by the intrinsic equation $s = k\sin\psi$ relative to an origin O, where the reference direction is horizontal and $s = 0$ when $\psi = 0$. Show that if a particle of mass m, moving from rest at a point P under the action of gravity alone, takes a time t_P to reach O, then t_P is independent of the position of P.

SOLUTION

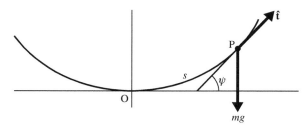

Resolving in the $\hat{\mathbf{t}}$-direction, we have

$$-mg \sin \psi = m\dot{v} = m \frac{d^2 s}{dt^2}$$

$$\Rightarrow \quad \frac{d^2 s}{dt^2} = -g \sin \psi$$

But, from the equation of the curve, we have $\sin \psi = \dfrac{s}{k}$, which gives

$$\frac{d^2 s}{dt^2} = -\frac{g}{k} s$$

This is the equation for simple harmonic motion with period $2\pi \sqrt{\dfrac{k}{g}}$. The period of SHM is independent of the amplitude of the motion. It follows that, whatever the position of P, the time taken for the particle to reach O is a quarter of the period of the oscillation. That is,

$$t_P = \frac{\pi}{2} \sqrt{\frac{k}{g}}$$

We can explore the curve in Example 8 by converting its intrinsic equation to parametric form, as follows.

We can see that for a small arc δs, corresponding to small increments δx and δy as shown, we have

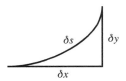

$$\delta s^2 \approx \delta x^2 + \delta y^2$$

$$\Rightarrow \quad \left(\frac{\delta s}{\delta x}\right)^2 \approx 1 + \left(\frac{\delta y}{\delta x}\right)^2$$

and so, in the limit,

$$\left(\frac{ds}{dx}\right)^2 = 1 + \left(\frac{dy}{dx}\right)^2$$

But $\dfrac{dy}{dx} = \tan \psi$. Hence, we have

$$\frac{ds}{dx} = \sqrt{1 + \tan^2 \psi} = \sec \psi$$

$$\Rightarrow \quad \frac{dx}{ds} = \cos \psi \qquad [10]$$

CHAPTER 8 MOTION OF A PARTICLE IN A PLANE

Similarly, we also have

$$\frac{dy}{ds} = \sin\psi \qquad [11]$$

For the curve in Example 8, we have

$$s = k\sin\psi \quad \Rightarrow \quad \frac{ds}{d\psi} = k\cos\psi$$

Hence, we obtain

$$\frac{dx}{ds} = \frac{dx}{d\psi}\frac{d\psi}{ds} = \frac{dx}{d\psi}\frac{1}{k\cos\psi}$$

And so, from equation [10], we arrive at

$$\frac{dx}{d\psi} = k\cos^2\psi$$

$$\Rightarrow \quad x = k\int \cos^2\psi\, d\psi$$

$$= \frac{k}{2}\int (1+\cos 2\psi)\, d\psi$$

$$= \frac{k}{4}(2\psi + \sin 2\psi) + c$$

As $x = 0$ when $\psi = 0$, we have $c = 0$, which gives

$$x = \frac{k}{4}(2\psi + \sin 2\psi)$$

Similarly, we obtain from equation [11]

$$\frac{dy}{d\psi} = k\sin\psi\cos\psi = \frac{k}{2}\sin 2\psi$$

$$\Rightarrow \quad y = \frac{k}{2}\int \sin 2\psi\, d\psi$$

$$= -\frac{k}{4}\cos 2\psi + c_1$$

As $y = 0$ when $\psi = 0$, we have $c_1 = \frac{k}{4}$, which gives

$$y = \frac{k}{4}(1 - \cos 2\psi)$$

Putting $2\psi = -\theta$, we have

$$x = \frac{k}{4}(\theta - \sin\theta) \quad \text{and} \quad y = \frac{k}{4}(1 - \cos\theta)$$

These are the standard parametric equations of the **cycloid**, the locus of a point on the circumference of a circle as it rolls without slipping along a straight line. (In this case, the cycloid is inverted.) Cycloids therefore have the so-called tautochrone property, such that particles descending from any point on the curve reach the lowest point in the same time. It can also be shown that, for a smooth particle, the cycloid is the curve of fastest descent between two points.

A common application of this curve, first proposed in the early part of the seventeenth century, is to gear teeth, whose flank profiles are often cut as arcs of cycloids to ensure rolling contact when the gears are in mesh.

The catenary

Consider a uniform chain, having weight per unit length w, hanging freely between two fixed points.

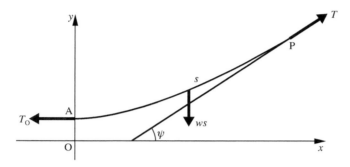

A section of the chain of arc length s between A, the lowest point, and a point P has weight ws. If the tensions in the chain at A and P are T_o and T respectively, as shown in the diagram, then resolving horizontally and vertically gives

$$T\cos\psi = T_o \qquad [12]$$

$$T\sin\psi = ws \qquad [13]$$

Dividing equation [13] by equation [12], we have

$$\tan\psi = \frac{ws}{T_o}$$

$$\Rightarrow \quad s = c\tan\psi \qquad [14]$$

where $c = \dfrac{T_o}{w}$.

This is the intrinsic equation of the curve, which is called the **catenary** (from the Latin *catena*, meaning a chain).

We can obtain the cartesian equation of a catenary as follows.

From equation [10], we have

$$\frac{ds}{dx} = \sec\psi = \sqrt{1 + \tan^2\psi}$$

$$\Rightarrow \quad \frac{ds}{dx} = \sqrt{1 + \frac{s^2}{c^2}} \quad \text{(from equation [14])}$$

CHAPTER 8 MOTION OF A PARTICLE IN A PLANE

This gives

$$\int \frac{c\,ds}{\sqrt{c^2+s^2}} = \int dx$$

$$\Rightarrow \quad c\,\operatorname{arsinh}\left(\frac{s}{c}\right) = x + C$$

When $x = 0$, $s = 0$, giving $C = 0$. Hence, we get

$$s = c\sinh\left(\frac{x}{c}\right)$$

But $\dfrac{dy}{dx} = \tan\psi$. So, from equation [14], we have

$$\frac{dy}{dx} = \frac{s}{c}$$

$$\Rightarrow \quad \frac{dy}{dx} = \sinh\left(\frac{x}{c}\right)$$

Integrating, we obtain

$$y = c\cosh\left(\frac{x}{c}\right) + C_1$$

Provided we choose our axes so that $y = c$ when $x = 0$, we have $C_1 = 0$, and so

$$y = c\cosh\left(\frac{x}{c}\right)$$

is the cartesian equation of the catenary.

Exercise 8D

1 A bead of mass $0.1\,\text{kg}$ moves at a constant speed of $4\,\text{m s}^{-1}$ along a smooth wire forming the curve with intrinsic equation $s = 4(1 - \cos\psi)$, lying in a horizontal plane. Find the horizontal component of the reaction between the bead and the wire when $\psi = \dfrac{\pi}{4}$.

2 A particle moves along the curve $s = 5\sin\psi$ so that $\dfrac{d\psi}{dt} = 0.5\,\text{rad s}^{-1}$. Initially, $\psi = 0$. Find, when $\psi = 1\,\text{rad}$,
   a) the speed of the particle
   b) the tangential and normal components of the acceleration of the particle.

3 A particle moves along the spiral $s = \psi^2$, starting from rest when $\psi = 0$, so that at time $t\,\text{s}$ its direction of motion is rotating at $0.2t\,\text{rad s}^{-1}$. Find the tangential and normal components of the particle's acceleration when $t = 2\,\text{s}$.

4 A particle of mass m slides from rest down a wire which is bent into the shape of a catenary $y = c \cosh\left(\dfrac{x}{c}\right)$ and fixed in a vertical plane with the vertex of the catenary at its lowest point. The intrinsic equation of the catenary is $s = c \tan \psi$, and the particle starts from the point where $\psi = \dfrac{\pi}{4}$.

  a) Show that the particle falls a vertical distance of $c\sqrt{2}$ in reaching the lowest point.
  b) Find the reaction of the wire on the particle when the particle reaches the lowest point.

5 A smooth tube is bent to form the cycloid $s = 4a \sin \psi$, and is fixed in a vertical plane with its vertex uppermost, as shown in the diagram. The points O and A correspond to $\psi = 0$ and $\psi = \dfrac{\pi}{2}$ respectively.

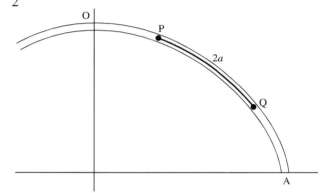

There are two particles, each of mass m, inside the tube and connected by means of a light, inextensible string of length $2a$. The particles are released from rest with the uppermost particle at O. Show that

a) the tension in the string has magnitude $\tfrac{1}{4}mg$ throughout the motion

b) the time taken for the lower particle to reach A is $2\sqrt{\dfrac{a}{3}}\,\text{arcosh } 3$.

Examination questions

Chapters 5 to 8

Chapter 5

1 A particle moving with speed u on a horizontal surface collides with a vertical straight wall and rebounds. The coefficient of restitution between the particle and the wall is $\frac{3}{4}$, and the angle between the direction of motion of the particle and the wall immediately before the collision is α. The particle's direction of motion immediately after the collision is perpendicular to its direction immediately before the collision.

a) Show that $\tan \alpha = \dfrac{2}{\sqrt{3}}$.

b) Find, in terms of u, the speed of the particle immediately after the collision.   (EDEXCEL)

2 A smooth uniform sphere is moving with speed 20 m s^{-1} on a smooth horizontal plane when it strikes a smooth vertical wall. After the impact the velocity of the sphere makes an angle θ with the wall where $\tan \theta = \frac{5}{12}$. Given that the coefficient of restitution between the sphere and the wall is $\frac{5}{9}$, find the speed of the sphere after the impact.   (EDEXCEL)

3 A snooker ball of mass 150 grams is struck by a snooker cue. The magnitude of the contact force, $F(t)$ newtons, between the cue and the ball can be modelled by the expression

$$F(t) = 12 \sin(20\pi t) \quad 0 \leqslant t \leqslant 0.05$$

where t seconds is the time from the start of the impact.

Given that the ball is initially at rest and assuming no spin is imparted to the ball, calculate the speed of the ball 0.05 seconds after the start of the impact.   (SQA/CSYS)

4 On a billiard table ABCD, $AB = 4a$ and $BC = 2a$. A smooth sphere P, of mass m, lies at rest at the centre of the table and a second smooth sphere Q, of equal radius and mass $2m$, is at rest at the mid-point of BC. Q is projected towards P with speed u and collides obliquely with P so that, after collision, P moves towards the pocket at A.

a) Explain why the direction of the line of centres at impact is AC.

b) Given that the coefficient of restitution is $\frac{1}{3}$, show that, after the impact, the speed of P is $\dfrac{16u}{9\sqrt{5}}$ and find the speed of Q.   (AEB 96)

5 Two smooth uniform spheres A and B, of equal radius, are moving on a smooth horizontal plane. Sphere A has mass 2 kg and velocity $(2\mathbf{i} + \mathbf{j})$ m s^{-1}, and sphere B has mass 3 kg and velocity $(-\mathbf{i} - \mathbf{j})$ m s^{-1}. The spheres collide when the line joining their centres is parallel to $\mathbf{j}$, as shown in the figure on the right.

Given that the coefficient of restitution between the spheres is $\frac{1}{2}$,

a) calculate the velocities of A and B immediately after the collision, giving your answers in vector form.

b) Calculate the loss in kinetic energy due to the collision. (EDEXCEL)

6 A collision occurs between two uniform smooth spheres A and B, of equal radius and unequal mass, which move on a smooth horizontal plane. Immediately before the collision A is moving with speed 12 m s^{-1} at an angle of 30° with the line of centres, and B is moving with speed 7 m s^{-1} perpendicular to the line of centres. Immediately after the collision the directions of motion of A and B make angles of 90° and 45° respectively with the line of centres (see diagram).

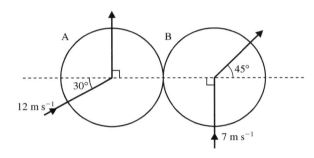

i) Show that the speed of B immediately after the collision is 9.9 m s^{-1} approximately.

ii) Calculate the coefficient of restitution.

iii) Given that the mass of A is 0.3 kg, calculate the magnitude of the impulse on A during the collision.

iv) It is given that the radius of each sphere is 5 cm and that after the collision each sphere moves in a straight line with constant speed. Calculate the time taken, from the collision, until the centres of the spheres are 110 cm apart. (OCR)

7 A uniform smooth sphere A of mass m is moving with speed u on a horizontal table when it collides with a stationary uniform sphere B of the same size as A and of mass $2m$. The direction of motion of A immediately before the impact makes an angle α with the line of centres of the spheres. The coefficient of restitution between the spheres is $\frac{2}{3}$.

a) Find the speed of B immediately after the impact.

Given that the speed of A immediately after the impact is $\frac{1}{3}u$,

b) show that $\sin^2 \alpha = \frac{1}{10}$. (EDEXCEL)

EXAMINATION QUESTIONS CHAPTERS 5 TO 8

Chapter 6

8 The diagram shows a particle P, of mass 0.4 kg, lying in equilibrium at a point B on a smooth horizontal table. The particle is attached to one end of a light spring, of natural length 0.5 m and modulus 20 N, and the other end of the spring is attached to a fixed point A.

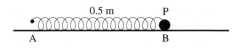

The particle is now displaced a distance 0.1 m from B in the sense AB and released from rest. In the subsequent motion, P is subject to a variable resistance of magnitude $4.8v$ N when its speed is v m s^{-1}. The displacement of P from B in the sense AB at time t s after release is denoted by x m.

a) Show that x satisfies the differential equation
$$\frac{d^2x}{dt^2} + 12\frac{dx}{dt} + 100x = 0$$

b) Find the general solution of the equation in part **a**.

c) Find, in terms of t, an expression for the extension of the spring. (WJEC)

9 O is a point on a straight line. A particle, P, of mass 0.1 kg moves along this line under the action of two forces. One force is equal to $0.16\overrightarrow{PO}$ newtons and the other force can be modelled by a dashpot of damping constant 0.08 N s m^{-1} joined between O and P.

i) Show that the particle's motion can be described by the differential equation
$$5\ddot{x} + 4\dot{x} + 8x = 0$$
where x is the displacement $\overrightarrow{OP}$, in metres, of the particle at time t seconds.

At time $t = 0$ the particle is at O and has a speed of 6 m s^{-1}.

ii) Find the greatest magnitude x will subsequently have. (NICCEA)

10 Two elastic strings AB and BC, each of natural length a and modulus of elasticity $\frac{1}{2}mg$, are joined at B. A particle of mass m is attached to the strings at B. The ends A and C of the strings are fixed at points which are a distance $4a$ apart on a smooth horizontal plane. The particle is set in motion and moves in a straight line along the line AC. Its displacement x is measured from the mid-point of the line AC. The particle is acted on by a resisting force of magnitude mnv, where v is the speed of the particle and $n = \sqrt{\left(\frac{g}{a}\right)}$.

a) Show that $\frac{d^2x}{dt^2} + n\frac{dx}{dt} + n^2 x = 0$.

Given that, when $t = 0$, $x = a$ and $\frac{dx}{dt} = 0$,

b) find an expression for x in terms of a, n and t. (EDEXCEL)

11 A basket and its contents have a total mass of M kilograms. The basket is attached to one end of a light spring. The other end of the spring is attached to a fixed point. The basket hangs in equilibrium.

Model the system as a perfect spring of modulus $2.5g$ N and natural length 0.5 m and a perfect dashpot of damping constant $2\sqrt{g}$ N s m^{-1}.

i) Find the extension of the spring in terms of M.

The basket is lifted a small distance vertically upwards and released from rest. Take x to be the displacement, in metres, of the basket from the equilibrium position at time t seconds.

ii) Obtain a differential equation which describes the motion of the basket and its contents.
iii) Find the condition for M if the system is to be strongly damped (over damped).
iv) Explain briefly the meaning of the term 'dashpot'   (NICCEA)

12

```
A              P     O                    B
●──────────────●─────────────────────────●
           ←─x─┤
```

The diagram shows an elastic string of natural length 2 m and modulus of elasticity 5 N stretched between two points A and B at a distance 4 m apart on a smooth horizontal plane. O is the mid-point of AB. A particle P of mass 1 kg, attached to the mid-point of the string, is pulled towards A through a distance of $\frac{1}{3}$ m and released from rest at time $t = 0$. A resistive force, equal in magnitude to twice the velocity, and a force given by $3e^{-t}$ in the direction towards A, also act on the particle during the subsequent motion.

Show that the equation of motion of the particle is

$$\frac{d^2x}{dt^2} + 2\frac{dx}{dt} + 10x = 3e^{-t}$$

where x is the displacement of the particle from O towards A.

Solve this equation to find x at time t.   (AEB 96)

Chapter 7

13 A particle P is projected with speed V from a point A on a plane which is inclined at an angle of $30°$ to the horizontal. The particle is projected up the plane at an angle θ above the plane and moves in a vertical plane through a line of greatest slope. P first strikes the plane at the point B and is then moving at right angles to the plane.

Show that $AB = \dfrac{4V^2}{7g}$.   (EDEXCEL)

14 A particle P is projected at time $t = 0$ with speed V from a point on a plane which is inclined at an angle of $45°$ to the horizontal. The particle is projected up the plane at an angle θ above the plane and moves in a vertical plane through a line of greatest slope.

a) Show that P first strikes the plane when $t = \dfrac{(2\sqrt{2})V\sin\theta}{g}$.

Given that P is moving horizontally when it first strikes the plane,

b) show that $\tan\theta = \frac{1}{3}$.

The coefficient of restitution between the particle and the inclined plane is $\frac{1}{2}$.

c) Show that the speed of the particle immediately after striking the plane for the first time is $\dfrac{V\sqrt{2}}{4}$.   (EDEXCEL)

15 The point O is on a plane inclined at an angle $\frac{\pi}{6}$ to the horizontal. A particle P is projected from O with speed V at an angle θ above the plane. P moves in a vertical plane through a line of greatest slope of the inclined plane and strikes the plane at a distance R from O, as shown in the figure on the right.

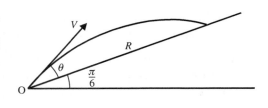

a) Show that $R = \frac{8V^2}{3g}\sin\theta\cos\left(\theta + \frac{\pi}{6}\right)$.

b) Show that R may be written in the form $R = \frac{2V^2}{3g}\left[2\sin\left(2\theta + \frac{\pi}{6}\right) - 1\right]$.

c) Hence, or otherwise, show that, as the angle of projection, θ, varies, the maximum possible range up the plane occurs when the direction of projection bisects the angle between the inclined plane and the upward vertical.   (EDEXCEL)

16 A particle P is projected at time $t = 0$ with speed V from a point O on a plane which is inclined at an angle α to the horizontal. The particle is projected up the plane at an angle θ above the plane and moves in a vertical plane through a line of greatest slope, first striking the plane at the point A.

a) Show that the magnitude of the component of the velocity of P perpendicular to the plane, when P strikes the plane at A, is the same as it was initially at O.

The plane is smooth and the coefficient of restitution between P and the plane is e. P strikes the plane for the second time at B.

b) Show that the times of flight from O to A and from A to B are in the ratio $1 : e$.

Given that B is the same point as O,

c) hence show that $\cot\theta\cos\alpha = (1 + e)$.   (EDEXCEL)

Chapter 8

17 Referred to the pole O and a fixed initial line, the curve C has polar equation $r = ae^{\theta\cot\alpha}$, where a is a positive constant and α is a constant, $0 \leqslant \alpha \leqslant \frac{\pi}{4}$. A particle P moves along C in such a way that OP rotates with constant angular speed ω. Show that

a) the speed v of P is $r\omega\operatorname{cosec}\alpha$

b) the magnitude of the acceleration of P is $\frac{v^2}{r}$.   (EDEXCEL)

18 A particle, of mass m, is attached to one end of an elastic string of natural length a and modulus mg. The other end of the string is fixed to a point, A, on a smooth horizontal surface.

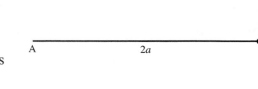

When the length of the string is $2a$, the particle is projected on the surface at right angles to the string with a speed of $\sqrt{\frac{27ag}{5}}$.

The motion of the particle is described using polar coordinates (r, θ), with A as pole.

a) Show that $r^2\dot{\theta}$ is constant and state its value.
b) By considering energy, **verify** that $3a$ is the maximum possible value of r.
c) Find the acceleration of the particle when r is at this maximum. (AEB 99)

19 Relative to the pole O and an initial line, the polar equation of the curve C is $r = ae^{k\theta}$, where a and k are positive constants. A particle P of mass m moves on C, which is fixed in a horizontal plane, so that OP rotates with constant angular speed ω.

a) Find in terms of m, r, k and ω the radial and transverse components of the horizontal resultant force acting on the particle.
b) Show that the horizontal resultant force has magnitude $(1 + k^2)mr\omega^2$. (EDEXCEL)

20 A smooth wire is in the shape of a cycloid. The wire is fixed in a vertical plane with its vertex, the origin O, uppermost. A bead of mass m is threaded on the wire. When $t = 0$ the bead is projected from O with speed u. At time t the bead is at the point P where the arc length OP $= s$ and the tangent to the curve at P makes an angle ψ with the horizontal, as shown in the figure on the right.

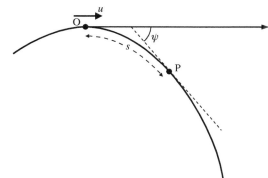

The intrinsic equation of the cycloid is $s = \dfrac{4k^2}{g} \sin \psi$, where k is a constant.

At P the bead has speed v.

a) Show that $v^2 = u^2 + \dfrac{g^2 s^2}{4k^2}$.

B is the point on the wire at which $\psi = \dfrac{\pi}{2}$.

Using the result in part **a**,

b) show that the bead reaches B after time $\dfrac{2k}{g}$ arsinh $\dfrac{2k}{u}$. (EDEXCEL)

21 A particle P moves with constant speed u along a catenary with intrinsic equation $s = c \tan \psi$. Show that the magnitude of the normal component of the acceleration of P, at time t after passing through the point where $\psi = 0$, is

$$\dfrac{cu^2}{c^2 + u^2 t^2}$$ (EDEXCEL)

9 Stability of equilibrium

A state of balance is attractive only when one is on a tightrope; seated on the ground there is nothing wonderful about it.
ANDRÉ GIDE

Centre of mass

In *Introducing Mechanics* (pages 221–39), we met the following results:

- The centre of mass, $\bar{\mathbf{r}}$, of a set of particles having masses $m_1, m_2, \ldots, m_n$ and placed at points with position vectors $\bar{\mathbf{r}}_1, \bar{\mathbf{r}}_2, \ldots, \bar{\mathbf{r}}_n$ is given by

$$\bar{\mathbf{r}} = \frac{\sum_{i=1}^{n} m_i \mathbf{r}_i}{\sum_{i=1}^{n} m_i}$$

- The centre of mass $(\bar{x}, \bar{y})$ of a lamina bounded by the curve $y = f(x)$, the x-axis and the lines $x = a$ and $x = b$ is given by

$$\bar{x} = \frac{\int_a^b \rho x f(x)\,dx}{\int_a^b \rho f(x)\,dx} \quad \text{and} \quad \bar{y} = \frac{\frac{1}{2}\int_a^b \rho (f(x))^2\,dx}{\int_a^b \rho f(x)\,dx}$$

In *Introducing Mechanics*, we took the density, ρ, to be constant. However, the same formulae are used when ρ is variable. In this chapter, we will meet examples where ρ is a function of x.

- The centre of mass of a solid of revolution formed by rotating about the x-axis the area bounded by $y = f(x)$, the x-axis and the lines $x = a$ and $x = b$ is $(\bar{x}, 0)$, where

$$\bar{x} = \frac{\int_a^b \rho x (f(x))^2\,dx}{\int_a^b \rho (f(x))^2\,dx}$$

Again, the density, ρ, need not be uniform.

Example 1 A lamina is bounded by the curve $y = \sin x$ and the x-axis between the values $x = 0$ and $x = \dfrac{\pi}{2}$. The density at the point (x, y) is $\rho = k \cos x$. Find the centre of mass of the lamina.

SOLUTION

For $\bar{x}$, we have

$$\int_0^{\frac{\pi}{2}} \rho x \mathrm{f}(x) \, \mathrm{d}x = \int_0^{\frac{\pi}{2}} kx \sin x \cos x \, \mathrm{d}x$$

$$= \frac{k}{2} \int_0^{\frac{\pi}{2}} x \sin 2x \, \mathrm{d}x$$

$$= \left[-\frac{k}{4} x \cos 2x \right]_0^{\frac{\pi}{2}} + \frac{k}{4} \int_0^{\frac{\pi}{2}} \cos 2x \, \mathrm{d}x \quad \text{(by parts)}$$

$$= \left[-\frac{k}{4} x \cos 2x + \frac{k}{8} \sin 2x \right]_0^{\frac{\pi}{2}}$$

$$= \frac{k\pi}{8}$$

For $\bar{y}$, we have

$$\int_0^{\frac{\pi}{2}} \rho (\mathrm{f}(x))^2 \, \mathrm{d}x = \int_0^{\frac{\pi}{2}} k \sin^2 x \cos x \, \mathrm{d}x$$

$$= \left[\frac{k}{3} \sin^3 x \right]_0^{\frac{\pi}{2}}$$

$$= \frac{k}{3}$$

And for $\bar{x}$ and $\bar{y}$, we have

$$\int_0^{\frac{\pi}{2}} \rho \mathrm{f}(x) \, \mathrm{d}x = \int_0^{\frac{\pi}{2}} k \sin x \cos x \, \mathrm{d}x$$

$$= \frac{k}{2} \int_0^{\frac{\pi}{2}} \sin 2x \, \mathrm{d}x$$

$$= \left[-\frac{k}{4} \cos 2x \right]_0^{\frac{\pi}{2}}$$

$$= \frac{k}{2}$$

Hence, we obtain

$$\bar{x} = \frac{\frac{k\pi}{8}}{\frac{k}{2}} = \frac{\pi}{4} \quad \text{and} \quad \bar{y} = \frac{\frac{1}{2} \times \frac{k}{3}}{\frac{k}{2}} = \frac{1}{3}$$

Example 2 A rod AB is 3 m long. The density of the material varies along its length, so that the density at a distance x m from A is $\dfrac{1}{1+x}$ kg m^{-1}. Find the distance of the centre of mass of the rod from A.

SOLUTION

We do not have a formula for this situation, but we can derive the result from first principles.

Consider a small section of rod, PQ, such that AP $= x$ and PQ $= \delta x$. Hence, we have

$$\text{Mass of PQ} \approx \frac{\delta x}{1+x}$$

which gives

$$\text{Moment of PQ about A} \approx \frac{x\delta x}{1+x}$$

If the distance of the centre of mass from A is $\bar{x}$, then we have

$$\bar{x} \approx \frac{\sum \dfrac{x\delta x}{1+x}}{\sum \dfrac{\delta x}{1+x}}$$

As $\delta x \to 0$, this becomes

$$\bar{x} = \frac{\int_0^3 \dfrac{x}{1+x}\,dx}{\int_0^3 \dfrac{1}{1+x}\,dx} = \frac{\int_0^3 \left(1 - \dfrac{1}{1+x}\right)dx}{\int_0^3 \dfrac{1}{1+x}\,dx}$$

$$\Rightarrow \quad \bar{x} = \frac{[x - \ln(1+x)]_0^3}{[\ln(1+x)]_0^3} = \frac{3 - \ln 4}{\ln 4} = 1.16$$

So, the centre of mass of the rod is 1.16 m from A.

Arcs and shells

When dealing with a lamina or a solid of revolution, we derive the formulae by considering an elemental section of width δx, which we then approximate as a rectangle or a disc respectively. This approach lets us down when we come to consider the arc of a curve or the surface of revolution (shell) generated by rotating the arc about the x-axis.

In these cases, we need to consider an element of arc of length δs.

If we are using cartesian coordinates, we have

$$\delta s \approx \sqrt{(\delta x)^2 + (\delta y)^2} = \sqrt{1 + \left(\frac{\delta y}{\delta x}\right)^2}\,\delta x$$

In some cases, the work is simplified by using polar coordinates. We obtain the expression for δs as follows.

We know that $x = r\cos\theta$ and $y = r\sin\theta$. These give

$$\delta x \approx \frac{dx}{d\theta}\delta\theta = \left(\frac{dr}{d\theta}\cos\theta - r\sin\theta\right)\delta\theta$$

and $\quad \delta y \approx \dfrac{dy}{d\theta}\delta\theta = \left(\dfrac{dr}{d\theta}\sin\theta + r\cos\theta\right)\delta\theta$

Substituting these into $\delta s \approx \sqrt{(\delta x)^2 + (\delta y)^2}$ and simplifying, we obtain

$$\delta s \approx \sqrt{r^2 + \left(\frac{dr}{d\theta}\right)^2}\,\delta\theta$$

Centre of mass of an arc of uniform density

Suppose the density (mass per unit length) of the arc is ρ. Then the mass of an element of arc length $\delta s \approx \rho\delta s$.

The moment of this element about the y-axis $\approx \rho x\delta s$.
So, if the x-coordinate of the centre of mass is $\bar{x}$, we have

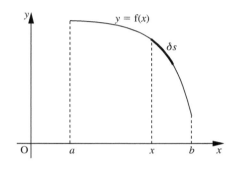

$$\bar{x} \approx \frac{\sum \rho x \delta s}{\sum \rho \delta s} = \frac{\sum \rho x \sqrt{1 + \left(\frac{\delta y}{\delta x}\right)^2}\,\delta x}{\sum \rho \sqrt{1 + \left(\frac{\delta y}{\delta x}\right)^2}\,\delta x}$$

Letting $\delta x \to 0$ and cancelling by ρ, we have

$$\bar{x} = \frac{\int_{x=a}^{x=b} x\,ds}{\int_{x=a}^{x=b} ds} = \frac{\int_a^b x\sqrt{1 + \left(\frac{dy}{dx}\right)^2}\,dx}{\int_a^b \sqrt{1 + \left(\frac{dy}{dx}\right)^2}\,dx} \qquad [1]$$

A similar process leads to

$$\bar{y} = \frac{\int_{x=a}^{x=b} f(x)\,ds}{\int_{x=a}^{x=b} ds} = \frac{\int_a^b f(x)\sqrt{1 + \left(\frac{dy}{dx}\right)^2}\,dx}{\int_a^b \sqrt{1 + \left(\frac{dy}{dx}\right)^2}\,dx} \qquad [2]$$

The corresponding result using polar coordinates is as follows.

The mass of an element of arc length $\delta s \approx \rho\delta s$.

The moment of this element about the y-axis $\approx \rho x\delta s$, which in polar coordinates becomes

$$\rho r\cos\theta \sqrt{r^2 + \left(\frac{dr}{d\theta}\right)^2}\,\delta\theta$$

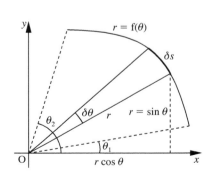

CHAPTER 9 STABILITY OF EQUILIBRIUM

If the x-coordinate of the centre of mass is $\bar{x}$, we have

$$\bar{x} \approx \frac{\sum \rho x \delta s}{\sum \rho \delta s} = \frac{\sum \rho r \cos\theta \sqrt{r^2 + \left(\frac{dr}{d\theta}\right)^2} \delta\theta}{\sum \rho \sqrt{r^2 + \left(\frac{dr}{d\theta}\right)^2} \delta\theta}$$

Letting $\delta\theta \to 0$ and cancelling by ρ, we obtain

$$\bar{x} = \frac{\int_{\theta_1}^{\theta_2} r \cos\theta \sqrt{r^2 + \left(\frac{dr}{d\theta}\right)^2} d\theta}{\int_{\theta_1}^{\theta_2} \sqrt{r^2 + \left(\frac{dr}{d\theta}\right)^2} d\theta} \qquad [3]$$

A similar process leads to

$$\bar{y} = \frac{\int_{\theta_1}^{\theta_2} r \sin\theta \sqrt{r^2 + \left(\frac{dr}{d\theta}\right)^2} d\theta}{\int_{\theta_1}^{\theta_2} \sqrt{r^2 + \left(\frac{dr}{d\theta}\right)^2} d\theta} \qquad [4]$$

These can easily lead to quite complex integrals, which can only be solved analytically in the simplest of cases.

Example 3 Find the coordinates of the centre of mass of a uniform wire bent into a semicircle of radius a.

SOLUTION

Taking the semicircle as shown, the polar equation is $r = a$ and the arc we require is for $-\frac{\pi}{2} \leq \theta \leq \frac{\pi}{2}$.

By symmetry, the y-coordinate of the centre of mass is zero.

As $r = a$, we have $\frac{dr}{d\theta} = 0$, and equation [3] gives

$$\bar{x} = \frac{\int_{-\frac{\pi}{2}}^{\frac{\pi}{2}} a^2 \cos\theta \, d\theta}{\int_{-\frac{\pi}{2}}^{\frac{\pi}{2}} a \, d\theta}$$

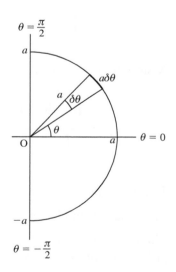

$$\Rightarrow \quad \bar{x} = \frac{\left[a^2 \sin\theta\right]_{-\frac{\pi}{2}}^{\frac{\pi}{2}}}{\left[a\theta\right]_{-\frac{\pi}{2}}^{\frac{\pi}{2}}} = \frac{2a^2}{\pi a} = \frac{2a}{\pi}$$

Hence, the centre of mass has coordinates $\left(\dfrac{2a}{\pi}, 0\right)$.

Centre of mass of a shell of uniform density

The shell, or surface of revolution, is formed by rotating a curve $y = f(x)$ about the x-axis.

By symmetry, $\bar{y} = 0$.

Let the density (mass per unit area) of the shell be ρ.

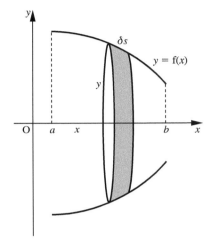

Consider an elemental section, shown shaded in the diagram. The external surface area of this is approximately that of a cylindrical shell of radius y and length δs, which is $2\pi y \, \delta s$.

Hence, its mass $\approx 2\pi \rho y \delta s$. Substituting for $y\delta s$, we obtain

$$2\pi\rho y \delta s = 2\pi\rho f(x)\sqrt{1 + \left(\frac{\delta y}{\delta x}\right)^2} \delta x$$

which gives

$$\text{Moment about } y\text{-axis} \approx 2\pi\rho x\, f(x)\sqrt{1 + \left(\frac{\delta y}{\delta x}\right)^2} \delta x$$

Hence, we have

$$\bar{x} \approx \frac{\sum 2\pi\rho x\, f(x)\sqrt{1 + \left(\frac{\delta y}{\delta x}\right)^2} \delta x}{\sum 2\pi\rho\, f(x)\sqrt{1 + \left(\frac{\delta y}{\delta x}\right)^2} \delta x}$$

Letting $\delta\theta \to 0$ and cancelling by $2\pi\rho$, we have

$$\bar{x} = \frac{\int_a^b x\, f(x)\sqrt{1 + \left(\frac{dy}{dx}\right)^2} dx}{\int_a^b f(x)\sqrt{1 + \left(\frac{dy}{dx}\right)^2} dx} \qquad [5]$$

In polar coordinates, the shell is formed by rotating the curve $r = f(\theta)$ about the x-axis. As before, the centre of mass is on the axis of symmetry.

Consider an elemental section, as shown. The external surface area of this is approximately that of a cylindrical shell of radius $r \sin \theta$ and length δs. Hence, we have

$$\text{Mass of section} \approx 2\pi \rho r \sin \theta \, \delta s = 2\pi \rho r \sin \theta \sqrt{r^2 + \left(\frac{dr}{d\theta}\right)^2} \, \delta\theta$$

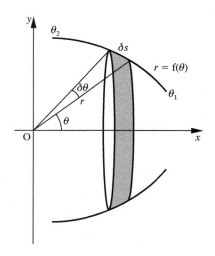

and

Moment of section about y-axis $\approx 2\pi \rho r^2 \sin \theta \cos \theta \, \delta s$

$$= \pi \rho r^2 \sin 2\theta \sqrt{r^2 + \left(\frac{dr}{d\theta}\right)^2} \, \delta\theta$$

which give

$$\bar{x} \approx \frac{\sum \pi \rho r^2 \sin 2\theta \sqrt{r^2 + \left(\frac{dr}{d\theta}\right)^2} \, \delta\theta}{\sum 2\pi \rho r \sin \theta \sqrt{r^2 + \left(\frac{dr}{d\theta}\right)^2} \, \delta\theta}$$

Letting $\delta\theta \to 0$ and cancelling $\pi\rho$, we have

$$\bar{x} = \frac{\int_{\theta_1}^{\theta_2} r^2 \sin 2\theta \sqrt{r^2 + \left(\frac{dr}{d\theta}\right)^2} \, d\theta}{\int_{\theta_1}^{\theta_2} 2r \sin \theta \sqrt{r^2 + \left(\frac{dr}{d\theta}\right)^2} \, d\theta} \qquad [6]$$

Example 4 Find the coordinates of the centre of mass of a hemispherical shell of radius a.

SOLUTION

As in Example 3, we have polar equation $r = a$, but this time for $0 \leq \theta \leq \frac{\pi}{2}$.

As $r = a$, $\frac{dr}{d\theta} = 0$, and so from equation [6], we have

$$\bar{x} = \frac{\int_0^{\frac{\pi}{2}} a^3 \sin 2\theta \, d\theta}{\int_0^{\frac{\pi}{2}} 2a^2 \sin \theta \, d\theta}$$

$$= \frac{\left[-\frac{1}{2}a^3 \cos 2\theta\right]_0^{\frac{\pi}{2}}}{\left[-2a^2 \cos \theta\right]_0^{\frac{\pi}{2}}}$$

$$= \frac{a^3}{2a^2} = \frac{a}{2}$$

So, the coordinates of the centre of mass are $\left(\frac{a}{2}, 0\right)$.

Exercise 9A

1. A straight rod AB of length $2a$ has density which varies such that, at a point which is a distance x from A, its mass per unit length is $m + \lambda x$. Find the distance of its centre of mass from A.

2. Find the position of the centre of mass of a uniform wire of length $2a\alpha$ which is bent to form an arc of a circle of radius a.

3. Find the position of the centre of mass of a uniform cap of a sphere of radius a which subtends an angle of 2α at the centre of the sphere.

4. A uniform wire is bent to form the portion of the curve $r = 1 - \cos\theta$ for $0 \leqslant \theta \leqslant \frac{\pi}{3}$. Find the coordinates of its centre of mass.

5. The portion of the curve $r = a\cos\theta$ for $0 \leqslant \theta \leqslant \frac{\pi}{4}$ is rotated about the x-axis. Find the coordinates of the centre of mass of the resulting uniform surface of revolution.

6. Derive an expression in cartesian coordinates for the position of the centre of mass of a uniform shell formed by rotating an arc of the curve $y = f(x)$ about the x-axis. Hence show that the centre of mass of the conical shell formed by rotating the line $x + y = a$ for $0 \leqslant x \leqslant a$ about the x-axis is at $(\frac{1}{3}a, 0)$.

7. A uniform lamina consists of the sector of the circle $r = a$ for $-\alpha \leqslant \theta \leqslant \alpha$. By treating the sector as being approximately composed of triangles of angle $\delta\theta$, find and evaluate an expression for the position of its centre of mass.

8. The sector in Question 7 may be regarded as a segment and a triangle. Find by integration the centre of mass of the segment and hence confirm your result for the centre of mass of the sector.

9. A uniform rope is suspended from two points, 4 m apart, so that its height above the ground is given by the catenary $y = \cosh x$ for $-2 \leqslant x \leqslant 2$. Show that the length of the rope is approximately 7.25 m and find the height of the centre of mass of the rope above the ground.

10. The portion of the curve $y = 2\sqrt{x}$ between $x = 0$ and $x = 3$ is rotated about the x-axis to form a uniform shell. Find the coordinates of its centre of mass.

11. Find the coordinates of the centre of mass of the uniform shell formed by rotating the curve $r = 1 - \cos\theta$ for $0 \leqslant \theta \leqslant \frac{\pi}{3}$ about the x-axis.

12. A uniform wire is bent to form the portion of the curve $y = x^2$ for $0 \leqslant x \leqslant 1$. Find the coordinates of its centre of mass. [**Hint** You may need to do a substitution for x in terms of $\sinh u$ to evaluate the integrals involved.]

Stability of equilibrium

Consider a uniform solid cone placed in equilibrium on a horizontal surface. To be in equilibrium, its centre of mass must be above a point of contact between the cone and the surface. In theory, there are three possible orientations of the cone for which this is true, as shown, although we would be hard put to realise **b** in practice.

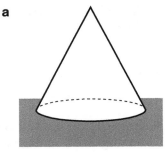

In position **a**, the cone is said to be in **stable equilibrium**. If the cone is displaced slightly from this position (by tipping it), it will move back towards this position when it is released.

In position **b**, the cone is said to be in **unstable equilibrium**. If the cone is displaced slightly from this position, it will continue to move further away.

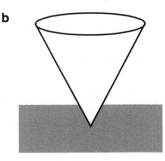

In position **c**, the cone is also in stable equilibrium as regards its being displaced by tipping it. However, in this position it can also be displaced by rolling it. If we do so, the cone will remain in its new position when it is released. It is said to be in **neutral equilibrium** with respect to this displacement.

We now need to explore the conditions for a system to be in equilibrium and the way in which we can distinguish the different types of equilibrium.

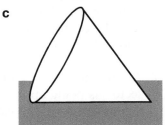

If a system is held at rest in a certain position, it has a total potential energy (gravitational plus elastic) of E. If on release it starts to move, it gains kinetic energy. As the principle of conservation of energy applies, this means that E must reduce.

In position **a** of the cone, displacing the cone from the equilibrium position raises the centre of mass and so increases the gravitational potential energy of the system. When it is released from this displaced position, it will move so as to decrease this potential energy, which means that it will move back towards the equilibrium position.

In position **b**, displacing the cone lowers the centre of mass and so decreases the gravitational potential energy of the system. When it is released from this displaced position, it cannot move back towards the equilibrium position, as this would involve an increase of potential energy. Instead, it will move away from the equilibrium position, as this further decreases the potential energy.

In position **c**, displacing the cone by tipping it raises the centre of mass, so it behaves in a similar fashion to position **a**. However, displacing it by rolling neither raises nor lowers the centre of mass. The potential energy is therefore constant and so on release the cone does not move.

We can see from the above discussion that a position of **stable equilibrium** corresponds to a **minimum potential energy** position. Any displacement increases the potential energy, and so on release the system moves back towards the minimum position.

Similarly, a position of **unstable equilibrium** corresponds to a **maximum potential energy** position. Any displaced position has lower potential energy and so the system cannot return to the equilibrium position when released.

If the potential energy is **constant** over a range of positions, the equilibrium is **neutral**.

Example 5 A ring of mass m is threaded onto a smooth circular hoop of radius a set in a vertical plane. The ring is attached to the lowest point on the hoop by means of a light spring of natural length a and modulus of elasticity mg. Find the positions of equilibrium of the system and examine their stability.

SOLUTION

Consider the ring at position P, as shown.

The length of the spring is $AP = 2a\cos\theta$.

The extension of the spring is $a(2\cos\theta - 1)$.

The height of the ring above the zero GPE level is
$$AC = a(1 + \cos 2\theta)$$

The potential energy of the system is

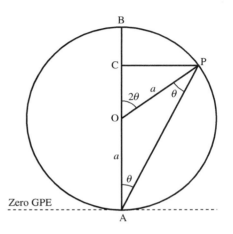

$$E = mga(1 + \cos 2\theta) + \frac{mga^2(2\cos\theta - 1)^2}{2a}$$
$$= mga[1 + \cos 2\theta + \tfrac{1}{2}(2\cos\theta - 1)^2]$$
$$\Rightarrow \quad \frac{dE}{d\theta} = mga[-2\sin 2\theta - 2\sin\theta(2\cos\theta - 1)]$$
$$= -2mga(2\sin 2\theta - \sin\theta)$$

For a position of equilibrium, E has either a maximum or a minimum value. That is, $\frac{dE}{d\theta} = 0$. Hence, we have
$$2\sin 2\theta - \sin\theta = 0$$
$$\Rightarrow \quad 4\sin\theta\cos\theta - \sin\theta = 0$$
$$\Rightarrow \quad \sin\theta(4\cos\theta - 1) = 0$$
$$\Rightarrow \quad \sin\theta = 0 \quad \text{or} \quad \cos\theta = 0.25$$
$$\Rightarrow \quad \theta = 0 \quad \text{or} \quad \pm 75.5°$$

There are, therefore, three positions of equilibrium: one at B when $\theta = 0$, and one on either side of the ring when AP makes an angle of 75.5° with the vertical.

We now consider
$$\frac{d^2 E}{d\theta^2} = -2mga(4\cos 2\theta - \cos\theta)$$

When $\theta = 0$, $\frac{d^2 E}{d\theta^2} < 0$, and so E is a maximum. This corresponds to unstable equilibrium.

When $\theta = \pm 75.5°$, $\frac{d^2 E}{d\theta^2} > 0$, and so E is a minimum. This corresponds to stable equilibrium.

The modelling assumptions would admit the possibility of an equilibrium position at A, but as this would involve the spring's being compressed to zero length, we can reasonably ignore it.

CHAPTER 9 STABILITY OF EQUILIBRIUM

Note The method used in Example 5 assumes that the potential energy function is continuous and differentiable. If this is not so, the position in which the potential energy takes its lowest value may not correspond to a stationary point.

Example 6 A solid, uniform cone has radius $2a$ and height $4a$. It rests on its base on a horizontal surface. Examine the stability of the equilibrium.

SOLUTION

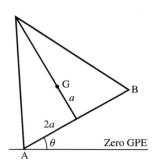

The diagram shows the cone tipped through an angle θ from the equilibrium position.

The centre of mass of a solid cone is a quarter of the way up its height, as shown.

If the mass of the cone is m, the potential energy function is

$$E = mga(2\sin\theta + \cos\theta)$$

As we allow θ to decrease, $(2\sin\theta + \cos\theta)$ also decreases. However, the function does not reach a minimum point when $\theta = 0$. This is because the function corresponds to a situation where the cone could continue to pivot about A as θ becomes negative. If this were the case, the potential energy would continue to fall, as shown in the first graph.

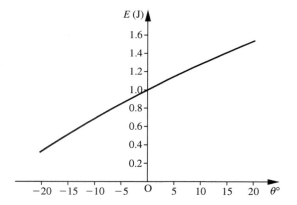

In practice, when B reaches the surface, the cone then starts to pivot about B, and the potential energy starts to increase. This means that the potential energy function is really

$$E = mga(2\sin|\theta| + \cos|\theta|)$$

as shown in the second graph. This function is not differentiable at $\theta = 0$, but the potential energy takes its lowest value at this point.

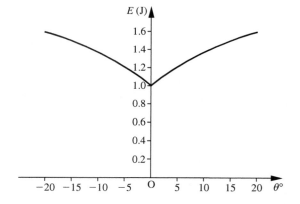

Exercise 9B

1. A uniform rod, PQ, of mass m and length $2l$, rests with its ends on two fixed, smooth, perpendicular planes whose line of intersection is horizontal. PQ lies in the plane ABC, which is perpendicular to the two fixed planes, as shown. AB is inclined to the horizontal at an angle α, and PQ is inclined at an angle θ to AB.

   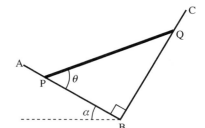

   a) Find the potential energy function for the rod.
   b) Show that the only position of equilibrium (other than when the rod lies on one of the two planes) is when the centre of the rod is vertically above the point B.
   c) Show that this position of equilibrium is unstable.

2. A child's toy consists of a solid hemispherical base of radius 3 cm, with a model dancer standing on its plane face. The base has a mass of 20 grams. The model has a mass of 5 grams, and its centre of mass is situated symmetrically 1 cm from the plane face of the base. The toy rests on a horizontal surface and is held with the plane face of the base making an angle θ with the horizontal.

   a) Find the potential energy function of the system.
   b) Show that when the toy stands upright it is in stable equilibrium.

3. A uniform pole, AB, of mass m and length $2l$, is smoothly hinged at A to a point on a vertical wall. An elastic rope, of natural length $2l$ and modulus of elasticity λ, is fastened to the end B and also to a point C on the wall, a distance of $2l$ vertically above A. The pole makes an angle of 2θ with the wall, as shown.

   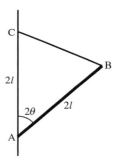

   a) Show that, provided $\lambda > mg$, the system has two equilibrium positions.
   b) Find the equilibrium positions for the case $\lambda = 3mg$, and establish whether or not they are stable.

4. ABCD is a rhombus made of freely jointed rods, each of length a and weight W. A and C are connected by means of a spring of natural length a and stiffness k. The structure is suspended from a fixed support at A.

   a) Find the potential energy function for the system in terms of θ, the angle between AB and the vertical.
   b) Find any positions of equilibrium and investigate their nature.

5. A rough, wooden block, of thickness $2b$ and weight W, rests across a fixed cylinder of radius r, so that the configuration is symmetrical, as shown. The friction is great enough so that pushing down on one end of the plank to incline it at a small angle, θ, to the horizontal does not cause it to slip.

   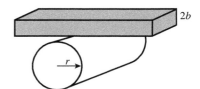

   a) Find the potential energy function in terms of θ.
   b) Show that the plank is in stable equilibrium when it is horizontal provided that $r > b$.

6 Particles of mass m are attached to either end and to the middle of a long, light, inextensible string of length $2a$. The string is placed symmetrically over two small, smooth pulleys, A and B, which are on the same level and a distance $2l$ apart. The string between the pulleys makes an angle θ with the vertical.

a) Find the potential energy function for the system.
b) Find the equilibrium position for the system and show that it is stable.

7 ABC is a uniform framework in the form of an equilateral triangle of side length a. D and E are two smooth pegs, a distance b apart on the same level. The framework is placed so that AB and AC rest on D and E respectively, as shown. Show that, provided $a > 4b$, there is a position of stable equilibrium in which AB and AC are not equally inclined to DE.

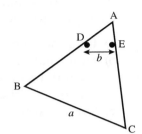

8 Two identical rods, AB and BC, each of length $2r$ and weight W, are freely jointed at B. They are placed symmetrically across a cylinder of radius r, as shown. Each rod makes an angle of θ with the horizontal.

a) Find the potential energy function for the system.
b) Show that the system is in stable equilibrium when $\cos^3\theta = \sin\theta$.

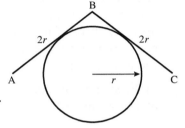

9 A ring of mass m is threaded on a smooth wire hoop of radius a, which is fixed in a vertical plane. The ring is attached by means of a light spring of natural length a and modulus $2mg$ to a point A on the hoop, level with its centre. Investigate the positions of equilibrium of the system and their stability.

10 Moment of inertia

'If everybody minded their own business,' said the Duchess in a hoarse growl, 'the world would go round a deal faster than it does.'
LEWIS CARROLL

Kinetic energy of rotation

You are already familiar with the fact that a moving body has kinetic energy, and that this energy only changes as a result of an external force acting on the body. If the energy increases, work is being done on the body. If the energy decreases, the body is doing work, typically against a resistance force or against gravity.

In the same way, a body rotating about a fixed axis has energy. We need to apply a torque to change the rate of rotation, and therefore the energy, of the body. Rotational energy can be made to do work. You may be familiar with toy cars which can travel a considerable distance using the energy stored in a rotating flywheel. There are even city transport systems powered by the energy stored in a large, floor-mounted flywheel, which is 'revved up' electrically at each stopping place.

The amount of energy stored in a particular body clearly depends on its rate of rotation. However, different bodies with the same rate of rotation can have different energies. The mass of the body is clearly one factor in this, but it turns out that the distribution of that mass in relation to the axis of rotation is also important. You may have observed this when watching an ice-skater performing a spin. If the skater holds a fixed attitude with arms outstretched, the rate of rotation remains constant. If, however, the arms are brought close to the body, or raised above the skater's head, the rate of rotation is seen to increase. As no external force has been applied, the rotational energy should be unchanged. What has happened is that the skater's mass distribution has altered, and this change has been compensated by a change in the rate of rotation so that the energy is constant.

Moment of inertia

We will now examine how the mass distribution affects the rotational kinetic energy by obtaining an expression for the kinetic energy of a rigid body rotating about a fixed axis with angular velocity $\dot{\theta}$.

The diagram on the right shows a plane section through the body such that the axis of rotation passes through point O and is perpendicular to the plane. The body can be modelled as a matrix of particles, P_i, of masses m_i, held together by internal forces. The particle P_i is at a distance r_i from O, as shown.

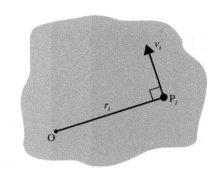

As the body is rotating about O, the velocity, v_i, of the particle P_i is perpendicular to OP_i, as shown. The kinetic energy of the particle is therefore $\frac{1}{2}m_i v_i^2$. But $v_i = r_i \dot{\theta}$, so the kinetic energy of the particle can be written as $\frac{1}{2}m_i r_i^2 \dot{\theta}^2$.

If we sum this energy for all the particles comprising the body, we obtain

$$\text{Total rotational KE} = \sum \tfrac{1}{2} m_i r_i^2 \dot{\theta}^2$$

As the angular velocity is the same at every point of a rigid body, we can write

$$\text{Total rotational KE} = \tfrac{1}{2}\left(\sum m_i r_i^2\right)\dot{\theta}^2 \qquad [1]$$

Compare this with the familiar formula for linear motion:

$$\text{Linear KE} = \tfrac{1}{2}mv^2 \qquad [2]$$

We see that the role of the linear velocity, v, in equation [2] has been taken by the angular velocity, $\dot{\theta}$, in equation [1].

In equation [2], the mass m is a measure of the body's 'linear inertia' or 'reluctance to move' – it requires a larger force to achieve a given acceleration in a more massive body. The counterpart of this in equation [1] is the quantity $\sum m_i r_i^2$, which depends both on the constituent masses and on their positions relative to the axis. This quantity determines the body's rotational inertia, and is called its **moment of inertia**. We write

$$\text{Moment of inertia, } I = \sum m_i r_i^2$$

The dimensions of I are ML^2. Its SI unit is the **kg m²**.

As the dimensions of $\dot{\theta}$ are T^{-1}, we can see from [1] that the dimensions of rotational kinetic energy are ML^2T^{-2}, the same as for linear kinetic energy. Hence, its SI unit is the **joule**.

On pages 209–16, we will see how, by the application of calculus, we can adapt our model of a rigid body as a set of discrete particles to find the moment of inertia of a continuous rigid body. For now, we will examine how in other rotational motion equations, the moment of inertia, I, takes a role analogous to that of mass in linear motion.

Equation of rotational motion

We take, once again, our particle model of a rigid body, and consider the acceleration of the particle P_i.

We saw on page 160 that the acceleration of a particle in relation to an origin O is given by

$$\ddot{\mathbf{r}} = (\ddot{r} - r\dot{\theta}^2)\hat{\mathbf{r}} + (r\ddot{\theta} + 2\dot{r}\dot{\theta})\hat{\mathbf{s}}$$

where $\hat{\mathbf{r}}$ and $\hat{\mathbf{s}}$ are unit vectors in the radial and transverse directions.

As P_i is constrained to move in a circle, r_i is constant. The acceleration of P_i is, therefore, given by

$$\ddot{\mathbf{r}}_i = -r_i\dot{\theta}^2\hat{\mathbf{r}} + r_i\ddot{\theta}\hat{\mathbf{s}}$$

as shown.

Let us suppose that the resultant of all forces, internal and external, acting on P_i is F_i, making an angle α_i to OP_i, as shown. The equation of motion of the particle in the transverse direction is

$$F_i \sin \alpha_i = m_i r_i \ddot{\theta}$$
$$\Rightarrow \quad F_i r_i \sin \alpha_i = m_i r_i^2 \ddot{\theta}$$

If we sum this for all the constituent particles, we obtain

$$\sum F_i r_i \sin \alpha_i = \sum m_i r_i^2 \ddot{\theta} \qquad [3]$$

Internal forces occur in equal and opposite pairs, so the left-hand side of equation [3] represents the total moment about O of all external forces acting on the body. That is, the torque or couple, C, say, acting on the body.

As the angular acceleration, $\ddot{\theta}$, is the same for all the particles, we can write equation [3] as

$$C = \left(\sum m_i r_i^2\right)\ddot{\theta}$$
$$\Rightarrow C = I\ddot{\theta} \qquad [4]$$

where I is the moment of inertia, as before.

If we compare equation [4] with the equation of linear motion, $F = ma$, we can again see how the role of the moment of inertia in rotational motion corresponds to that of mass in linear motion.

Angular momentum

We again model the rigid body as a collection of particles. Particle P_i has mass m_i and velocity $r_i\dot{\theta}$ perpendicular to OP_i. Its linear momentum is, therefore, $m_i r_i \dot{\theta}$.

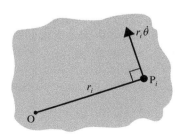

When examining rotational motion, we use the **moment of the momentum** of the particle about O. This gives

$$\text{Moment of momentum} = (m_i r_i \dot{\theta}) r_i = m_i r_i^2 \dot{\theta}$$

If we sum this for all the constituent particles, we obtain the total moment of momentum of the body. This is more usually called the **angular momentum**. Hence, we have

$$\text{Angular momentum} = \sum m_i r_i^2 \dot{\theta}$$

The value $\dot{\theta}$ is the same for all particles, so we have

$$\text{Angular momentum} = \left(\sum m_i r_i^2\right)\dot{\theta}$$

That is,

$$\text{Angular momentum} = I\dot{\theta} \qquad [5]$$

If we compare equation [5] with the expression mv for linear momentum, we can again see how the role of the moment of inertia in rotational motion corresponds to that of mass in linear motion.

The dimensions of angular momentum are ML^2T^{-1}. The SI unit is the **kg m² s⁻¹**, but this is rarely used explicitly, as the angular momentum is normally a stage on the way to calculating some other quantity.

Uniform angular acceleration

As with linear motion, there are formulae relating the parameters of angular motion in the case where the angular acceleration is constant.

We will use the following notation:

Initial angular velocity	ω
Final angular velocity	Ω
Angular displacement	θ
Angular acceleration	α
Time	t

We then have the following formulae:

$$\Omega = \omega + \alpha t$$
$$\theta = \tfrac{1}{2}(\omega + \Omega)t$$
$$\theta = \omega t + \tfrac{1}{2}\alpha t^2$$
$$\theta = \Omega t - \tfrac{1}{2}\alpha t^2$$
$$\Omega^2 = \omega^2 + 2\alpha\theta$$

Example 1 A flywheel has a moment of inertia of 300 kg m^2. It is rotating at 2 rad s^{-1}. A constant torque of 150 N m is applied to the flywheel for 10 seconds, in a sense such as to increase its rate of rotation. Find
a) the angular velocity at the end of this time
b) the angular displacement during this time
c) the increase in the kinetic energy of the flywheel.

SOLUTION

The equation of rotational motion is $C = I\ddot{\theta} = I\alpha$ for constant angular acceleration.

In this case, we have $C = 150\,\text{Nm}$ and $I = 300\,\text{kg}\,\text{m}^2$, which give

$$\alpha = \frac{150}{300} = 0.5\,\text{rad}\,\text{s}^{-2}$$

We also have initial angular velocity $\omega = 2\,\text{rad}\,\text{s}^{-1}$ and time $t = 10\,\text{s}$.

a) Using $\Omega = \omega + \alpha t$, we have

$$\Omega = 2 + 0.5 \times 10 = 7$$

So, the angular velocity at the end of the time is $7\,\text{rad}\,\text{s}^{-1}$.

b) Using $\theta = \omega t + \frac{1}{2}\alpha t^2$, we have

$$\theta = 2 \times 10 + \tfrac{1}{2} \times 0.5 \times 100 = 45$$

So, the angular displacement during this time is $45\,\text{rad}$.

c) Using rotational $\text{KE} = \frac{1}{2}I\dot{\theta}^2$, we have

$$\text{Initial KE} = \tfrac{1}{2} \times 300 \times 4 = 600\,\text{J}$$

$$\text{Final KE} = \tfrac{1}{2} \times 300 \times 49 = 7350\,\text{J}$$

So, the increase in $\text{KE} = 7350 - 600 = 6750\,\text{J}$.

Example 2 A gate has a moment of inertia about its hinges of $500\,\text{kg}\,\text{m}^2$. The gate is open and perpendicular to the gateway. It is given an impulse so that it starts to close with an initial angular velocity of $3\,\text{rad}\,\text{s}^{-1}$. The hinges provided a resisting torque of $400\,\text{N}\,\text{m}$. Find
a) the time taken for the gate to shut
b) the angular momentum of the gate at the instant that it shuts.

SOLUTION

Using $C = I\ddot{\theta}$, we have

$$\text{Acceleration, } \alpha = \frac{-400}{500} = -0.8\,\text{rad}\,\text{s}^{-2}$$

We also have initial velocity $\omega = 3\,\text{rad}\,\text{s}^{-1}$ and angular displacement $\theta = \dfrac{\pi}{2}\,\text{rad}$.

a) Using $\theta = \omega t + \frac{1}{2}\alpha t^2$, we have

$$\frac{\pi}{2} = 3t - 0.4t^2$$

$$\Rightarrow \quad 0.8t^2 - 6t + \pi = 0$$

Solving this quadratic equation, we obtain $t = 0.566$ or 6.934.

The first of these values is clearly the required one, so the gate takes $0.566\,\text{s}$ to shut.

b) Using $\Omega^2 = \omega^2 + 2\alpha\theta$, we have

$$\Omega^2 = 9 - 1.6 \times \frac{\pi}{2} = 6.488$$

$$\Rightarrow \quad \Omega = 2.547$$

Using angular momentum $= I\dot{\theta}$, we have

$$\text{Final angular momentum} = 500 \times 2.547 = 1273 \text{ kg m}^2 \text{ s}^{-1}$$

Example 3 A load of mass 20 kg is being raised on the end of a light, inextensible rope by means of a winch. The pulley of the winch has radius 0.2 m and moment of inertia 2.5 kg m². The pulley bearing is assumed to be smooth. The load starts from rest and accelerates at a rate of 0.4 m s⁻². Taking g $= 9.8$ m s⁻², find
a) the torque driving the pulley
b) the kinetic energy of the system after 3 seconds.

SOLUTION

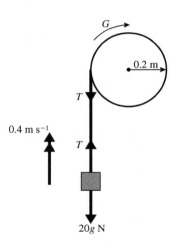

a) Resolving vertically for the load, we have

$$T - 20g = 20 \times 0.4$$
$$\Rightarrow \quad T = 204 \qquad [1]$$

Let the torque driving the pulley be G, as shown. The total moment acting on the pulley is then $G - 0.2T$.

The angular acceleration, α, of the pulley is such that a point on the rim of the pulley has a transverse acceleration of 0.4 m s⁻².

The transverse acceleration is given by

$$r\ddot{\theta} = 0.2 \times \alpha$$
$$\Rightarrow \quad 0.2\alpha = 0.4$$
$$\Rightarrow \quad \alpha = 2 \text{ rad s}^{-2}$$

Using $C = I\ddot{\theta}$, we have

$$G - 0.2T = 2.5 \times 2 \quad \Rightarrow \quad G = 0.2T + 5$$

Substituting from equation [1], we have

$$G = 0.2 \times 204 + 5 = 45.8$$

So, the driving torque is 45.8 N m.

b) After 3 seconds, the load is travelling at a speed of 1.2 m s⁻¹. Using $KE = \frac{1}{2}mv^2$, we have

$$\text{KE of load} = \frac{1}{2} \times 20 \times 1.2^2 = 14.4 \text{ J}$$

A point on the rim of the pulley also has speed 1.2 m s⁻¹. This speed is given by $r\dot{\theta}$, so we have

$$0.2\dot{\theta} = 1.2 \quad \Rightarrow \quad \dot{\theta} = 6$$

Using rotational KE = $\frac{1}{2}I\dot{\theta}^2$, we have

KE of pulley = $\frac{1}{2} \times 2.5 \times 36 = 45$ J

So, the total KE of the system is $45 + 14.4 = 59.4$ J

Example 4 The diagram shows two particles, each of mass m, connected by a light, inextensible string. One particle rests on a smooth, horizontal table. The other is suspended on the string, which passes over a pulley of radius r at the edge of the table. The moment of inertia of the pulley is $6mr^2$. The system is released from rest. Assuming that the pulley can rotate freely, and that the contact between the string and the pulley is rough, find the angular acceleration of the pulley.

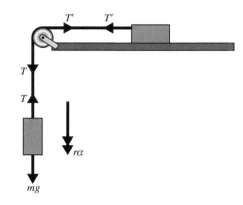

SOLUTION

Because the pulley is not light, the tensions in the two portions of the string are different. Call them T and T', as shown.

If the angular acceleration of the pulley is α, the linear acceleration of a point on its rim, and hence of the string and the particles, is $r\alpha$, as shown.

Using $F = ma$ for the particles, we have

$$mg - T = mr\alpha \qquad [1]$$

and

$$T' = mr\alpha \qquad [2]$$

The resultant torque acting on the pulley is $(T - T')r$. Using $C = I\ddot{\theta}$, we have

$$(T - T')r = 6mr^2\alpha \qquad [3]$$

Substituting from [1] and [2] into [3], we have

$$(mg - 2mr\alpha)r = 6mr^2\alpha$$
$$\Rightarrow \quad g - 2r\alpha = 6r\alpha$$
$$\Rightarrow \quad \alpha = \frac{g}{8r}$$

So, the angular acceleration of the pulley is $\frac{g}{8r}$.

Exercise 10A

1 A flywheel has a moment of inertia of 60 kg m^2.
   a) Find the angular acceleration of the flywheel if a torque of 150 N m is applied.
   b) Find the torque needed to achieve an angular acceleration of 1.6 rad s^{-2}.
   c) Find the kinetic energy of the flywheel if its angular momentum is $420 \text{ kg m}^2 \text{ s}^{-1}$.

CHAPTER 10 MOMENT OF INERTIA

2 A constant torque of 80 N m is applied to a rotating flywheel. As a result, the angular velocity of the flywheel is reduced from 9 rad s^{-1} to 3 rad s^{-1} in 12 seconds. Find
 a) the moment of inertia of the flywheel
 b) the loss of kinetic energy.

3 A flywheel having a moment of inertia of 600 kg m^2 is rotating at 300 rev min^{-1}. Find the frictional couple which must be applied to bring the flywheel to rest in 10 seconds.

4 A constant couple is applied to a wheel which has a moment of inertia of 75 kg m^2 and which can rotate on smooth bearings. The wheel starts from rest. After 150 revolutions, the wheel is rotating at 300 rev min^{-1}. Find the moment of the couple.

5 Particles A and B, of mass 5 kg and 2 kg respectively, are connected together by means of a light, inextensible string. The string passes over a pulley of radius 0.3 m, which has a moment of inertia of 20 kg m^2. Contact between the string and the pulley is rough, and the pulley is free to rotate on smooth bearings. The system is held at rest with the two particles level, and is then released.
 a) Find the accelerations of the system.
 b) Find the angular velocity of the pulley after particle A has fallen 2 m.
 c) Find the total kinetic energy of the system at this point, and confirm that it is equal to the loss of gravitational potential energy of the system.

6 A particle of mass 3 kg is attached to one end of a light, inextensible string. The string is wound round a cylindrical drum of radius 0.2 m and moment of inertia 12 kg m^2, which is mounted on a horizontal axle. The particle is suspended on the string and is released from rest. Find the acceleration of the particle if the drum is subject to a constant frictional couple of 2 N m.

7 The diagram shows particles of mass 10 kg and 20 kg. They are connected by means of a light, inextensible string, which passes over two pulleys which are mounted at either end of a horizontal bench. The pulleys each have a radius of 0.1 m and a moment of inertia of 0.2 kg m^2, and are free to rotate on smooth bearings. The system is released from rest. Assuming that the string does not slip on the pulleys, find the total kinetic energy of the system after 5 seconds.

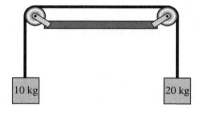

8 A uniform cylinder, of radius $2a$ and moment of inertia $2Ma^2$, is free to rotate about its horizontal axis. A light, inextensible string is wound round the cylinder and a particle of mass m is suspended on its free end. If the system is released from rest, find the acceleration of the particle.

9 Particles of mass M and m (where $M > m$) are attached to the ends of a light, inextensible string which passes over a pulley of radius a and moment of inertia I. The contact between the string and the pulley is rough, and the pulley is free to rotate on a smooth, horizontal axle. Find an expression for the angular acceleration of the pulley.

Calculating the moment of inertia

First, we will find the moment of inertia of certain simple objects. We will then use these to establish the moment of inertia for other objects.

Uniform ring

We will find the moment of inertia of a uniform ring, of mass M and radius a, about an axis through its centre and perpendicular to the plane of the ring. By this, we mean an idealised ring having no thickness, just mass concentrated along the circumference of a circle.

Suppose that a unit arc length has mass m. We can then regard a small element, δs, of the circumference of the circle as being approximately a particle of mass $m \delta s$ at a distance a from the axis. The moment of inertia of such a particle is $ma^2 \delta s$. Therefore, for the whole ring, we have

$$\text{Moment of inertia} \approx \sum ma^2 \delta s = ma^2 \sum \delta s$$

In the limit, as $\delta s \to 0$, this gives

$$\text{Moment of inertia} = ma^2 \int_0^{2\pi a} \mathrm{d}s = m^2 \left[s \right]_0^{2\pi a} = 2\pi ma^3$$

But $2\pi ma = M$, so we have

$$\text{Moment of inertia} = Ma^2$$

Uniform rod

We will consider a uniform, idealised rod of zero thickness, having mass M and length $2l$.

First, it is clear that if the axis about which we seek to find the moment of inertia is parallel to the rod, we can apply an argument similar to that for a ring in the previous section. If the axis is at a distance of a from the rod, the moment of inertia is Ma^2. (You might wish to fill in the detail of this.)

We will now examine the situation where the axis is through the centre of the rod and perpendicular to it, as shown.

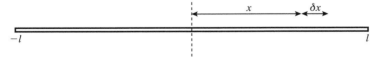

Consider a small element of length δx, at a distance of x from the axis. If the mass per unit length of the rod is m, the mass of this element is $m\delta x$.

The moment of inertia of this element is approximately $mx^2\delta x$. Therefore, for the whole rod, we have

$$\text{Moment of inertia} \approx \sum mx^2 \delta x = m \sum x^2 \delta x$$

In the limit, as $\delta x \to 0$, this gives

$$\text{Moment of inertia} = m \int_{-l}^{l} x^2 \, dx = m \left[\frac{x^3}{3}\right]_{-l}^{l} = \frac{2ml^3}{3}$$

But $2ml = M$, so we have

$$\text{Moment of inertia} = \frac{Ml^2}{3}$$

Uniform shells of revolution

We are now in a position to find the moments of inertia of uniform shells of revolution, which can be thought of as constructed of rings. Three examples are given below.

Example 5 Find the moment of inertia of a cylindrical shell of mass M, length l and radius a.

SOLUTION

The cylinder can be regarded as being made of small cylindrical elements of length δs, as shown.

Each element is approximately a ring of radius a. If the mass per unit area of the cylinder is m, the mass of this elemental ring is $2\pi ma\delta s$.

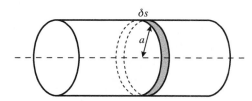

Hence, its moment of inertia is approximately $2\pi ma^3 \delta s$.

Therefore, for the whole cylinder, we have

$$\text{Moment of inertia} \approx \sum 2\pi ma^3 \delta s = 2\pi ma^3 \sum \delta s$$

In the limit, as $\delta s \to 0$, this gives

$$\text{Moment of inertia} = 2\pi ma^3 \int_0^l ds = 2\pi ma^3 \left[s\right]_0^l = 2\pi ma^3 l$$

But $2\pi mal = M$, so the moment of inertia of a cylindrical shell of mass M and radius a is Ma^2.

We can apply a similar argument to any hollow object which is a surface of revolution, as shown in Examples 6 and 7.

Example 6 Find the moment of inertia of an open hollow cone of mass M, base radius a and height h about its axis of symmetry.

SOLUTION

Let the mass per unit area of the cone be m.

The surface area of the cone is πal, where l is the slant height of the cone, as shown. So, $\pi mal = M$.

Let the semi-vertical angle of the cone be θ, as shown.

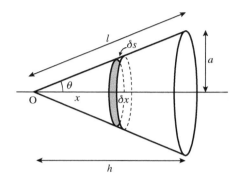

The diagram shows an elemental slice of thickness δx at a distance of x from the vertex, O. This approximates to a cylinder of height δs, where δs is the corresponding portion of the generating line, as shown. (Notice that we must use the 'arc' length δs rather than the thickness δx in such situations, otherwise the approximation is not good enough.)

The radius of the cylinder is $\dfrac{ax}{h}$, by similar triangles. Its surface area is, therefore, $\dfrac{2\pi ax\delta s}{h}$. But $\delta s = \dfrac{\delta x}{\cos\theta}$, so the surface area becomes $\dfrac{2\pi ax\delta x}{h\cos\theta}$.

The elemental slice is, therefore, approximately a ring of mass $\dfrac{2\pi amx\delta x}{h\cos\theta}$.

Its moment of inertia is approximately $\dfrac{2\pi a^3 mx^3\delta x}{h^3\cos\theta}$.

Therefore, for the whole cone, we have

$$\text{Moment of inertia} \approx \sum \frac{2\pi a^3 mx^3\delta x}{h^3\cos\theta} = \frac{2\pi a^3 m}{h^3\cos\theta}\sum x^3\delta x$$

In the limit, as $\delta x \to 0$, this gives

$$\text{Moment of inertia} = \frac{2\pi a^3 m}{h^3\cos\theta}\int_0^h x^3\,dx$$

$$= \frac{2\pi a^3 m}{h^3\cos\theta}\left[\frac{x^4}{4}\right]_0^h = \frac{\pi a^3 mh}{2\cos\theta}$$

But $\pi mal = M$, and $l = \dfrac{h}{\cos\theta}$. Hence, we have

$$\text{Moment of inertia} = \frac{Ma^2}{2}$$

CHAPTER 10 MOMENT OF INERTIA

Example 7 Find the moment of inertia of a spherical shell of radius a and mass M about an axis through its centre.

SOLUTION

Let the mass per unit area of the sphere be m.

The surface area of the sphere is $4\pi a^2$, so $4\pi m a^2 = M$.

We consider an elemental slice of the sphere perpendicular to the y-axis, as shown. The arc of the generating circle, δs, corresponding to this element subtends an angle $\delta \theta$ at the centre of the sphere. This gives $\delta s = a\, \delta \theta$.

The elemental slice is approximately a cylindrical shell with radius $a \cos \theta$ and height $a\, \delta \theta$. This has a surface area of $2\pi a^2 \cos \theta\, \delta \theta$, and hence its mass is $2\pi m a^2 \cos \theta\, \delta \theta$.

The moment of inertia of this element about the y-axis is approximately $2\pi m a^4 \cos^3 \theta\, \delta \theta$. Therefore, for the whole sphere, we have

$$\text{Moment of inertia} \approx \sum 2\pi m a^3 \cos^3 \theta\, \delta \theta = 2\pi m a^3 \sum \cos^3 \theta\, \delta \theta$$

In the limit, as $\delta \theta \to 0$, this gives

$$\text{Moment of inertia} = 2\pi m a^3 \int_{-\frac{\pi}{2}}^{\frac{\pi}{2}} \cos^3 \theta\, d\theta$$

$$= 2\pi m a^3 \int_{-\frac{\pi}{2}}^{\frac{\pi}{2}} \cos \theta\, (1 - \sin^2 \theta)\, d\theta$$

$$= 2\pi m a^3 \left[\sin \theta - \tfrac{1}{3} \sin^3 \theta \right]_{-\frac{\pi}{2}}^{\frac{\pi}{2}}$$

$$= \frac{8\pi m a^4}{3}$$

But $4\pi m a^2 = M$, so we have

$$\text{Moment of inertia} = \frac{2Ma^2}{3}$$

Uniform disc

We can also use the result for the moment of inertia of a uniform ring to find the moment of inertia of a uniform disc about the axis through its centre and perpendicular to the disc. From this, we can progress to dealing with solids of revolution.

Consider a uniform disc of radius a and mass M.

Let the mass per unit area of the disc be m. This gives $\pi m a^2 = M$.

We take an element of the disc forming an annulus (ring-shaped lamina) of width δx at a distance of x from the centre of the disc, as shown.

The element can be 'unrolled' to give an approximate rectangle with length $2\pi x$ and width δx. This therefore has mass $2\pi m x \delta x$.

The moment of inertia of the element about an axis through the centre of the disc and perpendicular to it is approximately that of a ring of mass $2\pi m x \delta x$ and radius x. This gives

$$\text{Moment of inertia of the element} \approx 2\pi m x^3 \delta x$$

Therefore, for the whole disc, we have

$$\text{Moment of inertia} \approx \sum 2\pi m x^3 \delta x = 2\pi m \sum x^3 \delta x$$

In the limit, as $\delta x \to 0$, this gives

$$\text{Moment of inertia} = 2\pi m \int_0^a x^3 \, dx$$

$$= 2\pi m \left[\frac{x^4}{4} \right]_0^a = \frac{\pi m a^4}{2}$$

But $\pi m a^2 = M$, so we have

$$\text{Moment of inertia of disc} = \frac{Ma^2}{2}$$

Uniform solids of revolution

Solids of revolution can be divided into disc-like elements, so we are now in a position to find the moment of inertia of a uniform solid of revolution about its axis of symmetry.

Example 8 Find the moment of inertia of a sphere of radius a and mass M about an axis through its centre.

SOLUTION

Let the mass per unit volume of the sphere be m.
This gives $\dfrac{4\pi m a^3}{3} = M$.

Let the sphere be generated by rotating the circle $x^2 + y^2 = a^2$ about the x-axis, and consider an element of thickness δx at distance of x from the centre, as shown.

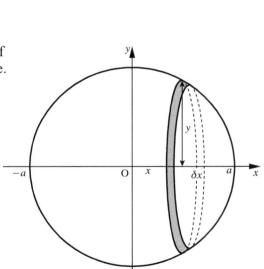

This element is approximately a disc of radius y and thickness δx. The mass of such a disc is $\pi m y^2 \delta x$, and its moment of inertia about the x-axis is $\dfrac{\pi m y^4 \delta x}{2}$.

Therefore, for the whole sphere, we have

$$\text{Moment of inertia} \approx \sum \frac{\pi m y^4 \delta x}{2} = \frac{\pi m}{2} \sum y^4 \delta x$$

In the limit, as $\delta x \to 0$, this gives

$$\begin{aligned}\text{Moment of inertia} &= \frac{\pi m}{2} \int_{-a}^{a} y^4 \, dx \\ &= \frac{\pi m}{2} \int_{-a}^{a} (a^2 - x^2)^2 \, dx \\ &= \frac{\pi m}{2} \left[a^4 x - \tfrac{2}{3} a^2 x^3 + \tfrac{1}{5} x^5 \right]_{-a}^{a} \\ &= \frac{8 \pi m a^5}{15}\end{aligned}$$

But $\dfrac{4\pi m a^3}{3} = M$, so we have

$$\text{Moment of inertia} = \frac{2 M a^2}{5}$$

Uniform laminae

We can also make use of our earlier result for a uniform rod to find the moments of inertia of uniform laminae which can be divided into rod-like elements.

Example 9 Find the moment of inertia of a rectangular lamina ABCD of mass M, where $AB = 2a$ and $BC = 2b$, about the axis of symmetry through the mid-point of AB.

SOLUTION

Let the mass per unit area of the rectangle be m. This gives $4mab = M$.

Consider an element of width δx, as shown. This approximates to a rod of length $2a$ and mass $2ma\delta x$.

Using the result for a rod, the moment of inertia of this element about the x-axis is $\dfrac{2ma^3 \delta x}{3}$.

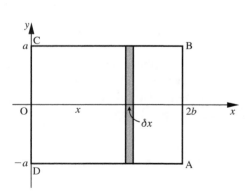

Therefore, for the whole rectangle, we have
$$\text{Moment of inertia} = \sum \frac{2ma^3 \delta x}{3} = \frac{2ma^3}{3} \sum \delta x$$

In the limit, as $\delta x \to 0$, we have
$$\text{Moment of inertia} = \frac{2ma^3}{3} \int_0^{2b} \mathrm{d}x$$
$$= \frac{2ma^3}{3} \left[x\right]_0^{2b} = \frac{4ma^3 b}{3}$$

But $4mab = M$, so we have
$$\text{Moment of inertia} = \frac{Ma^2}{3}$$

Example 10 Find the moment of inertia of triangular lamina ABC of mass M, where AB = $2a$, BC = $2b$ and angle B = $90°$, about the side BC.

SOLUTION

Let the mass per unit area of the triangle be m. This gives $2mab = M$.

Take BA and BC to be the x- and y-axes, as shown.

The equation of the line AC is
$$y = 2b - \frac{bx}{a}$$

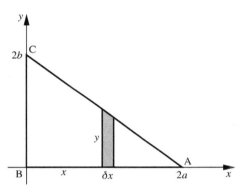

Consider an element of width δx, as shown. This approximates to a rod of mass $my\delta x$ at a distance of x from the line BC. The approximate moment of inertia of this element about BC is $mx^2 y \delta x$.

Therefore, for the whole triangle, we have
$$\text{Moment of inertia} \approx \sum mx^2 y \delta x = m \sum x^2 y \delta x$$

In the limit, as $\delta x \to 0$, this gives
$$\text{Moment of inertia} = m \int_0^{2a} x^2 y \, \mathrm{d}x$$
$$= m \int_0^{2a} x^2 \left(2b - \frac{bx}{a}\right) \mathrm{d}x$$
$$= m \left[\frac{2bx^3}{3} - \frac{bx^4}{4a}\right]_0^{2a}$$
$$= \frac{4ma^3 b}{3}$$

But $2mab = M$, so we have
$$\text{Moment of inertia} = \frac{2Ma^2}{3}$$

CHAPTER 10 MOMENT OF INERTIA

Non-uniform bodies

So far, we have dealt only with uniform bodies. It is possible to find the moment of inertia for some non-uniform bodies, provided that the density of the body varies with position in some convenient and well-defined way.

Example 11 Repeat Example 10 in the case where the mass per unit area of the lamina is directly proportional to the distance from BC.

SOLUTION

Let the mass per unit area at a distance x from BC be mx. Hence, the mass of the element shown is now approximately $mxy\delta x$.

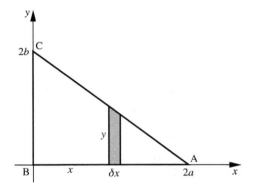

The mass of the triangle is, therefore, approximately $\sum mxy\delta x$.

As $\delta x \to 0$, this gives

$$M = m \int_0^{2a} xy \, dx$$

$$= m \int_0^{2a} x\left(2b - \frac{bx}{a}\right) dx$$

$$= m \left[bx^2 - \frac{bx^3}{3a} \right]_0^{2a}$$

$$\Rightarrow \quad M = \frac{4ma^2 b}{3}$$

The moment of inertia of the element about BC is approximately $mx^3 y \delta x$.

The moment of inertia of the triangle is, therefore, approximately $\sum mx^3 y \delta x$.

As $\delta x \to 0$, this gives

$$\text{Moment of inertia} = m \int_0^{2a} x^3 y \, dx$$

$$= m \int_0^{2a} x^3 \left(2b - \frac{bx}{a}\right) dx$$

$$= m \left[\frac{bx^4}{2} - \frac{bx^5}{5a} \right]_0^{2a}$$

$$= \frac{8ma^4 b}{5} = \frac{4ma^2 b}{3} \times \frac{6a^2}{5}$$

But as $\dfrac{4ma^2 b}{3} = M$, we have

$$\text{Moment of inertia} = \frac{6Ma^2}{5}$$

Exercise 10B

1. Show that the moment of inertia of a uniform rod of mass M and length a about an axis through one end and perpendicular to the rod is $\dfrac{Ma^2}{3}$.

2. Show that the moment of inertia of a uniform rod of mass M and length $2l$ about an axis through its centre and making an angle of α to the rod is $\dfrac{Ml^2 \sin^2 \alpha}{3}$.

3. Find the moment of inertia of a rod, AB, of mass M and length πa, which is bent into a semicircle, about AB.

4. Find the moment of inertia of a solid cylinder of mass M and radius a about its axis.

5. Find the moment of inertia of a lamina of mass M, consisting of an equilateral triangle of side length $2a$, about **a)** its axis of symmetry and **b)** a side.

6. Find the moment of inertia of an annulus of mass M, consisting of a circular lamina with centre O and radius $2a$ from which a concentric hole of radius a has been removed, about an axis through O and perpendicular to the plane of the annulus.

7. Prove that the moment of inertia of a uniform solid cone of mass M and base radius a about its axis of symmetry is $\dfrac{3Ma^2}{10}$.

8. A shell comprises the curved surface of a frustum of a cone. The two circular ends have radii of a and $2a$, and the distance between them is h. The mass of the shell is M. Find its moment of inertia about its axis of symmetry.

9. A lamina of mass M comprises an isosceles trapezium ABCD. The parallel sides are AB and CD, of length a and $2a$ respectively, and the distance between them is h. Find the ratio of the moments of inertia of the lamina about AB and CD.

10. A uniform solid is formed by rotating about the x-axis the region bounded by the curve $y^2 = x$, the line $x = a$ and the x-axis. If the solid has mass M, find its moment of inertia about the x-axis.

11. Find the moment of inertia of a lamina in the form of quadrant of a circle, having mass M and radius a, about one of its straight edges.

CHAPTER 10 MOMENT OF INERTIA

Composite bodies

Consider an object formed of two parts, A and B. Let the moments of inertia of the whole object about some axis and of A and B about the same axis be I, I_1 and I_2 respectively. If the object is rotating with angular velocity $\dot\theta$ about the axis, the kinetic energy of rotation of the whole object is the sum of the energies of A and B. That is,

$$\tfrac{1}{2}I\dot\theta^2 = \tfrac{1}{2}I_1\dot\theta^2 + \tfrac{1}{2}I_2\dot\theta^2$$
$$\Rightarrow \quad I = I_1 + I_2$$

So, in general, the moment of inertia of a composite body about some axis is the sum of the moments of inertia of its component parts about the same axis.

Example 12 A placard consists of a rectangular board, PQRS, of mass 2.4 kg, where PQ = 0.8 m and QR = 0.4 m, attached to a pole, TU, of mass 1.5 kg and length 1.8 m. The pole is attached so that T is at the centre of PQRS and passes through the mid-point of PQ, as shown. Find the moment of inertia of the placard about the axis XY, where X and Y are the mid-points of QR and SP respectively.

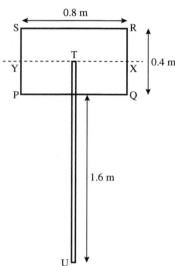

SOLUTION

We model the pole as a uniform rod and the board as a uniform lamina.

We saw in Exercise 10B, Question **1**, that the moment of inertia of a rod of mass M and length a about a perpendicular axis through its end is $\dfrac{Ma^2}{3}$.

We also saw in Example 9 (page 214) that the moment of inertia of a rectangular lamina ABCD of mass M, where AB = $2a$ and BC = $2b$, about the axis of symmetry through the mid-point of AB, is $\dfrac{Ma^2}{3}$.

Applying these results in this case, we have

$$\text{For PQRS:} \quad \text{Moment of inertia about XY} = \dfrac{2.4 \times 0.2^2}{3}$$
$$= 0.032 \,\text{kg m}^2$$

$$\text{For TU:} \quad \text{Moment of inertia about XY} = \dfrac{1.5 \times 1.8^2}{3}$$
$$= 1.62 \,\text{kg m}^2$$

So, the moment of inertia of the placard about XY = $0.032 + 1.62 = 1.652 \,\text{kg m}^2$.

Parallel and perpendicular axes theorems

Parallel axis theorem

We now meet the first of two theorems which enable us to derive the moment of inertia of some bodies about a desired axis from already known moments about related axes.

The **parallel axis theorem** states:

> If I_G is the moment of inertia of a body of mass M about an axis through G, the centre of mass of the body, and I_O is the moment of inertia about a parallel axis at a distance d from G, then
> $$I_O = I_G + Md^2$$

Proof

Take the centre of mass, G, to be the origin and the z-axis to be that axis about which we are taking moments.

Let the parallel axis be the line $x = x_o$, $y = y_o$, which is at a distance d from G, where
$$d^2 = x_o^2 + y_o^2$$

Let a constituent particle, P, of mass m, be at the point $P(x, y, z)$.

The moment of inertia of P about the axis through G is $m(x^2 + y^2)$.

The moment of inertia of P about the parallel axis is
$$m[(x - x_o)^2 + (y - y_o)^2]$$

So, we have
$$I_G = \sum m(x^2 + y^2)$$
and
$$I_O = \sum m[(x - x_o)^2 + (y - y_o)^2]$$
$$= \sum m(x^2 + y^2) + \sum m(x_o^2 + y_o^2) - 2x_o \sum mx - 2y_o \sum my$$

But as G is the centre of mass, we have $\sum mx = \sum my = 0$.

Also, we have $x_o^2 + y_o^2 = d^2$ and $\sum m = M$.

Hence, we find
$$I_O = I_G + Md^2$$
as required.

CHAPTER 10 MOMENT OF INERTIA

Example 13 Find the moment of inertia of the uniform L-shaped lamina shown, which has mass 3 kg, about the line AB.

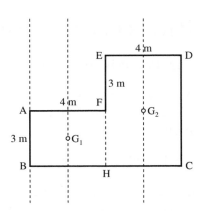

SOLUTION

The lamina is composed of rectangles ABHF and EHCD, of mass 1 kg and 2 kg respectively. Let the centres of mass of these rectangles be G_1 and G_2, as shown.

Using the standard result for a rectangular lamina (page 215), we find that the moment of inertia of ABHF about the dashed axis through G_1 is

$$I_{G_1} = \frac{1 \times 2^2}{3} = 1\tfrac{1}{3} \text{ kg m}^2$$

Similarly, for EHCD, we find

$$I_{G_2} = \frac{2 \times 2^2}{3} = 2\tfrac{2}{3} \text{ kg m}^2$$

Using the parallel axis theorem, we have

For ABHF: Moment of inertia about AB $= I_{G_1} + 1 \times 2^2 = 5\tfrac{1}{3} \text{ kg m}^2$
For EHCD: Moment of inertia about AB $= I_{G_2} + 2 \times 6^2 = 74\tfrac{2}{3} \text{ kg m}^2$

So, the moment of inertia of the whole lamina about AB is $5\tfrac{1}{3} + 74\tfrac{2}{3} = 80 \text{ kg m}^2$.

Perpendicular axes theorem

The parallel axis theorem applies to **any** body. This second result is **only applicable to plane objects**. It **cannot** be used for three-dimensional bodies.

The **perpendicular axes theorem** states:

If I_x and I_y are the moments of inertia of a plane object about two perpendicular axes Ox and Oy in the plane, then its moment of inertia, I_z, about an axis Oz perpendicular to the plane is given by

$$I_z = I_x + I_y$$

Proof

Let P be a constituent particle, of mass m, at the point $P(x, y)$, a distance r from Oz.

The moment of inertia of P about Oz is mr^2.

But $r^2 = x^2 + y^2$, so this becomes $m(x^2 + y^2)$.

Hence, for the whole body, we have

$$I_z = \sum m(x^2 + y^2) = \sum mx^2 + \sum my^2$$

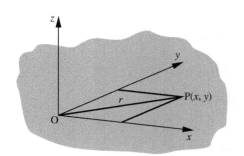

But $\sum mx^2 = I_y$ and $\sum my^2 = I_x$, so we have

$$I_z = I_x + I_y$$

as required.

Example 14 Find the moment of inertia of a ring of mass M and radius a about a diameter.

SOLUTION

Consider axes as shown, with Ox and Oy as two perpendicular diameters of the ring. The moments of inertia of the ring about the axes are I_x, I_y and I_z respectively.

We know from the standard result (page 209) that $I_z = Ma^2$.

By symmetry, $I_x = I_y$, and by the perpendicular axes theorem, $I_z = I_x + I_y$. Hence, we have

$$I_x = I_y = \tfrac{1}{2}Ma^2$$

So, the moment of inertia of the ring about a diameter is $\tfrac{1}{2}Ma^2$.

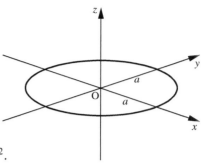

Example 15 ABCD is a uniform lamina of mass M. $AB = 2a$ and $BC = 2b$. Find the moment of inertia of the lamina about an axis through A and perpendicular to the plane ABCD.

SOLUTION

Consider axes as shown, where Ox and Oy are the lines of symmetry of the rectangle. The moments of inertia of the ring about the axes are I_x, I_y and I_z respectively.

We know from the standard result (page 215) that

$$I_x = \frac{Ma^2}{3} \quad \text{and} \quad I_y = \frac{Mb^2}{3}$$

By the perpendicular axes theorem, we have

$$I_z = \frac{Ma^2}{3} + \frac{Mb^2}{3} = \frac{M(a^2 + b^2)}{3}$$

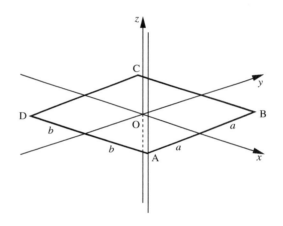

Let I_A be the required moment of inertia about the axis shown through A. As O is the centre of mass of the lamina, we have, from the parallel axis theorem,

$$I_A = I_z + M \times OA^2$$
$$= \frac{M(a^2 + b^2)}{3} + M(a^2 + b^2)$$
$$\Rightarrow \quad I_A = \frac{4M(a^2 + b^2)}{3}$$

Radius of gyration

The moment of inertia of a body of mass M about a given axis can be expressed in the form Mk^2. This is equivalent to the moment of inertia of a particle of mass M placed at a distance of k from the axis. The quantity k is called the **radius of gyration** of the body. The moment of inertia of a body is sometimes given by stating the radius of gyration.

For example, we know (page 215) that the moment of inertia of a rectangular lamina of length $2a$ about the line of symmetry perpendicular to its length is $\dfrac{Ma^2}{3}$. Hence, the radius of gyration for this body and axis is $\dfrac{a}{\sqrt{3}}$.

Example 16 Find the radius of gyration of a cylindrical shell of mass M and radius a about an axis on the surface of the cylinder and parallel to its axis.

SOLUTION

We know from the standard result (Example 5, page 210) that the moment of inertia of the cylinder shell about its axis is Ma^2.

By the parallel axis theorem, the required moment of inertia is given by

$$I = Ma^2 + Ma^2 = 2Ma^2$$

So, the radius of gyration of the cylinder about the required axis is $a\sqrt{2}$.

Summary of standard results

Most examination boards provide a list of standard moments of inertia. From these, it is possible to calculate the moment of inertia of compound bodies, or of the standard bodies about other axes.

Uniform body, mass M	Axis	Moment of inertia
Rod of length $2a$	Through centre of rod and perpendicular to it	$\dfrac{Ma^2}{3}$
Rod of any length	Parallel to rod and at a distance a	Ma^2
Rectangle of length $2a$	Line of symmetry perpendicular to length	$\dfrac{Ma^2}{3}$
Ring of radius a, centre O	Through O, perpendicular to ring	Ma^2
Disc of radius a, centre O	Through O, perpendicular to disc	$\dfrac{Ma^2}{2}$
Hollow cylinder of radius a	Axis of cylinder	Ma^2
Hollow sphere of radius a	A diameter	$\dfrac{2Ma^2}{3}$
Solid cylinder of radius a	Axis of cylinder	$\dfrac{Ma^2}{2}$
Solid sphere of radius a	A diameter	$\dfrac{2Ma^2}{5}$

Exercise 10C

1 A rectangle, ABCD, has mass 5 kg. AB = 6 m and BC = 4 m. Find its moment of inertia about

  a) a line through E in the plane ABCD and perpendicular to AB, where E is on AB and AE = 1 m

  b) a line through F in the plane ABCD and perpendicular to BC, where BF = 1 m.

2 a) Find the moment of inertia of a ring of mass M and radius a about AB, where A and B are the ends of a diameter of the ring.

  b) Particles of mass M are attached to the ring at A and B.
   i) 'Find the moment of inertia of the system about an axis through the centre of the ring and perpendicular to it.
   ii) State the moment of inertia of the system about AB.
   iii) Hence find the moment of inertia about the diameter perpendicular to AB.

3 A shovel is modelled as a uniform rod, AB, of mass 1 kg and length 1 m, attached to a coplanar uniform lamina, CDEF, of mass 2 kg, where CD = 20 cm, DE = 30 cm and B is the mid-point of CD, with AB perpendicular to CD. Find the moment of inertia of the shovel about **a)** CD, **b)** EF and **c)** an axis through A and perpendicular to the plane CDEF.

4 a) Find the moment of inertia of a square lamina of mass m and side length a about the axis through the centre of the square and perpendicular to it.

  b) Deduce the moment of inertia of a cube of mass M and side length a about an axis through the mid-points of two opposite faces.

  c) Hence find the moment of inertia of the cube in part **b** about an edge.

5 Find the radius of gyration of a triangle, comprising three equal uniform rods of length $2a$, about the axis through its centre of mass and perpendicular to the plane of the triangle.

6 Find the moment of inertia of a semicircular lamina of mass M and radius a about **a)** its straight edge and **b)** its line of symmetry.

7 Find the moment of inertia about its axis of a straight, hollow pipe of mass M and length l, whose external and internal radii are a and b respectively.

8 A uniform disc of radius $3a$ and centre O has a circular hole of radius a and centre A cut in it, where OA = a. Find its radius of gyration about

  a) OA
  b) the axis through O in the plane of the lamina and perpendicular to OA
  c) the axis through O perpendicular to the plane of the lamina.

9 A hollow sphere has external radius R and internal radius r. Show that its radius of gyration about a diameter is $\sqrt{\dfrac{2(R^5 - r^5)}{5(R^3 - r^3)}}$.

11 Rotation about a fixed axis

Round the world for ever and aye.
MATTHEW ARNOLD

We have already established the following results (pages 201–4) for the rotation about a given fixed axis of a body whose moment of inertia about that axis is I.

The equation of rotational motion is

$$C = I\ddot{\theta}$$

where C is the resultant torque acting on the body.

The rotational kinetic energy is given by

$$\text{Rotational KE} = \tfrac{1}{2}I\dot{\theta}^2$$

where $\dot{\theta}$ is the angular velocity about the axis.

The angular momentum (moment of momentum) is given by

$$\text{Angular momentum} = I\dot{\theta}$$

We will now use these to explore the mechanics of a body rotating about a fixed axis, first by examining the work–energy principle in relation to such rotational motion.

Total energy of a rotating body

Suppose that a body of mass M is rotating about a fixed axis. Its moment of inertia about the axis is I. Point P on the axis is such that the axis is perpendicular to the plane containing P and G, the centre of mass of the body. Unless the axis is vertical, or G and P coincide, G undergoes vertical displacement as the body rotates, and so the gravitational potential energy (GPE) of the body changes.

Consider the simplest situation in which the axis is horizontal, so that the plane containing P and G is vertical. We take the zero potential energy position to
be when G is vertically below P.

Let PG $= h$. In a general position, with PG making an angle θ to the downward vertical, as shown, we have

$$\text{GPE} = Mgh(1 - \cos\theta)$$

The total energy of the body is, therefore, given by

$$\text{Total energy} = \tfrac{1}{2}I\dot{\theta}^2 + Mgh(1 - \cos\theta)$$

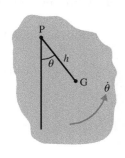

Work done by a couple

Suppose that a couple C is applied to the body. This is equivalent to equal and opposite forces, F, a distance a apart, where $C = Fa$. Without loss of generality, we can position the first of these forces so that its line of action passes through the axis, as shown. This force then does no work as the body rotates.

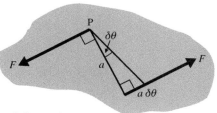

Suppose that the body rotates through a small angle $\delta\theta$. The second force is displaced by approximately $a\delta\theta$. The work done is given by

$$\delta W \approx Fa\delta\theta$$

But $C = Fa$, so we have

$$\frac{\delta W}{\delta\theta} \approx C$$

which, in the limit, as $\delta\theta \to 0$, gives

$$\frac{dW}{d\theta} = C$$

Hence, as the body rotates through an angle α, the work done by the couple is

$$W = \int_0^\alpha C\,d\theta$$

In particular, if the couple is constant, we have $W = C\alpha$.

Relationship between work and energy

We will continue to consider a body rotating about a horizontal axis, although the results apply more generally.

Consider the body shown, rotating about an axis through P. G is the centre of mass. The moment of inertia about P is I. Let the body be subjected to a couple C, as shown.

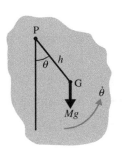

The total torque acting on the body is $C - Mgh\sin\theta$. So, by the equation of rotational motion, we have

$$C - Mgh\sin\theta = I\ddot\theta \qquad [1]$$

But $\ddot\theta = \dfrac{d\dot\theta}{dt} = \dfrac{d\theta}{dt}\dfrac{d\dot\theta}{d\theta} = \dot\theta\dfrac{d\dot\theta}{d\theta}$, so equation [1] becomes

$$C - Mgh\sin\theta = I\dot\theta\frac{d\dot\theta}{d\theta} \qquad [2]$$

CHAPTER 11 ROTATION ABOUT A FIXED AXIS

Suppose the body rotates from a position θ_1 to a position θ_2, where it has angular velocities $\dot\theta_1$ and $\dot\theta_2$ respectively. From equation [2], we have

$$\int_{\theta_1}^{\theta_2} (C - Mgh \sin\theta) \, d\theta = \int_{\dot\theta_1}^{\dot\theta_2} I\dot\theta \, d\dot\theta$$

$$\Rightarrow \quad \int_{\theta_1}^{\theta_2} C \, d\theta + \Big[Mgh \cos\theta\Big]_{\theta_1}^{\theta_2} = \Big[\tfrac{1}{2} I\dot\theta^2\Big]_{\dot\theta_1}^{\dot\theta_2}$$

$$\Rightarrow \quad \int_{\theta_1}^{\theta_2} C \, d\theta = \tfrac{1}{2} I(\dot\theta_2^2 - \dot\theta_1^2) + Mgh(\cos\theta_1 - \cos\theta_2) \qquad [3]$$

But $\int_{\theta_1}^{\theta_2} C \, d\theta$ is the work done by the applied couple. Also, $h(\cos\theta_1 - \cos\theta_2)$ is the amount by which G has been raised. Equation [3] therefore shows that

> Work done = Change of KE + Change of GPE
> = Change of total energy

Thus, the principle, familiar from our study of linear motion, that work done equals change of energy, also applies to rotational motion.

It follows that if no external couple is applied to the body, its total mechanical energy remains constant. That is, the **principle of conservation of mechanical energy** applies to problems involving rotational motion.

Example 1 A uniform rod, AB, of length a and mass m, is free to rotate in a vertical plane about a horizontal axis through A. The rod is gently displaced from rest with B vertically above A.

a) Find the angular velocity of the rod when B reaches a point vertically below A.
b) If, when B is vertically below A, the angular velocity is in fact half that found in part **a**, find the magnitude of the constant frictional couple at the axis.

SOLUTION

Let the zero GPE level be at A, as shown. Let the final angular velocity be ω.

The moment of inertia of the rod about A is $\tfrac{1}{3} ma^2$.

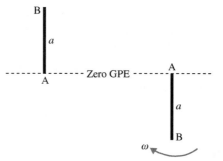

a) The centre of mass of the rod is at the mid-point of AB. So, at the start, we have

$$KE = 0$$
$$GPE = \tfrac{1}{2} mga$$

In the second position, we have

$$KE = \tfrac{1}{2} \times \tfrac{1}{3} ma^2 \times \omega^2 = \tfrac{1}{6} ma^2 \omega^2$$
$$GPE = -\tfrac{1}{2} mga$$

No external couple is applied and there are no sudden changes, so mechanical energy is conserved. Hence, we have

$$\tfrac{1}{2}mga = \tfrac{1}{6}ma^2\omega^2 - \tfrac{1}{2}mga$$
$$\Rightarrow \quad \omega = \sqrt{\frac{6g}{a}}$$

b) The final angular velocity is now $\sqrt{\dfrac{3g}{2a}}$, giving a final kinetic energy of $\tfrac{1}{4}mga$. The potential energies are as in part **a**. We therefore have

Work done by friction = Change of energy
$$= (\tfrac{1}{4}mga - \tfrac{1}{2}mga) - \tfrac{1}{2}mga$$
$$= -\tfrac{3}{4}mga$$

If C is the magnitude of the frictional couple, the work done by friction is $-C\pi$. Hence, we have

$$-C\pi = -\tfrac{3}{4}mga$$
$$\Rightarrow \quad C = \frac{3mga}{4\pi}$$

Example 2 A constant tangential force of magnitude 12 N is applied to the rim of a stationary, uniform, circular flywheel of mass 100 kg and radius 0.5 m. Find the speed at which the flywheel is rotating after it has completed 25 revolutions.

SOLUTION

We will use energy methods, although the problem could equally well be solved using the constant acceleration equations.

Let the final angular velocity be ω.

The moment of inertia is

$$\frac{100 \times 0.5^2}{2} = 12.5 \,\text{kg}\,\text{m}^2$$

The final kinetic energy is, therefore, $\tfrac{1}{2} \times 12.5\omega^2$ J.

The work done by the applied force is

$$12 \times 25 \times 2\pi \times 0.5 = 300\pi \,\text{J}$$

The potential energy is constant, as the axis is through the centre of mass. Therefore, we have

$$\tfrac{1}{2} \times 12.5\omega^2 = 300\pi$$
$$\Rightarrow \quad \omega = \sqrt{48\pi} = 12.3 \,\text{rad s}^{-1}$$

Example 3 A uniform cylinder can rotate freely about its axis, which is horizontal. The cylinder has radius 20 cm and mass 5 kg. A particle of mass 2 kg is attached by means of a light, inextensible string to a point on the cylinder and the string is wound onto the cylinder. The particle is held below the cylinder with the string vertical and taut, and is released from rest. Find the speed at which the particle is travelling after the cylinder has made two complete revolutions. (Assume the string is still partially wound on the cylinder at this stage.)

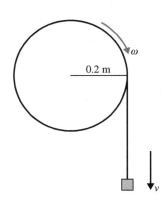

SOLUTION

The moment of inertia of the cylinder is

$$\tfrac{1}{2} \times 5 \times 0.2^2 = 0.1 \, \text{kg m}^2$$

If at any stage of the motion the particle is falling with speed $v \, \text{m s}^{-1}$ and the cylinder is rotating with angular speed $\omega \, \text{rad s}^{-1}$, we have

$$v = 0.2\omega \quad \Rightarrow \quad \omega = 5v$$

since the string can be assumed not to slip.

The potential energy of the cylinder does not change during its motion, since the axis passes through the centre of mass.

If we take the zero level for the potential energy of the particle to be at its initial position, we have

$$\text{Initial KE} = \text{Initial GPE} = 0 \, \text{J}$$

After two revolutions, the particle has fallen a distance

$$2 \times 2\pi \times 0.2 = 0.8\pi \, \text{m}$$

We therefore have

$$\text{Final KE of particle} = \tfrac{1}{2} \times 2v^2 = v^2 \, \text{J}$$
$$\text{Final KE of cylinder} = \tfrac{1}{2} \times 0.1\omega^2 = \tfrac{1}{2} \times 0.1 \times (5v)^2 = 1.25v^2 \, \text{J}$$
$$\text{Final GPE of particle} = -2g \times 0.8\pi = -1.6\pi g \, \text{J}$$

There are no external forces and no sudden changes, so energy is conserved. We therefore have

$$v^2 + 1.25v^2 - 1.6\pi g = 0$$
$$\Rightarrow \quad v = \sqrt{\frac{1.6\pi g}{2.25}} = 4.68$$

So, the particle is falling at a speed of $4.68 \, \text{m s}^{-1}$.

Example 4 A uniform disc of mass M and radius a can rotate about a horizontal axis which is tangential to the disc. The disc is held in a horizontal position and released from rest. If the angular velocity of the disc when it reaches a vertical position is $\sqrt{\dfrac{g}{5a}}$, find the average magnitude of the frictional couple exerted at the axis.

SOLUTION

Let the zero GPE level be at the axis, as shown.

In the vertical position, the centre of mass, G, has fallen a distance a, so we have

$$\text{Initial GPE} = 0$$
$$\text{Final GPE} = -Mga$$

The moment of inertia of a disc about an axis through its centre and perpendicular to the disc is $\frac{1}{2}Ma^2$.

Therefore, the moment of inertia about a diameter is $\frac{1}{4}Ma^2$, by the perpendicular axes theorem.

Hence, the moment of inertia about the given axis is

$$\tfrac{1}{4}Ma^2 + Ma^2 = \tfrac{5}{4}Ma^2$$

by the parallel axis theorem.

We therefore have

$$\text{Initial KE} = 0$$
$$\text{Final KE} = \tfrac{1}{2} \times \tfrac{5}{4}Ma^2 \times \frac{g}{5a} = \frac{Mga}{8}$$

If the average frictional couple is C, the work done by friction is $-C \times \frac{1}{2}\pi$.

Using Work done = Change of energy, we have

$$-\tfrac{1}{2}C\pi = \frac{Mga}{8} - Mga = -\frac{7Mga}{8}$$

$$\Rightarrow \quad C = \frac{7Mga}{4\pi}$$

Exercise 11A

1. A uniform rod, AB, of length 2 m and mass 1 kg, has a particle of mass 1 kg attached at A. The rod can rotate freely about a horizontal axis through its mid-point. The rod is held in a horizontal position and released from rest. Find its angular velocity when it reaches a vertical position.

2. A shop sign of mass 2 kg is in the form of a rectangular lamina, ABCD, where AB = 1 m and BC = 1.6 m. It is suspended from a horizontal axis AB and is hanging at rest with its plane vertical. It is then given an initial angular velocity of 4 rad s^{-1}. The sign comes instantaneously to rest when it reaches an angle of 60° to the downward vertical.

   a) Find the magnitude of the frictional couple, assumed constant.
   b) Does the sign return to the vertical position? If so, with what angular velocity is it moving when it does so?

3 A uniform cylinder can rotate freely about its axis, which is horizontal. The radius of the cylinder is 0.3 m and its mass is 4 kg. A light rope of length 6 m is wound completely onto the cylinder and is then pulled with a constant force of 20 N. Assuming that the rope continues to exert the same force on the cylinder until the moment that it loses contact, find the final angular speed of the cylinder.

4 A ring of mass M and radius a can rotate freely about an axis through a point on its circumference and perpendicular to the plane of the ring. Initially, the ring is hanging at rest. It is then set in motion with angular velocity ω. Find the value of ω if the ring just makes a complete revolution.

5 A uniform circular flywheel of mass 8 kg and radius 30 cm is rotating at 40 rev min^{-1} about a smooth, horizontal axis through its centre. A brake pad is applied to its rim and the wheel comes to rest in 5 revolutions. If the coefficient of friction between the brake pad and the wheel is 0.8, find the normal force which has been applied to the brake pad.

6 A uniform circular flywheel has mass 50 kg and diameter 1.2 m. It is fixed in a vertical plane to the end of a horizontal axle, of radius 0.1 m and negligible mass. The whole is being driven by a motor, and is winching up a particle of mass 6 kg by means of a light rope wound around the axle. The flywheel is turning at a rate of 12 rad s^{-1} when the motor is disconnected. Find how much further the particle will be raised before the system comes instantaneously to rest.

7 A uniform rod, AB, of length 2 m and mass 1 kg, has a particle of mass 1 kg attached to it at B. It can rotate freely about a horizontal axis through A. The end B is attached by means of an elastic string of natural length 1 m to a point a distance 2 m above A. The system is held at rest with AB horizontal and is then released. If the rod just reaches the vertical position, find the modulus of elasticity of the string.

8 The diagram shows the striking mechanism for a large clock. The striker consists of a uniform rod, of length 55 cm and mass 0.5 kg, and a uniform spherical head, of mass 1.5 kg and radius 5 cm. It can rotate freely about a horizontal axis through the end of the rod. The striker is raised to an angle of 30° to the downward vertical and is then released from rest. It strikes the bell when the rod is vertical. Find the angular speed of the striker at the moment that it hits the bell.

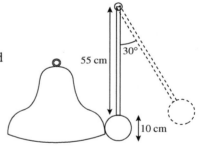

9 A uniform circular flywheel has mass 5 kg and radius 0.4 m. A particle of mass 3 kg is attached to the rim of the flywheel. The flywheel can rotate freely about an axis through its centre and perpendicular to its plane. The axis is positioned so that the plane of the flywheel makes an angle of 45° to the vertical. The system is at rest with the particle at the highest position, and is then slightly displaced. Find the angular velocity when the particle reaches the lowest position.

10 A uniform cylinder of mass 3 kg and radius 20 cm can rotate freely about its axis, which is horizontal. A particle of mass 1 kg is placed on the curved surface of the cylinder at its highest point. The coefficient of friction between the cylinder and the particle is 0.6. The system is displaced slightly from rest, and the cylinder and the particle rotate together.

  a) Find the angular speed of the system when the cylinder has turned through an angle θ.
  b) Investigate whether the particle slips before it loses contact with the cylinder, and find the value of θ at which it starts to move relative to the cylinder.

Compound pendulum

In *Introducing Mechanics* (pages 405–8), we considered the motion of an idealised **simple pendulum**, consisting of a point mass suspended on a light, inextensible string or rod. We found that, provided the oscillations were small, the motion was simple harmonic. We now extend this to cover the more general case of a body oscillating on a horizontal axis which does not pass through its centre of mass. Such an arrangement is called a **compound pendulum**.

The diagram shows such a body, of mass M. The horizontal axis passes through point P in the plane perpendicular to the axis and containing the centre of mass, G. Let $PG = h$, and let the moment of inertia about the axis be I.

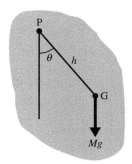

The equation of rotational motion is then

$$Mgh \sin \theta = -I\ddot{\theta}$$

$$\Rightarrow \quad \ddot{\theta} = -\frac{Mgh}{I} \sin \theta$$

If we confine the problem to small oscillations, we can use the approximation $\sin \theta \approx \theta$. The approximate equation of motion is then

$$\ddot{\theta} = -\frac{Mgh}{I} \theta$$

This is the equation of simple harmonic motion. In the standard notation, we have

$$\omega^2 = \frac{Mgh}{I}$$

The period of the oscillation is given by

$$T = 2\pi \sqrt{\frac{I}{Mgh}}$$

It might appear that the period of the oscillation depends on the mass of the body. If true, this would be different from the situation of a simple pendulum. However, the moment of inertia, I, depends on the mass, and so the period turns out to be independent of the mass.

We can see this more readily if we work in terms of the radius of gyration.

Let the radius of gyration of the body about a horizontal axis through G be k. The moment of inertia about this axis is Mk^2. The moment of inertia about the parallel axis through P is then given by

$$I = M(k^2 + h^2)$$

This gives the period of oscillation in the form

$$T = 2\pi \sqrt{\frac{k^2 + h^2}{gh}}$$

which is then seen to be independent of the mass.

CHAPTER 11 ROTATION ABOUT A FIXED AXIS

A simple pendulum of length l has period $T = 2\pi\sqrt{\dfrac{l}{g}}$. A simple pendulum of length $l = \dfrac{k^2 + h^2}{h}$ would have the same period as the compound pendulum. This expression is called the length of the **equivalent simple pendulum**.

Example 5 A uniform rod, AB, of length $2a$ and mass $2m$, has midpoint C. A particle of mass m is attached to B. The rod is suspended from a horizontal axis through P, a point on the rod between A and C and where $PC = x$. It is made to perform small oscillations.

a) Find the period of the oscillations.
b) Find the value of x for which the period is a minimum.
c) Find the length of the equivalent simple pendulum which would have this minimum period.

SOLUTION

a) The moment of inertia of the rod about C is $\dfrac{2ma^2}{3}$.

The moment of inertia of the rod about P is, therefore,
$$\frac{2ma^2}{3} + 2mx^2 = \frac{2m(a^2 + 3x^2)}{3}$$

Hence, the moment of inertia of the whole system about P is
$$\frac{2m(a^2 + 3x^2)}{3} + m(a + x)^2 = \frac{m(5a^2 + 6ax + 9x^2)}{3}$$

The equation of rotational motion about P is
$$-2mgx \sin\theta - mg(a + x) \sin\theta = \frac{m(5a^2 + 6ax + 9x^2)}{3}\ddot\theta$$
$$\Rightarrow \ddot\theta = -\frac{3g(a + 3x)}{5a^2 + 6ax + 9x^2}\sin\theta$$

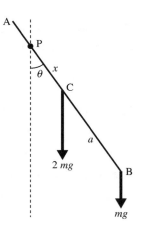

When the oscillations are small, we can write this as
$$\ddot\theta = -\left[\frac{3g(a + 3x)}{5a^2 + 6ax + 9x^2}\right]\theta$$

This is simple harmonic motion with period
$$T = 2\pi\sqrt{\frac{5a^2 + 6ax + 9x^2}{3g(a + 3x)}} \qquad [1]$$

b) The period is a minimum when $y = \dfrac{5a^2 + 6ax + 9x^2}{a + 3x}$ is a minimum.

Differentiating and equating to zero, we obtain
$$\frac{dy}{dx} = \frac{27x^2 + 18ax - 9a^2}{(a + 3x)^2} = 0$$
$$\Rightarrow \quad 3x^2 + 2ax - a^2 = 0$$
$$\Rightarrow \quad (3x - a)(x + a) = 0$$
$$\Rightarrow \quad x = \tfrac{1}{3}a \quad \text{or} \quad x = -a$$

As P was located between A and C, we have $x = \tfrac{1}{3}a$, which gives a minimum period of

$$T = 2\pi\sqrt{\frac{4a}{3g}}$$

c) From equation [1], the length of the equivalent simple pendulum is

$$\frac{5a^2 + 6ax + 9x^2}{3(a + 3x)}$$

When $x = \tfrac{1}{3}a$ this gives a length of $\tfrac{4}{3}a$.

Note In part **b**, there was a second root, $x = -a$. As can be seen in the first graph below, this corresponds to a maximum point of the function, and furthermore a point for which T^2 would be negative.

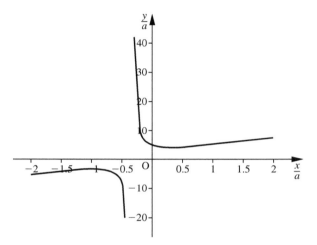

However, the function $\dfrac{5a^2 + 6ax + 9x^2}{a + 3x}$ can be written as $\dfrac{4a^2 + (a + 3x)^2}{a + 3x}$, which has rotational symmetry about the point $\left(-\dfrac{a}{3}, 0\right)$, corresponding to the centre of mass of the system.

Suspending the system from an axis through B, but still taking x as positive in the direction CA, gives

$$T = 2\pi\sqrt{\frac{-(5a^2 + 6ax + 9x^2)}{3g(a + 3x)}}$$

and hence the second, reflected graph, which is shown at the top of the next page.

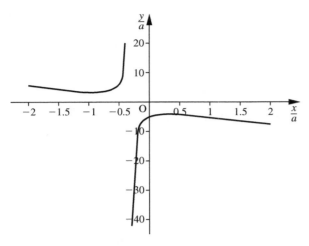

This indicates that, were the system made to oscillate about an axis through B, the period would be the same, minimum period. This is an example of a more general fact: if a body oscillates about an axis through point P in the plane through G perpendicular to the axis, oscillations about a point Q on PG produced would have the same period provided PQ is the length of the equivalent simple pendulum.

Exercise 11B

1 A uniform rod, AB, of length 2 m and mass m, has a particle of mass m attached to B. It is made to perform small oscillations about a horizontal axis through A. Find the period of the oscillations and the length of the equivalent simple pendulum.

2 A uniform disc of radius a is free to oscillate about an horizontal axis through a point A on its circumference. T_1 is the period when the axis is tangential to the disc. T_2 is the period when the axis is perpendicular to the plane of the disc. Find the ratio $T_1:T_2$.

3 A uniform, solid hemisphere of mass M and radius 1 m performs small oscillations about a horizontal axis which forms a diameter of its plane face. Find the period of the oscillations.

4 A uniform, square lamina of mass M and side length $2a$ performs small oscillations about a horizontal axis through a point on its diagonal and perpendicular to the plane of the lamina. Find the distance of the axis from the centre of the lamina if the period of the oscillations is a minimum.

5 A pendulum consists of a uniform rod of mass M and length $2a$ attached to the rim of a circular ring, also of mass M and having radius a, such that the rod is a normal to the circle. The pendulum oscillates in the plane of the ring about the free end of the rod.
   a) Find the period of the oscillation.
   b) Find the percentage change in the period if the ring were replaced by a disc of the same size and mass.

6 A body of mass M has moment of inertia Mk^2 about a horizontal axis through G, its centre of mass. The body is made to perform small oscillations about a parallel axis through a point P in the plane through G perpendicular to the axis, where $PG = h$.

   a) Find the length, l, of the equivalent simple pendulum.
   b) Show that, if Q is the point on PG produced such that $PQ = l$, the period of oscillation about a parallel axis through Q is the same as that about the axis through P.
   c) Show that the period of oscillation about the axis through P is a minimum when $h = k$.

7 A uniform rod, AB, of length $2a$ and mass M, hangs vertically from a horizontal axis through A. The mid-point of AB is fixed rigidly to the mid-point of a second, identical, rod, CD, which is horizontal. The whole is made to perform small oscillations in the plane of the rods.

   a) Show that the period is the same as if the rod AB were oscillating alone.
   b) Show that there is a second point on AB to which the rod CD could be symmetrically fixed without affecting the period of the oscillation.

8 Show that the minimum period of small oscillations of a uniform disc of radius a about a horizontal axis perpendicular to the disc is $2\pi\sqrt{\dfrac{a\sqrt{2}}{g}}$.

9 Two identical laminae, each of mass M and radius of gyration k about an axis through the centre of mass perpendicular to the lamina, hang at rest from the same horizontal axis perpendicular to their planes. The axis is a distance h_1 from the centre of mass of the first lamina and a distance h_2 from that of the second. The laminae are then fixed together in this position and made to perform small oscillations.

   a) Find an expression for the period of the oscillations.
   b) Hence show that this period lies between those of the laminae if they were allowed to oscillate separately.

Reaction at the axis

When a body rotates about a fixed axis, there will in general be a reaction force at the axis to keep that point of the body stationary. This reaction force will change in magnitude and direction as the body rotates. We will confine ourselves to the case of a horizontal axis.

Consider a body of mass M rotating about a horizontal axis through a point P in the plane containing the centre of mass G and perpendicular to the axis. Let $PG = h$. At a given instant, PG makes an angle θ with the downward vertical.

The force exerted on the body by the axis is most conveniently represented by components X and Y, perpendicular and parallel to PG, as shown in the right-hand diagram at the top of the next page.

CHAPTER 11 ROTATION ABOUT A FIXED AXIS

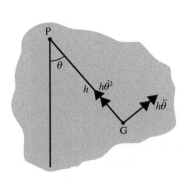
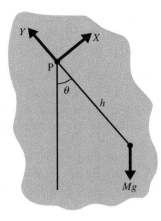

The components of acceleration of the centre of mass are $h\ddot{\theta}$ perpendicular to PG and $h\dot{\theta}^2$ in the direction GP, as shown in the left-hand diagram.

Resolving the forces in these two directions and applying Newton's second law, we have

$$X - Mg\sin\theta = Mh\ddot{\theta}$$
$$Y - Mg\cos\theta = Mh\dot{\theta}^2$$

Assuming we are able to establish both $\dot{\theta}$ and $\ddot{\theta}$, we can solve these equations for X and Y.

Example 6 A uniform disc, centre A, has mass 5 kg and radius 2 m. It is free to rotate about an axis through point P on its circumference, perpendicular to the plane of the disc. The disc is hanging at rest when it is set in motion with initial angular velocity 3 rad s^{-1}. Find the force on the axis when the disc has rotated through $\dfrac{\pi}{6}$ rad.

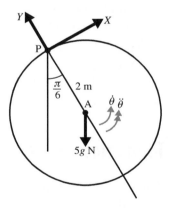

SOLUTION

The moment of inertia of the disc about an axis through A is

$$\frac{5 \times 2^2}{2} = 10 \text{ kg m}^2$$

By the parallel axis theorem, the moment of inertia about the axis through P is, therefore,

$$10 + 5 \times 2^2 = 30 \text{ kg m}^2$$

The equation of rotational motion about P is

$$-5g \sin\left(\frac{\pi}{6}\right) \times 2 = 30\ddot{\theta}$$

$$\Rightarrow \quad \ddot{\theta} = -\frac{g}{6} \text{ rad s}^{-2}$$

If we take the initial position of the disc to be the zero GPE level, we have

$$\text{Initial GPE} = 0\,\text{J}$$
$$\text{Initial KE} = \tfrac{1}{2} \times 30 \times 3^2 = 135\,\text{J}$$
$$\text{Final GPE} = 5g \times 2\left[1 - \cos\left(\frac{\pi}{6}\right)\right] = 5g(2-\sqrt{3})\,\text{J}$$
$$\text{Final KE} = \tfrac{1}{2} \times 30 \times \dot{\theta}^2 = 15\dot{\theta}^2\,\text{J}$$

There is no external torque acting on the body and there are no sudden changes, so energy is conserved. Hence, we have

$$15\dot{\theta}^2 + 5g(2-\sqrt{3}) = 135$$
$$\Rightarrow \dot{\theta}^2 = 9 - \frac{g(2-\sqrt{3})}{3}$$

The force on the axis has components X and Y, as shown. Resolving perpendicular and parallel to PA, we obtain

$$X - 5g\sin\left(\frac{\pi}{6}\right) = 5 \times 2 \times \left(\frac{-g}{6}\right)$$
$$Y - 5g\cos\left(\frac{\pi}{6}\right) = 5 \times 2 \times \left[9 - \frac{g(2-\sqrt{3})}{3}\right]$$

which give

$$X = \frac{5g}{6} = 8.17\,\text{N} \quad \text{and} \quad Y = 90 + \frac{5g(7\sqrt{3}-8)}{6} = 123.7\,\text{N}$$

The force on the axis is, therefore,

$$\sqrt{X^2 + Y^2} = 124\,\text{N}$$

making an angle with AP of

$$\arctan\left(\frac{8.17}{124}\right) = 3.78°$$

Exercise 11C

1. A uniform rod, AB, of mass 3 kg and length 2 m, is able to rotate freely about a horizontal axis through A. The rod is held at rest with B vertically above A, and is then slightly displaced. Find the force on the axis when the rod has rotated **a)** 180°, **b)** 90° and **c)** 120°.

2. A uniform, circular ring, of mass 5 kg and radius 1 m, is free to rotate about a horizontal axis tangential to the ring. Initially, the ring hangs at rest. It is then set in motion with angular velocity ω.

   a) Find the value of ω when the ring can just complete a circle about the axis.
   b) For this value of ω, find the force on the axis when the plane of the ring is first horizontal.

CHAPTER 11 ROTATION ABOUT A FIXED AXIS

3 A uniform, circular lamina, centre O and radius a, has mass M. It is free to rotate about a horizontal axis through O perpendicular to the lamina. A particle of mass M is attached to the lamina at a point P on its rim. The lamina is released from rest with OP horizontal. Find the force on the axis when the lamina has rotated through $60°$.

4 A uniform rod, AB, of length $3a$ and mass M, is free to rotate about a horizontal axis through a point P on the rod, where $AP = a$. The rod is held in a horizontal position and released from rest. Show that when the rod becomes vertical, the reaction on the axis is $\dfrac{3Mg}{2}$.

5 A uniform disc, centre O, has mass M and radius a. It is free to rotate about a horizontal axis perpendicular to its plane, passing through a point P on its rim. The disc is held with OP horizontal and is released from rest.
   a) Find the vertical and horizontal components of the force on the axis when the disc has rotated through an angle θ.
   b) Find the minimum and maximum magnitude of the force on the axis during the motion.

6 A uniform rod, AB, of length $2\,\text{m}$ and mass $4\,\text{kg}$, is pivoted about a horizontal axis through A. It is held with B vertically above A and is set in motion with an initial angular velocity of $3\,\text{rad s}^{-1}$. Its motion is opposed by a frictional torque of magnitude $\dfrac{4g}{\pi}\,\text{N m}$. Find the force on the pivot when the rod reaches the horizontal position.

7 A uniform, square lamina, ABCD, of mass M and side length $4a$, lies on a rough horizontal table with AB parallel to the edge of the table and at a distance a from it, so that three quarters of the lamina overhangs the edge of the table. The lamina is held in this position and is released from rest. If the coefficient of friction between the lamina and the table is 0.5, find the angle through which the lamina rotates before it starts to slip.

8 A uniform rod, AB, of mass M has a particle of mass M attached to B. It is laid on a rough table perpendicular to the table edge, with its mid-point at the edge and B projecting out. The rod is released from rest, and starts to slip when it has turned through an angle of $\arctan(0.25)$. Find the coefficient of friction between the rod and the table.

Impulse and momentum

The equation of motion of a rigid body rotating about a fixed axis is
$$C = I\ddot{\theta}$$
where C is the resultant torque acting on the body.

Integrating with respect to t, we obtain
$$\int_{t_1}^{t_2} C\,dt = \left[I\dot{\theta}\right]_{t_1}^{t_2} \qquad [4]$$

The quantity $I\dot{\theta}$ is the angular momentum of the body, so the right-hand side of equation [4] represents the change in angular momentum.

If, during a short period δt, the torque C is approximately constant and equivalent to a force P acting at a distance a from the axis, we have

$$C\delta t \approx a \times P\delta t$$

But $P\delta t$ is the impulse of the force P. Hence, $C\delta t$ is the moment of that impulse about the axis.

The limit of the sum of these small impulses leads to the conclusion that the left-hand side of equation [4] represents the moment of the impulse of C over the given time period.

Hence, we have

> Moment of impulse about the axis = Change of angular momentum

Example 7 A circular turntable, of radius 0.5 m and mass 2 kg, is rotating in a horizontal plane about a smooth axis through its centre at a rate of $10 \,\text{rad s}^{-1}$. A constant tangential braking force, P, is applied to its rim, and the turntable comes to rest in 3 seconds. Find the value of P.

SOLUTION

The moment of inertia of the turntable is

$$\frac{2 \times 0.5^2}{2} = 0.25 \,\text{kg m}^2$$

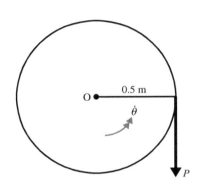

We therefore have

$$\text{Initial angular momentum} = 0.25 \times 10 = 2.5 \,\text{kg m}^2 \,\text{s}^{-1}$$
$$\text{Final angular momentum} = 0$$

The moment of impulse of the force P is

$$-P \times 0.5 \times 3 = -1.5P$$

Hence, we have

$$-1.5P = 0 - 2.5$$
$$\Rightarrow \quad P = 1\tfrac{2}{3} \,\text{N}$$

Example 8 A uniform flywheel, of radius 2 m and mass 40 kg, is rotating at a rate of $8 \,\text{rad s}^{-1}$. A variable braking force of $(10 - 2t)\,\text{N}$ is applied to its rim for a period of 2 s. Find the angular speed of the flywheel at the end of this time.

SOLUTION

Let the final angular speed be ω.

The moment of inertia of the flywheel is

$$\frac{40 \times 2^2}{2} = 80 \,\text{kg m}^2$$

The moment of the applied force is $-2(10 - 2t)\,\text{N m}$.

Using Moment of impulse = Change of angular momentum, we have

$$-2\int_0^2 (10-2t)\,dt = 80\omega - 80 \times 8$$

$$\Rightarrow \quad -2\Big[10t - t^2\Big]_0^2 = 80\omega - 640$$

$$\Rightarrow \quad -32 = 80\omega - 640$$

$$\Rightarrow \quad \omega = 7.6$$

So, the final rate of rotation is $7.6\,\text{rad s}^{-1}$.

Example 9 A uniform rod, AB, has length 4 m and mass 6 kg. It is free to rotate about a horizontal axis through A. It is held at rest with B vertically above A and then slightly displaced. When B reaches its lowest point, it strikes a fixed peg. If the rod next comes to rest with AB horizontal, find the magnitude of the impulse exerted by the peg on the rod.

SOLUTION

We need to find the angular velocities, ω_1 and ω_2, of the rod immediately before and after its encounter with the peg.

The moment of inertia of the rod is

$$\frac{6 \times 4^2}{3} = 32\,\text{kg m}^2$$

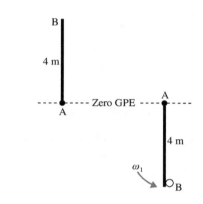

Take the zero GPE level as shown in the diagram.

Initially, we have

$$\text{KE} = 0\,\text{J}$$
$$\text{GPE} = 6g \times 2 = 12g\,\text{J}$$

Immediately before the peg is struck, we have

$$\text{KE} = \tfrac{1}{2} \times 32 \times \omega_1^2 = 16\omega_1^2\,\text{J}$$
$$\text{GPE} = 6g \times (-2) = -12g\,\text{J}$$

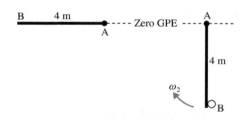

Energy is conserved, so we also have

$$16\omega_1^2 - 12g = 12g$$

$$\Rightarrow \quad \omega_1 = \sqrt{\frac{3g}{2}}\,\text{rad s}^{-1}$$

Immediately after the peg is struck, we have

$$\text{KE} = \tfrac{1}{2} \times 32 \times \omega_2^2 = 16\omega_2^2\,\text{J}$$
$$\text{GPE} = -12g\,\text{J}$$

When the rod reaches the horizontal position, we have

$$\text{KE} = 0 \quad \text{and} \quad \text{GPE} = 0$$

Again, energy is conserved, so we have

$$16\omega_2{}^2 - 12g = 0$$

$$\Rightarrow \quad \omega_2 = \frac{\sqrt{3g}}{2} \text{ rad s}^{-1}$$

Let the impulse exerted by the peg be J, as shown. Hence, the moment of the impulse about A is $4J$.

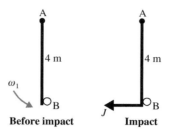
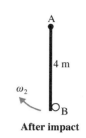

Before impact   Impact   After impact

Using Moment of impulse = Change of angular momentum, we obtain

$$4J = 32 \times \frac{\sqrt{3g}}{2} + 32 \times \sqrt{\frac{3g}{2}}$$

$$\Rightarrow \quad J = 4\sqrt{3g}\left(\sqrt{2} + 1\right) \text{ N s}$$

If an impulse is applied to a rigid body which is pivoted about a fixed axis, there is in general an impulsive reaction at the axis. The magnitude of this impulsive reaction depends on the position of the point on the body at which the external impulse is applied. This will be familiar to anyone who has played cricket. If the ball strikes the bat in the right spot, the hands hardly feel the impact, whereas at other points the hands feel significant jarring.

There is a point, called the **centre of percussion**, at which an applied impulse will cause zero impulsive reaction at the axis.

Example 10 A uniform rod, AB, of mass M and length $2a$, is able to turn freely about a horizontal axis through A. The rod is hanging at rest. A horizontal impulse J is applied to the rod at a point a distance x below A. Find the impulsive reaction at the axis and hence the value of x for which the impulsive reaction is zero.

SOLUTION

Let the impulsive reaction at the axis be X, as shown.

The moment of inertia of the rod about A is $\dfrac{4Ma^2}{3}$.

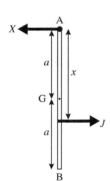

Immediately after the impulse, the rod is turning about A with angular velocity ω. This means that the centre of mass, G, is moving with velocity $v = a\omega$.

The total impulse on a body in any given direction gives the change of momentum in that direction of its centre of mass. This gives

$$J - X = Mv - 0 = Ma\omega \qquad [1]$$

The total moment of impulse about the axis at A gives the change of angular momentum. Hence, we have

$$Jx = \frac{4Ma^2\omega}{3} \qquad [2]$$

From equations [1] and [2], we have

$$Jx = \frac{4a(J-X)}{3}$$

$$\Rightarrow \quad X = \frac{(4a-3x)J}{4a}$$

Therefore, the impulsive reaction at the axis is zero when $x = \frac{4a}{3}$, and this point is the centre of percussion of the system.

Conservation of angular momentum

We have seen that

Change of angular momentum = Moment of impulse

It follows that if the moment of impulse about the given axis is zero, the angular momentum of the system about that axis is constant. This is the **principle of conservation of angular momentum**:

If, about a given axis, there is no resultant torque acting on a system, the angular momentum of the system about that axis is constant.

Example 11 A uniform, circular turntable, of radius 0.2 m and mass 2 kg, is rotating in a horizontal plane about a smooth axis with angular velocity 12 rad s^{-1}. At a certain instant, a particle of mass 1 kg, at rest, is caught on a small peg on the rim of the turntable and moves with it. Find the angular velocity of the system after this has occurred.

SOLUTION

Let the final angular velocity be ω.

The moment of inertia of the turntable is

$$\frac{2 \times 0.2^2}{2} = 0.04 \text{ kg m}^2$$

Before becoming attached to the turntable, the particle has zero momentum. Hence, the angular momentum of the system is initially

$$0.04 \times 12 = 0.48 \text{ kg m}^2 \text{ s}^{-1}$$

The moment of inertia of the turntable plus the particle is

$$0.04 + 1 \times 0.2^2 = 0.08 \text{ kg m}^2$$

The final angular momentum of the system is therefore 0.08ω.

There are no external forces, so angular momentum is conserved. Hence, we have

$$0.08\omega = 0.48 \quad \Rightarrow \quad \omega = 6$$

So, the final angular velocity is 6 rad s^{-1}.

Example 12 A uniform rod, AB, of mass 2 kg and length 30 cm, is rotating on the surface of a smooth table about a vertical axis through A. Its angular velocity is 10 rad s^{-1}. A particle of mass 0.5 kg, moving on the surface of the table at a speed of 2 m s^{-1}, collides with the rod, striking it at B. At the moment of impact, B and the particle are moving towards each other and AB is perpendicular to the direction of motion of the particle. If the coefficient of restitution between the rod and the particle is 0.5, find the velocity of the particle after impact.

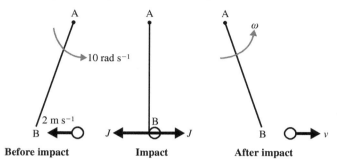

SOLUTION

After impact, the rod moves with angular velocity ω and the particle with velocity v, as shown.

The moment of inertia of the rod about A is

$$\frac{2 \times 0.3^2}{3} = 0.06 \text{ kg m}^2$$

The angular momentum of the rod about A prior to impact is, therefore,

$$0.06 \times 10 = 0.6 \text{ kg m}^2 \text{ s}^{-1}$$

The angular momentum of the particle about A is the moment about A of its linear momentum. That is,

$$-0.5 \times 2 \times 0.3 = -0.3 \text{ kg m}^2 \text{ s}^{-1}$$

Hence, the total angular momentum of the system about A before impact is

$$0.6 - 0.3 = 0.3 \text{ kg m}^2 \text{ s}^{-1}$$

Similarly, the total angular momentum about A after impact is

$$0.06\omega + 0.15v$$

At impact, the rod and the particle suffer impulses J and $-J$, as shown. As the total impulse is zero, angular momentum is conserved. Hence, we have

$$0.06\omega + 0.15v = 0.3 \qquad [1]$$

To obtain a second equation, we apply Newton's law of restitution to the point B and the particle. The speed of B prior to impact is $10 \times 0.3 = 3$ m s^{-1}, and after impact is 0.3ω. Thus, we have

Speed of approach: $\quad 3 + 2 = 5$
Speed of separation: $\quad v - 0.3\omega$

Hence, we have
$$\frac{v - 0.3\omega}{5} = 0.5$$
$$\Rightarrow \quad v - 0.3\omega = 2.5 \qquad [2]$$

Solving equations [1] and [2] for v, we obtain $v = 2.286$.

So, the particle moves with a speed of $2.29 \, \text{m s}^{-1}$ after the collision.

Exercise 11D

1. A uniform, square lamina, of mass 0.5 kg and side length 60 cm, is mounted with its plane horizontal on a vertical axis. The lamina is rotating at a rate of $4 \, \text{rad s}^{-1}$ when a torque of magnitude 6 N m is applied for 2 s in the same sense as the lamina's rotation. Find the angular speed of the lamina at the end of this time.

2. A uniform cylinder, of radius 0.5 m and mass 20 kg, is rotating about its geometrical axis at a rate of $8 \, \text{rad s}^{-1}$. Find the magnitude of the constant torque needed to double the kinetic energy of the cylinder in 5 s.

3. A uniform flywheel, of radius 0.4 m and mass 60 kg, is rotating at a rate of $300 \, \text{rev min}^{-1}$. A brake pad is pressed against the rim of the wheel and its speed is halved in 10 s. If the coefficient of friction between the wheel and the brake pad is 0.8, find the normal force with which the pad is pressed onto the wheel.

4. A uniform flywheel, of radius 2 m and mass 50 kg, is rotating at a rate of $8 \, \text{rad s}^{-1}$. It is brought to rest by a torque whose magnitude at time t s is $(10t + 120)$ N m. Find the length of time before the flywheel stops.

5. A uniform sphere, of radius a and mass M, is free to rotate about a diameter. The sphere is initially at rest. A torque, whose magnitude at time t is $\dfrac{Mga}{1+t}$, is applied to the sphere for 9 s. Find the angular speed at which the sphere is then rotating.

6. A uniform rod, AB, of length $2a$ and mass M, is free to rotate about a fixed horizontal axis through A. The rod is hanging at rest when a horizontal impulse J is applied to B and the rod just reaches the horizontal position.

    a) Find the magnitude of the impulse.
    b) Find the impulsive reaction at A.

7. A uniform rod, AB, of length $2a$ and mass M, has a particle of mass M attached to it at C, where $AC = \frac{1}{2}a$. The rod is free to rotate about a fixed horizontal axis through A. The rod is hanging at rest when a horizontal impulse J is applied to it at a point P, a distance x below A. Find the impulsive reaction in the axis, and hence find the value of x for this impulsive reaction to be zero.

8. A uniform rod, AB, of length $2a$ and mass M, is free to rotate about a fixed horizontal axis through A. The rod is held at rest with B vertically above A, and is then slightly displaced. When B reaches the point immediately below A, it collides with a fixed peg and rebounds. If the rod is next instantaneously at rest when AB is horizontal, find the coefficient of restitution between the rod and the peg.

9 A uniform turntable, of radius a and mass M, is rotating at a rate of ω. A particle of mass M, initially at rest, is lowered onto the surface of the turntable at a distance of $\tfrac{1}{2}a$ from its centre, and moves with the turntable without slipping. Find the new rate of rotation of the turntable.

10 A pulley can be regarded as a uniform disc of radius 20 cm and mass 2 kg. It is free to rotate about a fixed horizontal axis. A light, inextensible string is wound round the pulley, leaving 2 m free, and the end of the string is attached to a particle of mass 1 kg. This particle is held close to the pulley and level with its centre, and is then dropped from rest. Find the angular velocity with which the pulley starts to move.

11 A uniform rod, AB, of length $2a$ and mass M, is free to rotate about a fixed horizontal axis through A. The rod is held horizontal and then set in motion with initial angular speed ω, so that B commences moving downwards. When the rod becomes vertical, B collides with a stationary particle of mass M, which adheres to it. In the subsequent motion, the rod is next instantaneously at rest in the horizontal position. Show that $\omega = \sqrt{\dfrac{33g}{a}}$.

12 A uniform rod, AB, of length $2a$ and mass $2M$, is rotating with angular speed ω on the surface of a smooth table about a fixed vertical axis through A. B is connected by means of a light, inextensible string to a particle P of mass M, which is at rest on the table. The string becomes taut at a point where B is moving away from P and BP is perpendicular to AB. Find the speed of the particle immediately after the string becomes taut.

13 A uniform rod, AB, of length $2a$ and mass M, is free to rotate on the surface of a smooth table about a fixed vertical axis through A. Initially, it is rotating with angular velocity ω. A particle, P, of mass m, moving on the surface of the table with speed $6a\omega$, collides directly with the rod, striking it at right angles at a point C, where $BC = \tfrac{1}{2}a$. The coefficient of restitution between the particle and the rod is 0.8. If the rod is brought to rest by the collision, find m and the new speed of the particle.

12 General motion of a rigid body in two dimensions

This wheel's on fire, rolling down the road.
BOB DYLAN

In the previous chapter, we studied the motion of bodies able to rotate about a fixed axis. We now turn our attention to bodies undergoing both translational and rotational motion. The general study of this would take us beyond the remit of this book – indeed, the tumbling motion of some astronomical bodies could take us into the realms of chaos theory. We will restrict ourselves to the motion of a lamina moving in its own plane, extending in a limited way to bodies which can be regarded as being the sum of such laminae.

Particle model

We will start by modelling the lamina as being composed of particles P_i, having masses m_i. Hence, the mass of the lamina is $M = \Sigma m_i$.

The particle P_i has position vector $\mathbf{r}_i = x_i\mathbf{i} + y_i\mathbf{j}$ relative to a fixed origin, O.

The centre of mass, G, of the lamina has position vector $\bar{\mathbf{r}} = \bar{x}\mathbf{i} + \bar{y}\mathbf{j}$.

The position vector of P_i relative to G is

$$\mathbf{r}'_i = x'_i\mathbf{i} + y'_i\mathbf{j} = (x_i - \bar{x})\mathbf{i} + (y_i - \bar{y})\mathbf{j}$$

Each particle is subject to forces, both internal and external. Let the resultant of the external forces on P_i be

$$\mathbf{F}_i = X_i\mathbf{i} + Y_i\mathbf{j}$$

Let the resultant of the internal forces on P_i be

$$\mathbf{F}_{Ii} = X_{Ii}\mathbf{i} + Y_{Ii}\mathbf{j}$$

The equation of motion for the particle P_i is, therefore,

$$\mathbf{F}_i + \mathbf{F}_{Ii} = m_i\ddot{\mathbf{r}}_i \qquad [1]$$

By Newton's third law, the internal forces occur in balancing pairs. Hence, when we sum equation [1] over all the particles, $\Sigma \mathbf{F}_{Ii} = \mathbf{0}$. It follows that

$$\Sigma \mathbf{F}_i = \Sigma m_i\ddot{\mathbf{r}}_i = \frac{\mathrm{d}}{\mathrm{d}t}(\Sigma m_i\dot{\mathbf{r}}_i) \qquad [2]$$

We can state equation [2] as follows:

> The rate of change of the linear momentum of the system in any direction is equal to the component of the resultant of the external forces in that direction.

In particular,

> When the component of the resultant of the external forces in some direction is zero, the linear momentum in that direction is constant.

This confirms the **principle of linear momentum**.

Motion of the centre of mass

By definition, we have
$$M\bar{\mathbf{r}} = \Sigma m_i \mathbf{r}_i$$
Differentiating, we obtain
$$M\dot{\bar{\mathbf{r}}} = \Sigma m_i \dot{\mathbf{r}}_i$$
which shows that the linear momentum of the body is the same as if it were a single particle of mass M located at the centre of mass.

Differentiating again, we have
$$M\ddot{\bar{\mathbf{r}}} = \Sigma m_i \ddot{\mathbf{r}}_i$$
But from equation [2], we have
$$\Sigma \mathbf{F}_i = \Sigma m_i \ddot{\mathbf{r}}_i$$
$$\Rightarrow \quad \Sigma \mathbf{F}_i = M\ddot{\bar{\mathbf{r}}}$$

We can state this as follows:

> The centre of mass moves as if the whole body were a particle situated there and all external forces acted there.

This provides some justification for modelling bodies as particles in many mechanics problems.

Translational and rotational motion of the lamina

Rotational motion

We now examine the moments of the forces acting on the lamina.

First, we remind ourselves that if a particle P at the point with position vector $\mathbf{r}$ is acted upon by a force $\mathbf{F}$, the moment of the force about the origin is given by the vector product of the position vector and the force vector (see page 27). That is, we have
$$\mathbf{r} \times \mathbf{F} = \mathbf{C}$$

CHAPTER 12 GENERAL MOTION OF A RIGID BODY IN TWO DIMENSIONS

Taking $\mathbf{F}_i$ and $\mathbf{F}_{Ii}$ to be the resultant external and internal forces acting on the particle P_i of our lamina, as before, and taking moments about the origin, we have

$$C_i \mathbf{k} = \mathbf{r}_i \times (\mathbf{F}_i + \mathbf{F}_{Ii})$$
$$= \mathbf{r}_i \times \mathbf{F}_i + \mathbf{r}_i \times \mathbf{F}_{Ii} \quad [3]$$

Again, as the internal forces occur in balancing pairs, $\Sigma \mathbf{r}_i \times \mathbf{F}_{Ii} = \mathbf{0}$. Hence, when we sum equation [3] over all the particles, we obtain

$$C\mathbf{k} = \Sigma \mathbf{r}_i \times \mathbf{F}_i \quad [4]$$

From equation [1], we have

$$\mathbf{r}_i \times (\mathbf{F}_i + \mathbf{F}_{Ij}) = \mathbf{r}_i \times m_i \ddot{\mathbf{r}}_i = m_i (\mathbf{r}_i \times \ddot{\mathbf{r}}_i)$$

Summing this over all the particles, and applying $\Sigma \mathbf{r}_i \times \mathbf{F}_{Ii} = \mathbf{0}$ as before, we have

$$\Sigma \mathbf{r}_i \times \mathbf{F}_i = \Sigma m_i (\mathbf{r}_i \times \ddot{\mathbf{r}}_i) \quad [5]$$

Consider $\mathbf{r}_i \times \dot{\mathbf{r}}_i = (x_i \dot{y}_i - y_i \dot{x}_i) \mathbf{k}$. If we differentiate this, we obtain

$$\frac{d}{dt}(x_i \dot{y}_i - y_i \dot{x}_i) \mathbf{k} = (x_i \ddot{y}_i - y_i \ddot{x}_i) \mathbf{k} = \mathbf{r}_i \times \ddot{\mathbf{r}}_i$$

We can therefore write equation [5] as

$$\Sigma \mathbf{r}_i \times \mathbf{F}_i = \frac{d}{dt}[\Sigma m_i (\mathbf{r}_i \times \dot{\mathbf{r}}_i)] \quad [6]$$

The right-hand side of equation [6] is the rate of change of the moment of momentum (the angular momentum) of the system. Hence, we can see that:

> The sum of the moments of the external forces about a fixed axis is equal to the rate of change of the angular momentum of the system about that axis.

In particular,

> When the sum of the moments of the external forces about some axis is zero, the angular momentum of the system about that axis is constant.

> This confirms the **principle of angular momentum**.

Translational and rotational motion

We stated earlier that the position vector of P_i relative to G is $\mathbf{r}'_i$. Hence, $\mathbf{r}_i = \bar{\mathbf{r}} + \mathbf{r}'_i$. Using this, equation [5] becomes

$$\Sigma (\bar{\mathbf{r}} + \mathbf{r}'_i) \times \mathbf{F}_i = \Sigma m_i [(\bar{\mathbf{r}} + \mathbf{r}'_i) \times (\ddot{\bar{\mathbf{r}}} + \ddot{\mathbf{r}}'_i)]$$

Expanding, we have

$$\Sigma (\bar{\mathbf{r}} \times \mathbf{F}_i) + \Sigma (\mathbf{r}'_i \times \mathbf{F}_i) = \Sigma (m_i \bar{\mathbf{r}} \times \ddot{\bar{\mathbf{r}}}) + \Sigma (m_i \bar{\mathbf{r}} \times \ddot{\mathbf{r}}'_i) + $$
$$+ \Sigma (m_i \mathbf{r}'_i \times \ddot{\bar{\mathbf{r}}}) + \Sigma (m_i \mathbf{r}'_i \times \ddot{\mathbf{r}}'_i)$$
$$= \Sigma (m_i \bar{\mathbf{r}} \times \ddot{\bar{\mathbf{r}}}) + \bar{\mathbf{r}} \times (\Sigma m_i \ddot{\mathbf{r}}'_i) + $$
$$+ (\Sigma m_i \mathbf{r}'_i) \times \ddot{\bar{\mathbf{r}}} + \Sigma (m_i \mathbf{r}'_i \times \ddot{\mathbf{r}}'_i)$$

But $\Sigma(\bar{\mathbf{r}} \times \mathbf{F}_i) = \Sigma(m_i \bar{\mathbf{r}} \times \ddot{\mathbf{r}})$. Also, by the definition of centre of mass, we have $\Sigma m_i \mathbf{r}'_i = \Sigma m_i \dot{\mathbf{r}}'_i = \Sigma m_i \ddot{\mathbf{r}}'_i = \mathbf{0}$. Hence, we find that

$$\Sigma(\mathbf{r}'_i \times \mathbf{F}_i) = \Sigma(m_i \mathbf{r}'_i \times \ddot{\mathbf{r}}'_i)$$

$$\Rightarrow \quad \Sigma(\mathbf{r}'_i \times \mathbf{F}_i) = \frac{\mathrm{d}}{\mathrm{d}t}(\Sigma(m_i \mathbf{r}'_i \times \dot{\mathbf{r}}'_i)) \qquad [7]$$

We can state equation [7] as follows:

> The moment of the external forces about the centre of mass is equal to the rate of change of angular momentum about the centre of mass.

This means that we may treat the centre of mass as a fixed point as far as the rotational motion of the body is concerned. The motion of the body may therefore be regarded as being composed of two independent components:

- the linear motion of the body regarded as a particle situated at the centre of mass, and
- the rotational motion of the body about the centre of mass as if it were fixed.

Angular momentum about a fixed point

The angular momentum of the lamina about the origin, O, is given by

$$\begin{aligned}
\text{Angular momentum} &= \Sigma m_i (\mathbf{r}_i \times \dot{\mathbf{r}}_i) \\
&= \Sigma m_i [(\bar{\mathbf{r}} + \mathbf{r}'_i) \times (\dot{\bar{\mathbf{r}}} + \dot{\mathbf{r}}'_i)] \\
&= \Sigma m_i (\bar{\mathbf{r}} \times \dot{\bar{\mathbf{r}}}) + \Sigma m_i (\bar{\mathbf{r}} \times \dot{\mathbf{r}}'_i) + \Sigma m_i (\mathbf{r}'_i \times \dot{\bar{\mathbf{r}}}) + \Sigma m_i (\mathbf{r}'_i \times \dot{\mathbf{r}}'_i) \\
&= \Sigma m_i (\bar{\mathbf{r}} \times \dot{\bar{\mathbf{r}}}) + \bar{\mathbf{r}} \times (\Sigma m_i \dot{\mathbf{r}}'_i) + (\Sigma m_i \mathbf{r}'_i) \times \dot{\bar{\mathbf{r}}} + \Sigma m_i (\mathbf{r}'_i \times \dot{\mathbf{r}}'_i)
\end{aligned}$$

As before, $\Sigma m_i \mathbf{r}'_i = \Sigma m_i \dot{\mathbf{r}}'_i = \mathbf{0}$, so we have

$$\begin{aligned}
\text{Angular momentum} &= \Sigma m_i (\bar{\mathbf{r}} \times \dot{\bar{\mathbf{r}}}) + \Sigma m_i (\mathbf{r}'_i \times \dot{\mathbf{r}}'_i) \\
&= M(\bar{\mathbf{r}} \times \dot{\bar{\mathbf{r}}}) + \Sigma m_i (\mathbf{r}'_i \times \dot{\mathbf{r}}'_i) \qquad [8]
\end{aligned}$$

The first term is the angular moment about the origin of the body regarded as a particle situated at the centre of mass. The second is the angular momentum of the body about its centre of mass. Hence, we have

> The angular momentum of a body of mass M and centre of mass G about a fixed axis through O is the angular momentum of a particle of mass M at G plus the angular momentum of the body about G.

Kinetic energy

The kinetic energy of particle P_i is

$$\begin{aligned}
\tfrac{1}{2} m_i \dot{\mathbf{r}}_i \cdot \dot{\mathbf{r}}_i &= \tfrac{1}{2} m_i (\dot{\bar{\mathbf{r}}} + \dot{\mathbf{r}}'_i) \cdot (\dot{\bar{\mathbf{r}}} + \dot{\mathbf{r}}'_i) \\
&= \tfrac{1}{2} m_i \dot{\bar{\mathbf{r}}} \cdot \dot{\bar{\mathbf{r}}} + m_i \dot{\bar{\mathbf{r}}} \cdot \dot{\mathbf{r}}'_i + \tfrac{1}{2} m_i \dot{\mathbf{r}}'_i \cdot \dot{\mathbf{r}}'_i
\end{aligned}$$

Summing over all the particles, we have

$$\text{KE} = \tfrac{1}{2} \Sigma m_i \dot{\bar{\mathbf{r}}} \cdot \dot{\bar{\mathbf{r}}} + \dot{\bar{\mathbf{r}}} \cdot \Sigma (m_i \dot{\mathbf{r}}'_i) + \tfrac{1}{2} \Sigma m_i \dot{\mathbf{r}}'_i \cdot \dot{\mathbf{r}}'_i$$

But $\Sigma m_i \dot{\mathbf{r}}'_i = \mathbf{0}$, as before. Also, $\dot{\mathbf{r}}'_i$ is the velocity of P_i relative to G.

For a rigid body, we have $\dot{\mathbf{r}}'_i = r'_i \dot{\theta} \hat{\mathbf{s}}$, where $\dot{\theta}$ is the angular velocity of the body about G and $\hat{\mathbf{s}}$ is the transverse unit vector. Hence, $\dot{\mathbf{r}}'_i \cdot \dot{\mathbf{r}}'_i = r'^2_i \dot{\theta}^2$, and the kinetic energy is given by

$$\text{KE} = \tfrac{1}{2}\Sigma m_i \dot{\mathbf{r}} \cdot \dot{\mathbf{r}} + \tfrac{1}{2}\Sigma m_i r'^2_i \dot{\theta}^2$$

$$\Rightarrow \quad \text{KE} = \tfrac{1}{2} M v^2 + \tfrac{1}{2} I \dot{\theta}^2$$

where v is the speed of the centre of mass, and I is the moment of inertia of the body about its centre of mass.

We can state the above equation as follows:

> The total kinetic energy of a body of mass M and centre of mass G is equal to the kinetic energy of a particle of mass M situated at G plus the rotational kinetic energy of the body about G.

Example 1 A uniform cylinder, of mass 3 kg and radius 0.2 m, rolls without slipping down a rough plane inclined at 20° to the horizontal.

a) Find the acceleration of its centre of mass.
b) Use energy methods to find the speed of the centre of mass after it has moved 4 m.

SOLUTION

a) Let the angular velocity of the cylinder about its axis be $\dot{\theta}$ rad s^{-1}.

Let the acceleration of the centre of mass be a.

As the point in instantaneous contact with the plane is at rest, the centre of mass of the cylinder is moving with speed

$$v = 0.2\dot{\theta}$$

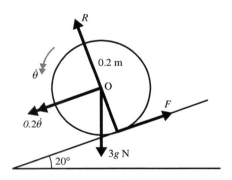

Differentiating, we obtain

$$a = 0.2\ddot{\theta} \quad \Rightarrow \quad \ddot{\theta} = 5a$$

The moment of inertia of the cylinder about its axis is

$$I = \tfrac{1}{2} \times 3 \times 0.2^2 = 0.06 \text{ kg m}^2$$

Resolving down the plane, we have

$$3g \sin 20° - F = 3a \quad [1]$$

Taking moments about O, we have

$$0.2F = I\ddot{\theta} = 0.06 \times 5a$$
$$\Rightarrow \quad F = 1.5a \quad [2]$$

Substituting from [2] into [1], we obtain

$$4.5a = 3g \sin 20°$$
$$\Rightarrow \quad a = 2.235 \text{ m s}^{-2}$$

The centre of mass therefore accelerates at a rate of 2.235 m s^{-2}.

Note It is worth comparing this value with the acceleration of $g \sin 20° = 3.35 \, \text{m s}^{-2}$ for a mass sliding down a smooth slope having the same inclination. Although the friction force does no work (the point of contact is stationary), the acceleration and therefore the speed of the two systems differ because some of the initial potential energy possessed by the cylinder is converted into rotational rather than linear kinetic energy, whereas for a sliding mass it all provides linear kinetic energy. We see this in part **b** of the example.

b) Let the speed of the centre of mass after 4 m be v and its angular speed about O be ω.

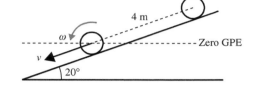

Take the final position of the centre of mass to be the zero potential energy level, as shown.

The initial energy situation is

  Linear KE = 0
  Rotational KE = 0
  GPE = $3g \times 4 \sin 20° = 40.22 \, \text{J}$

The final energy situation is

  Linear KE = $\frac{1}{2} \times 3 \times v^2 = 1.5v^2$
  Rotational KE = $\frac{1}{2} \times I\omega^2 = 0.03\omega^2$
  GPE = 0

There is no work done by external forces, so energy is conserved. Hence, we have

$$1.5v^2 + 0.03\omega^2 = 40.22 \qquad [3]$$

But, as the cylinder is not slipping, we have

$$v = 0.2\omega \quad \Rightarrow \quad \omega = 5v$$

Equation [3] then becomes

$$2.25v^2 = 40.22$$
$$\Rightarrow \quad v = 4.23$$

So, the centre of mass is travelling at $4.23 \, \text{m s}^{-1}$.

(You should verify that this is the same value as that given using the acceleration from part **a**.)

Example 2 A cylinder has a mass of 0.5 kg and a radius of 0.1 m. A light, inextensible string is wound around the cylinder.

a) The free end of the string is held in a fixed position and the cylinder is allowed to fall so that the string unwinds. Assuming that the unwound string remains vertical and the axis of the cylinder remains horizontal, calculate
  i) the acceleration of the centre of mass of the cylinder
  ii) the tension in the string.

CHAPTER 12 GENERAL MOTION OF A RIGID BODY IN TWO DIMENSIONS

b) If instead the free end of the string were pulled upwards so that the centre of mass of the cylinder remained at rest, calculate the acceleration of the free end of the string.

SOLUTION

a) The moment of inertia of the cylinder is

$$\frac{0.5 \times 0.1^2}{2} = 0.0025 \text{ kg m}^2$$

At some time t, the displacement of the centre of mass, O, of the cylinder below the fixed point, P, is x, as shown.

The angular velocity of the cylinder is $\dot{\theta}$, and as the point on the cylinder at which the string loses contact is instantaneously at rest, we have

$$\dot{x} = 0.1\dot{\theta}$$

and differentiating, we obtain

$$\ddot{x} = 0.1\ddot{\theta} \qquad [1]$$

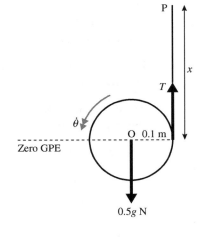

i) Resolving vertically (down positive), we have

$$0.5g - T = 0.5\ddot{x} \qquad [2]$$

Taking moments about O, we have

$$0.1T = 0.0025\ddot{\theta} \qquad [3]$$

Substituting from [1] into [3], we obtain

$$0.1T = 0.025\ddot{x} \qquad [4]$$

Substituting [4] into [2], we obtain

$$0.5g - 0.25\ddot{x} = 0.5\ddot{x}$$
$$\Rightarrow \quad \ddot{x} = 6.53$$

So, the centre of mass of the cylinder accelerates downwards at a rate of 6.53 m s^{-2}.

ii) Substituting back into [4], we have $T = 1.63$.

So, the tension in the string is 1.63 N.

b) Suppose the point P is given an upward acceleration of a, resulting in the point O remaining stationary.

Resolving vertically, we have

$$T_1 - 0.5g = 0 \qquad [5]$$

Taking moments about O, we have

$$0.1T_1 = 0.0025\ddot{\theta} \qquad [6]$$

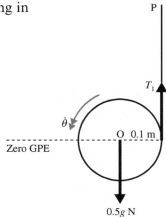

The point at which the string loses contact with the cylinder now has an upward acceleration of $0.1\ddot{\theta}$, and this must equal a. Hence, equation [6] becomes

$$0.1T_1 = 0.025a \qquad [7]$$

From equations [5] and [7], we have

$$0.05g = 0.025a$$
$$\Rightarrow \quad a = 2g$$

So, the free end of the string must be accelerated upwards at a rate of $2g\,\text{m s}^{-2}$.

Example 3 A uniform rod, AB, of length 4 m and mass 3 kg, is lying at rest on a smooth horizontal surface. Its centre of mass is G. It is given a horizontal impulse of magnitude 12 N s perpendicular to the rod at a point C, where CG = 1.5 m. Find

a) the speed of G
b) the angular speed of the rod
c) the point O on the rod about which the rod begins to turn.

SOLUTION

The moment of inertia of the rod is

$$\frac{3 \times 2^2}{3} = 4\,\text{kg m}^2$$

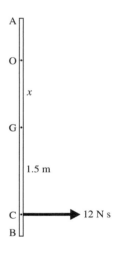

a) Let u be the speed of G immediately after the impulse. Then, using Impulse = Change of momentum, we have

$$12 = 3u \quad \Rightarrow \quad u = 4$$

So, the centre of mass starts to move with speed $4\,\text{m s}^{-1}$ in the direction of the impulse.

b) Let ω be the angular speed of the rod after the impulse. Then, using Moment of impulse = Change of angular momentum, we have

$$12 \times 1.5 = 4\omega \quad \Rightarrow \quad \omega = 4.5$$

So, the rod starts to turn with an angular speed of $4.5\,\text{rad s}^{-1}$.

c) O is the point on the rod at a distance x from G, as shown. When the rod starts to move, O moves relative to G at a speed of $4.5x\,\text{m s}^{-1}$ in the direction opposite to that in which G moves.

The overall initial speed of O is, therefore, $4 - 4.5x\,\text{m s}^{-1}$.

If O is the instantaneous centre of rotation, its initial velocity is zero. Hence, we have

$$4 - 4.5x = 0 \quad \Rightarrow \quad x = 0.889$$

So, the rod begins to rotate about a point 0.889 m from the centre of mass.

Example 4 A sphere, of radius $5a$ and mass m, is rolling along a rough horizontal plane with speed v. It strikes the rough edge of a step of height a, with friction being sufficient to prevent slipping. The impact is perfectly inelastic. Find

a) the angular velocity with which the sphere begins to turn about the edge of the step
b) the maximum and minimum values of v if the sphere is to mount the step and continue rolling without losing contact with the step.

SOLUTION

The moment of inertia of the sphere about its centre, O, is given by

$$\frac{2m \times (5a)^2}{5} = 10ma^2$$

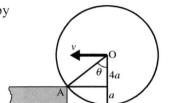

a) As the sphere is rolling without slipping, its angular velocity before meeting the step is $\dfrac{v}{5a}$.

Let A be the edge of the step. Immediately prior to contact with the step, the total angular momentum of the sphere about A is

$$10ma^2 \times \frac{v}{5a} + mv \times 4a = 6mva$$

The moment of inertia of the sphere about A is

$$10ma^2 + m \times (5a)^2 = 35ma^2$$

Let the angular velocity with which the sphere begins to turn about A be ω. Hence, the angular momentum of the sphere about A immediately after impact is $35ma^2\omega$.

As the impulse involved in the collision has no moment about A, angular momentum about A is conserved. Hence, we have

$$35ma^2\omega = 6mva \quad \Rightarrow \quad \omega = \frac{6v}{35a}$$

b) The problem of maximum velocity is a complex one. The angular velocity about A decreases as the centre of mass rises after the impact, so it would be reasonable to assume that if contact is lost, it happens immediately after impact. However, the greater the angular velocity, the smaller the normal reaction and so the greater the likelihood that slipping will occur. In practice, at speeds near the critical one, the behaviour of the system will be difficult to predict. We will make the modelling assumption that no slipping ever occurs and that loss of contact occurs immediately after impact.

Let the normal reaction at A be R. While in contact with the edge of the step, the centre, O, of the sphere is moving in a circle about A. Resolving in the direction OA immediately after contact, we have

$$mg\cos\theta - R = ma\omega^2$$

which gives

$$R = mg\cos\theta - ma\omega^2$$
$$\Rightarrow R = mg \times \frac{4}{5} - ma \times \left(\frac{6v}{35a}\right)^2$$
$$\Rightarrow R = \frac{4mg}{5} - \frac{36mv^2}{1225a}$$

The sphere stays in contact with the edge of the step provided R is positive. That is, provided

$$\frac{4mg}{5} > \frac{36mv^2}{1225a}$$
$$\Rightarrow v^2 < \frac{245ga}{9}$$

Hence, the maximum velocity consistent with remaining in contact with the step is $\frac{7\sqrt{5ga}}{3}$.

To find the minimum velocity needed, we consider the energy situation. Immediately after impact, the total kinetic energy of the sphere is

$$\tfrac{1}{2} \times 35ma^2 \times \omega^2 = \frac{35ma^2}{2} \times \left(\frac{6v}{35a}\right)^2 = \frac{18mv^2}{35}$$

To get all the way up the step, the sphere needs to gain potential energy of mga. Hence, we have

$$\frac{18mv^2}{35} > mga$$
$$\Rightarrow v^2 > \frac{35ga}{18}$$

So, the minimum velocity consistent with successfully mounting the step is $\frac{\sqrt{70ga}}{6}$.

Small oscillations of a system

On pages 231–4, we examined the motion of a compound pendulum. This is just one example of a situation in which a system makes small oscillations about a position of stable equilibrium.

In general, suppose a system has a position, O, of stable equilibrium in which its potential energy is E_0, by definition a minimum value. If we now displace the system by a small amount to a position A, where its potential energy is E_1, and release it from rest, it will return to O.

However, provided there are no dissipative forces, when the system reaches O it has kinetic energy $E_1 - E_0$, and so will continue beyond O until it reaches a point B where its potential energy is again E_1. Subsequently, it will oscillate between A and B.

In our analysis of such situations, we will assume that all displacements and velocities are small, so that higher powers may be neglected.

Example 5 A uniform, solid cylinder, of mass 2 kg and radius 0.1 m, can roll without slipping inside a fixed, hollow cylinder of radius 1 m. The axes of both cylinders are parallel and horizontal. Find the period of small oscillations about the lowest position.

SOLUTION

The small cylinder has moment of inertia

$$I = 0.5 \times 2 \times 0.1^2 = 0.01 \text{ kg m}^2$$

Suppose that, at some stage in its motion, the small cylinder has angular speed ω about its centre G, and that the point P on its circumference is in contact with the large cylinder, as shown.

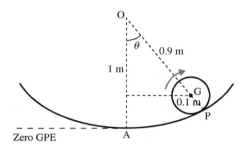

Hence, we have

Velocity of G $= 0.9\dot{\theta}$

which gives

Velocity of P $= 0.9\dot{\theta} - 0.1\omega$

As there is no slipping, the instantaneous velocity of P is zero. Therefore, we have

$$0.9\dot{\theta} - 0.1\omega = 0$$
$$\Rightarrow \quad \omega = 9\dot{\theta}$$

For the energy situation, we have

Translational kinetic energy $= 0.5 \times 2 \times (0.9\dot{\theta})^2 = 0.81\dot{\theta}^2$

Rotational kinetic energy about G $= 0.5 \times 0.01\omega^2 = 0.405\dot{\theta}^2$

Potential energy in position shown $= 2g(1 - 0.9\cos\theta)$

The system is conservative, so we have

$$1.215\dot{\theta}^2 + 2g(1 - 0.9\cos\theta) = \text{Constant}$$

Differentiating with respect to t, we obtain

$$2.43\dot{\theta}\ddot{\theta} + 1.8g\dot{\theta}\sin\theta = 0$$

As θ is small, we have $\sin\theta \approx \theta$. Hence, we find

$$\ddot{\theta} \approx \frac{1.8g}{2.43}\theta = -\frac{196}{27}\theta$$

This is simple harmonic motion with period $T = 2\pi\sqrt{\dfrac{27}{196}} = \dfrac{3\pi\sqrt{3}}{7}$ s.

Example 6 A particle of mass 1 kg rests on a smooth horizontal table. It is attached by means of identical elastic strings to each of points A, B, C and D, which form a square whose diagonals are of length 4 m. The natural length of each string is 1 m and each has modulus of elasticity 40 N. The particle is initially at rest at the centre of the square. It is then displaced slightly in the direction of A and is released from rest. Find the period of oscillation of the subsequent motion.

SOLUTION

The diagram shows the particle in a general position, P, displaced by an amount x from the centre.

From the diagram, we see that $AP = 2 - x$, $CP = 2 + x$, and $BP = DP = (4 + x^2)^{\frac{1}{2}}$.

For the energy situation, we have

$$\text{Kinetic energy at P} = 0.5 \times 1 \times \dot{x}^2 = 0.5\dot{x}^2$$

$$\text{Total potential energy at P} = \frac{40(1+x)^2}{2} + \frac{40(1-x)^2}{2} + \frac{2 \times 40[(4+x^2)^{\frac{1}{2}} - 2]^2}{2}$$

$$= 360 + 80x^2 - 160(4+x^2)^{\frac{1}{2}}$$

As the system is conservative, we have

$$0.5\dot{x}^2 + 360 + 80x^2 - 160(4+x^2)^{\frac{1}{2}} = \text{Constant}$$

Differentiating with respect to t, we obtain

$$\dot{x}\ddot{x} + 160\dot{x}x - 160x\dot{x}(4+x^2)^{-\frac{1}{2}} = 0$$

$$\Rightarrow \quad \ddot{x} = -160x + 160x(4+x^2)^{-\frac{1}{2}}$$

$$\Rightarrow \quad \ddot{x} = -160x + 80x(1 - \tfrac{1}{8}x^2 \ldots) \quad \text{by the binomial theorem}$$

If x is small, we can ignore x^3 and higher powers. Hence, we have

$$\ddot{x} \approx -80x$$

This is simple harmonic motion with period $T = \dfrac{\pi}{\sqrt{80}} = 0.351$ s.

Exercise 12

1 A uniform solid sphere, of mass 5 kg and radius 0.4 m, is rolling down a rough plane inclined at an angle of 20° to the horizontal.

   a) Assuming that no slipping occurs, find the acceleration of its centre of mass.
   b) Show that if no slipping is to occur, the coefficient of friction between the sphere and the plane must be at least 0.104.

2 A uniform disc and a circular hoop, each of mass 2 kg and radius 0.5 m, are rolled simultaneously from rest down a rough slope of length 20 m, inclined at 10° to the horizontal. Assuming that no slipping occurs and that both objects follow a straight path down the line of greatest slope, find which reaches the bottom first and by what distance.

3 A uniform solid cylinder, of radius 0.4 m, rolls without slipping down a rough plane inclined at 30° to the horizontal. If the cylinder starts from rest, find the velocity of it centre of mass
a) after 4 s, and b) after 4 m.

4 A light, inextensible string is wound around a uniform, solid cylindrical reel of mass m and radius a. The end of the string is held stationary and the reel falls from rest with its axis horizontal so that the string unwinds. Find the acceleration of the centre of the reel and the tension in the string.

5 A uniform rod, of length $2l$ and mass m, is held in a vertical position with one end resting on a smooth horizontal plane. It is then slightly disturbed from this position and allowed to fall.

a) Explain why the path followed by its centre of mass is a vertical straight line.

b) Show that when the rod reaches the horizontal position it has angular velocity $\sqrt{\dfrac{3g}{2l}}$.

6 A large beam, of mass M, is being pushed on level ground with the help of three cylindrical rollers, each of mass m and radius a. No slipping takes place between the beam and the rollers or between the rollers and the ground. If the beam is moving at a speed v, show that the total kinetic energy of the system is $\dfrac{(8M + 9m)v^2}{16}$.

7 A uniform, solid cylinder, of radius a, is rolling without slipping up a rough plane inclined at an angle α to the horizontal. If at some instant the angular velocity of the cylinder is ω, show that it will travel a further distance of $\dfrac{3a^2\omega^2}{4g \sin \alpha}$ up the slope before coming instantaneously to rest.

8 A uniform hoop, of mass m and radius a, has a particle of mass m attached to a point on its circumference. The hoop is held in a vertical plane and in contact with a rough horizontal surface. Initially, the particle is at the topmost point of the hoop, which is then displaced slightly from rest and allowed to roll in a straight line. Show that when the hoop has rolled through an angle θ, the speed of its centre of mass is

$$\sqrt{\dfrac{ga(1 - \cos \theta)}{2 + \cos \theta}}$$

9 A solid, uniform, spherical ball, of radius a and centre A, is held at rest at the top of a rough fixed sphere of radius b and centre B. It is displaced slightly and rolls from rest down the surface of the fixed sphere. At a subsequent moment, the line AB makes an angle of θ with the upward vertical.

a) Show that the ball has rotated by an amount $\dfrac{(a + b)\theta}{a}$, and hence show that

$$\dot{\theta} = \sqrt{\dfrac{10g(1 - \cos \theta)}{7(a + b)}} \quad \text{and} \quad \ddot{\theta} = \dfrac{5g \sin \theta}{7(a + b)}$$

b) Given that the ball has mass M, find expressions for the friction force and the normal reaction force at the instantaneous point of contact between the ball and the sphere.

c) Deduce that, whatever the coefficient of friction, the ball must start slipping before it loses contact with the sphere. Find the value of θ at which this occurs if the coefficient of friction is 0.6.

10 A uniform rod, AB, of length 2 m and mass 3 kg, lies at rest on a smooth horizontal surface. The end A is struck a horizontal blow perpendicular to the rod. The magnitude of the impulse it receives is 15 N s. Find the initial velocities of A and B.

11 A uniform rod, AB, of length $2a$ and mass m, lies at rest on a smooth horizontal surface. A horizontal impulse of magnitude J is applied perpendicular to the rod at a point P, where $PB = x$. If the initial velocity of the end A is zero, find the value of x.

12 A uniform, circular disc, of mass m and radius a, is spinning freely about its stationary centre with angular velocity ω, with its plane horizontal and resting on a smooth horizontal surface. A point on its rim is suddenly fixed.

   a) Find the new motion of the disc.
   b) Show that the ratio of the kinetic energy of the system before and after the occurrence is $3:1$.

13 Two particle, A and B, each have mass 2 kg. They are connected by means of a light rod of length 1 m and lie at rest on a smooth horizontal surface. A third particle, C, also of mass 2 kg, is travelling on the surface perpendicular to AB at a speed of $5\,\mathrm{m\,s^{-1}}$. It strikes the particle A. If particle C is brought to rest by the collision, find the velocities of the particles A and B immediately after the collision.

14 A uniform, circular disc, of radius a, is rolling without slipping along a rough horizontal plane at speed v. It strikes a kerb of height $\dfrac{2a}{5}$. Find the angular velocity with which the disc starts to rotate about the edge of the kerb.

15 Two equal, uniform rods, AB and BC, each of length $2a$ and mass m, are freely jointed at B. They lie at rest in a straight line on a smooth horizontal surface. A horizontal impulse J is applied to A perpendicular to AB. Find the ratio of the initial speeds of the centres of mass of the rods.

16 Three particles, A, B and C, each have mass m. A and B are connected by means of a light rod of length $2a$. C is connected to B by means of a light, inextensible string. All three particles lie at rest on a smooth horizontal surface, with C next to B. C is then projected horizontally away with speed v from B so that, when the string becomes taut, the angle ABC is $60°$. Find the velocities of the three particles immediately after the string becomes taut.

17 The particles, A and B, and the light rod described in Question **16** are replaced by a uniform rod of length $2a$ and mass $2m$. Examine the motion of this system when the string becomes taut.

18 A uniform rod, AB, of length $4a$ and mass m, lies at rest on a smooth horizontal plane. A particle of mass m, moving on the plane with a speed v, strikes the rod at right angles at a point P, where $AP = a$. The particle becomes attached to the rod. Find the initial velocities of the ends A and B immediately after the collision.

19 A uniform rod of length $2r$ rests symmetrically on a fixed, rough horizontal cylinder of radius r. The rod is displaced slightly without slipping and is released from rest. At a general point in its motion, it makes an angle θ with the horizontal, as shown in the diagram. Show that, provided we can ignore all powers and products of θ, $\dot{\theta}$ and $\ddot{\theta}$ of order three or higher, the rod performs simple harmonic motion with period $2\pi\sqrt{\dfrac{r}{3g}}$.

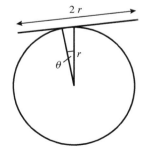

20 A smooth hoop of radius a has a bead of mass m threaded onto it. The hoop is fixed in a vertical position. One end of a light elastic string of natural length a and modulus of elasticity $2mg$ is attached to the bead. The other end of the string is fixed to a point a distance $2a$ vertically below the centre of the hoop. The bead is displaced slightly from the lowest point of the hoop and released from rest. Find the period of oscillation of the system.

21 A heavy, uniform chain of length $5a$ and mass M hangs over a rough pulley. The pulley is a uniform disc of mass m and radius a, and is free to turn about a smooth horizontal axis through its centre. A particle of mass m is fixed to a point on the rim of the pulley. Initially the system is in equilibrium with equal lengths of chain on either side of the pulley and the particle at the lowest point of the pulley.

a) Show that the equilibrium is stable provided $M < 2.5m$.
b) Find, in the case where $M = 2m$, the period of oscillation of the system if it is displaced slightly and released.

Examination questions

Chapters 9 to 12

Chapter 9

1 ABCD is a rhombus consisting of four freely jointed uniform rods, each of mass m and length $2a$. The rhombus is freely suspended from A and is prevented from collapsing by two light springs, each of natural length a and modulus of elasticity $2mg$. One spring joins A and C and the other joins B and D, as shown in the figure on the right.

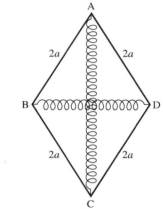

a) Show that when AB makes an angle θ with the downward vertical, the potential energy V of the system is given by

$$V = -8mga(\sin\theta + 2\cos\theta) + \text{constant}$$

b) Hence find the value of θ, in degrees to one decimal place, for which the system is in equilibrium.

c) Determine whether this position of equilibrium is stable or unstable. (EDEXCEL)

2 A uniform rod AB has mass m and length $2a$. One end A is freely hinged to a fixed point. One end of a light elastic string, of natural length a and modulus $\frac{1}{2}mg$, is attached to the other end B of the rod. The other end of the string is attached to a small ring C which can move freely on a smooth horizontal wire fixed at a height of $3a$ above A and in the vertical plane through A, as shown in the figure on the right.

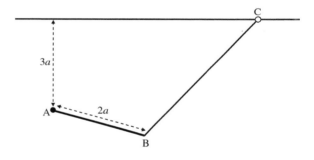

a) Explain why, when the system is in equilibrium, the elastic string is vertical.

b) Show that, when BC is vertical and the rod AB makes an angle θ with the downward vertical, the potential energy, V, of the system is given by

$$V = mga(\cos^2\theta + \cos\theta) + \text{constant}$$

c) Hence find the values of θ, $0 \leqslant \theta \leqslant \pi$, for which the system is in equilibrium.

d) Determine whether each position of equilibrium is stable or unstable. (EDEXCEL)

3 A small ring R, of mass m, is free to slide on a smooth wire in the shape of a circle with radius a. The wire is fixed in a vertical plane. A light inextensible string has one end attached to R and passes over a small smooth pulley at P, where P is one end of the horizontal diameter of the wire. The other end of the string is attached to a mass $km(k < 1)$ which hangs freely, as shown in the figure on the right.

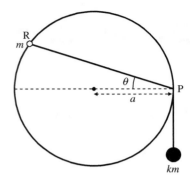

PR makes an angle θ with the horizontal.

a) Show that the potential energy of the system, V, is given by
$$V = mga(\sin 2\theta + 2k\cos\theta) + \text{constant}$$

Given that $k = \frac{1}{2}$,

b) find, in radians to three decimal places, the values of θ, $-\frac{\pi}{2} \leqslant \theta \leqslant \frac{\pi}{2}$, for which the system is in equilibrium.

c) Determine whether each of the positions of equilibrium is stable or unstable.

(EDEXCEL)

4 A system consists of four equal rods, each of length $2a$, freely jointed to form a rhombus ABCD, and an elastic string connecting B to D. The rod AB is fixed in a horizontal position, with ABCD hanging in a vertical plane, as shown in the figure on the right. Each rod has mass m and the modulus of elasticity of the string is mg. The string has natural length a.

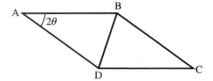

In a general configuration with the string taut, $\angle \text{BAD} = 2\theta$, $\theta < \frac{1}{4}\pi$.

a) Show that the length of the string is $4a\sin\theta$.

b) Show that when the string is taut the potential energy of the system is
$$-4mga\sin 2\theta + \tfrac{1}{2}mga(4\sin\theta - 1)^2 + c$$
where c is a constant.

c) Show that if $\theta = \alpha$ is a position of equilibrium, then
$$2\sin 2\alpha - \cos\alpha - 2\cos 2\alpha = 0$$

d) Given that this equation has a root in the range $\beta < \alpha < \frac{1}{4}\pi$, where $\sin\beta = \frac{1}{4}$, show that this position of equilibrium is stable. (EDEXCEL)

Chapter 10

5 Show, by integration, that the moment of inertia of a uniform rod, of mass m and length $2a$, about an axis passing through the centre of the rod and making an angle θ with the rod is $\frac{1}{3}ma^2\sin^2\theta$. (EDEXCEL)

6 Three identical uniform rods AB, BC and CA, each of mass m and length $2a$, are joined together at their ends to form a plane rigid framework in the form of an equilateral triangle as shown in the diagram on the right. Find the moment of inertia of this triangle about an axis through A and perpendicular to its plane. (NEAB)

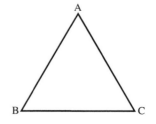

7 A bucket of mass $3m$ is attached to one end of a rope and moves in a vertical line. The rope passes vertically up from the bucket and is wrapped round a cylindrical drum of radius a and mass m. The other end is attached to a fixed point on the rim of the drum. The drum is free to rotate about a fixed smooth horizontal axis, as shown in the figure on the right. The bucket is released from rest with the rope taut and falls vertically downwards. The bucket is modelled as a particle, the drum as a uniform cylinder, the rope as light, thin and inextensible, and air resistance is assumed to be negligible.

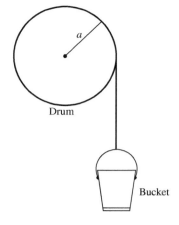

Find the initial acceleration of the bucket, giving your answer in terms of g.   (EDEXCEL)

8 A spacecraft is rotating about a fixed axis with angular speed $0.2\,\text{rad s}^{-1}$. By firing rocket motors, a couple of constant moment is applied to the spacecraft and, after 5 complete revolutions, the angular speed of the spacecraft is $0.5\,\text{rad s}^{-1}$. Find the angular acceleration of the spacecraft.   (OCR)

9 The diagram on the right shows a uniform rod AB of mass 6 kg and length 1.2 m. A particle of mass 1.5 kg is attached to the rod at A and a similar particle is attached at B.

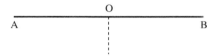

   a) Show that the moment of inertia of the combined system of the rod and particles about an axis through the centre O and perpendicular to the rod is $1.8\,\text{kg m}^2$.
   b) The rod is mounted on a fixed vertical axis through O and can rotate in a horizontal plane about this axis. The rotational motion is resisted by a constant frictional couple $C\,\text{Nm}$. Given that when the rod is set in motion with an initial angular speed of $240\,\text{rad s}^{-1}$ it comes to rest in 36 seconds, find the value of C.   (NEAB)

Chapter 11

10 The uniform solid cuboid shown in the figure opposite has mass m kg, length $12a$ metres, breadth $4a$ metres and depth $2a$ metres.

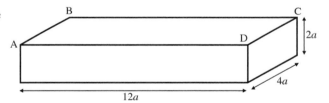

   i) Find the moment of inertia of the cuboid about the edge AB.

   Initially ABCD is in a horizontal plane and the cuboid is released from rest. Its rotation about the fixed axis AB is resisted by a constant couple of magnitude k N m. When the face ABCD becomes vertical the cuboid has an angular speed of $\sqrt{\dfrac{6g}{37a}}$ radians per second.

   ii) Find k in terms of m, g, a and π.   (NICCEA)

11 Two flywheels are rotating, in the same direction, about the same fixed axis. One wheel has moment of inertia $3.6\,\text{kg m}^2$ about the axis, and angular speed $45\,\text{rad s}^{-1}$. The other has moment of inertia $1.4\,\text{kg m}^2$ about the axis, and angular speed $75\,\text{rad s}^{-1}$. A clutch mechanism then locks the two flywheels together so that they rotate with the same angular speed $\omega\,\text{rad s}^{-1}$. Find ω.   (OCR)

12 a) You may assume that the moment of inertia of a disc, of mass m and radius a, about an axis through its centre perpendicular to the plane of the disc is $\frac{1}{2}ma^2$.

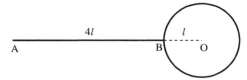

Use integration to show that the moment of inertia of a solid sphere, of mass m and radius r, about an axis through its centre is $\frac{2}{5}mr^2$.

b) A rod AB, of length $4l$ and mass $3m$, has a solid sphere, of mass m and radius l, attached to one end, B, as shown in the diagram.

ABO is a straight line where O is the centre of the sphere.

The rod is attached to a smooth pivot at A so that it can move freely in a vertical plane. AB is held horizontally and then released.

Show that the maximum angular velocity of the system in the resulting motion is $\sqrt{\dfrac{110g}{207l}}$.

(AEB 99)

13 The door, shown in the diagram on the right, is opened so that it makes an angle $\dfrac{\pi}{2}$ with the door frame. It is released from rest and closes by means of a mechanism which exerts a constant couple of magnitude 144 N m about the line of the hinges, which is vertical. The door is of mass 72 kg and width 1 m and may be modelled as a uniform rectangular lamina.

a) Write down the work done by the closing mechanism in closing the door.

b) Hence, by applying the work-energy principle, or otherwise, find the angular speed of the door at the instant it reaches the closed position. (NEAB)

14 A uniform rod AB, of mass m and length $2a$, is free to rotate in a vertical plane about a fixed horizontal axis passing through the point C on the rod, where $AC = \frac{2}{3}a$. Find the moment of inertia of the rod about this axis.

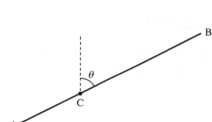

Initially the rod is vertical with B above C. It is slightly disturbed from rest in this position. At an instant during the subsequent motion, AB makes an angle θ with the upward vertical (see diagram).

i) Show that the angular speed of the rod at this instant is $\sqrt{\left(\dfrac{3g}{2a}(1-\cos\theta)\right)}$.

ii) Find the angular acceleration of the rod at this instant.

At this instant, the force acting on the rod at C has components R along $\overrightarrow{AB}$ and S perpendicular to $\overrightarrow{AB}$. Show that $R = \frac{1}{2}mg(3\cos\theta - 1)$ and find an expression for S in terms of m, g and θ. (OCR)

15 A uniform rod, of mass m and length $2a$ with centre G, is free to rotate about a smooth horizontal axis passing through a point O on the rod, where $GO = \frac{1}{3}a$. When the rod is vertical, with G below O, it has angular speed $\dfrac{3}{2}\sqrt{\left(\dfrac{2g}{a}\right)}$. Show that, when the rod is horizontal,

a) the angular acceleration of the rod has magnitude $\dfrac{3g}{4a}$,

b) the angular speed of the rod is $\sqrt{\left(\dfrac{3g}{a}\right)}$.

c) Find the magnitude of the force exerted by the rod on the axis when the rod is horizontal.

16 A uniform rod AB has mass m and length $2a$. A particle of mass m is attached to the end B. The loaded rod is free to rotate about a fixed smooth horizontal axis L, perpendicular to the rod and passing through a point O of the rod, where AO $= \frac{1}{2}a$.

a) Show that the moment of inertia of the loaded rod about L is $\dfrac{17ma^2}{6}$.

The loaded rod is held at rest in a vertical position with B above O and slightly disturbed.

b) Find the angular acceleration of the loaded rod when it is horizontal.

c) Hence find the magnitude of the vertical component of the force exerted on the loaded rod by the axis when the rod is horizontal.   (EDEXCEL)

Chapter 12

17 A yo-yo is modelled as a uniform circular disc of mass m and radius a. One end of a light inextensible string is fixed to a point on the rim of the disc. The string is wrapped several times round the circumference. The disc is held in a vertical plane, with the other end of the string held fixed. The disc is projected vertically downwards with speed $3\sqrt{(ag)}$ so that the string, as it unwinds, remains vertical. Given that the string does not fully unwind, find the speed of the centre of the disc after it has travelled a distance of $12a$.   (EDEXCEL)

18 A uniform rod, of mass m and length $2a$, is at rest on a smooth horizontal plane. A horizontal impulse of magnitude $4mv$ is applied perpendicular to the rod at a point at a distance of $\frac{1}{4}a$ from its centre O.

a) i) Show that, immediately after the impulse has been applied, the angular speed of the rod is $\dfrac{3v}{a}$.

ii) Express the speed of O, immediately after the impulse has been applied, in terms of v.

b) Show that the kinetic energy of the rod when it starts moving is $\dfrac{19}{2}mv^2$.   (NEAB)

19 A uniform rod AB, of mass m and length $2a$, lies at rest on a smooth horizontal table.

A horizontal impulse of magnitude J, whose direction makes an angle α, $0 < \alpha < \dfrac{\pi}{2}$, with AB, is applied to the end A, as shown in the figure on the right.

a) Show that the kinetic energy generated by the impulse is

$$\dfrac{J^2(1 + 3\sin^2\alpha)}{2m}$$

Immediately after the impulse is applied, the end B moves in a direction which makes an angle θ with AB produced.

b) Show that $\tan\theta = 2\tan\alpha$.   (EDEXCEL)

20 Four light rods each of length 2a are freely hinged at their ends to form a rhombus ABCD which is suspended at A from a fixed point. A light spring of natural length 2a and stiffness $\frac{3mg}{a}$ connects the points A and C. A particle of mass $\frac{3m}{2}$ is attached at the point C and each of the rods AB and AD carries a particle of mass m at its mid-point. The arrangement is shown on the right.

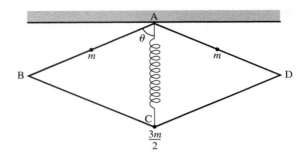

The point C can move freely in the vertical line through A. The angle between AB and the downward vertical AC is θ.

i) Show that the potential energy V of the system relative to the zero level through A is given by
$$V = mga(24\cos^2\theta - 32\cos\theta + 6)$$

ii) Deduce that there are two positions of equilibrium and show that only one of these is stable.

The system now performs small oscillations about the position of stable equilibrium where $\theta = \alpha$.

iii) Show that the kinetic energy T is given by $T = ma^2\dot{\theta}^2(13 - 12\cos^2\theta)$.

iv) By putting $\theta = \alpha + \phi$ and assuming both ϕ and $\dot{\phi}$ remain very small, show that
$$T = \frac{23}{3}ma^2\dot{\phi}^2.$$

You are now **given** that, near the equilibrium position, $V \approx \frac{mga}{3}(-14 + 40\phi^2)$.

v) Find the approximate period of small oscillations about the position of stable equilibrium.

(MEI)

21 A uniform solid spherical ball rolls up a line of greatest slope of a rough inclined plane. The plane is inclined at an angle α to the horizontal, where $\sin\alpha = \frac{1}{10}$. The centre of mass of the ball has an initial speed of $5\,\text{m s}^{-1}$ and the plane is sufficiently rough to prevent the ball from slipping.

Find, to the nearest metre, how far the ball rolls from its starting point before it comes to instantaneous rest.   (EDEXCEL)

22 A log of mass 20 kg rolls without slipping, with its axis horizontal, down a rough plane inclined at an angle of 30° to the horizontal. The log is modelled as a uniform cylinder of radius 10 cm.

The coefficient of friction between the log and the plane is μ. Show that $\mu \geq \frac{\sqrt{3}}{9}$.

(EDEXCEL)

Answers

Exercise 1A

1 a) 7 units **b)** $\frac{2}{7}\mathbf{i} + \frac{3}{7}\mathbf{j} - \frac{6}{7}\mathbf{k}$ **c)** $-3\mathbf{i} - 4.5\mathbf{j} + 9\mathbf{k}$ **d)** 73.4°, 64.4°, 149° **2 a)** 105° **b)** $-5\mathbf{i} + 2\mathbf{j} + \mathbf{k}$ **c)** 156°
3 $\pm(16\mathbf{i} - 16\mathbf{j} + 28\mathbf{k})$ **4** 46.7° or 133.3°, $\mathbf{V} = 3.86\mathbf{i} + 2.05\mathbf{j} \pm 4.11\mathbf{k}$ **5** $\pm 4\sqrt{3}(\mathbf{i} + \mathbf{j} + \mathbf{k})$ **6** $(10.4\mathbf{i} + 27.3\mathbf{j} + 27.3\mathbf{k})$ N

Exercise 1B

1 52.2 N, 5.39° **2 a)** $1.3\mathbf{i} + 0.5\mathbf{j} + 0.4\mathbf{k}$ **3 a)** $\dfrac{(4k+1)\mathbf{i} + (2k-1)\mathbf{j} + (k+2)\mathbf{k}}{k+1}$, $\dfrac{(4k+2)\mathbf{i} + (2k+3)\mathbf{j} + (k-3)\mathbf{k}}{k+1}$ **c)** $1:(k+1)$
4 a) $\mathbf{r} = (3\lambda + 2)\mathbf{i} + (\lambda + 2)\mathbf{j} + (2\lambda + 1)\mathbf{k}$, $\dfrac{x-2}{3} = y - 1 = \dfrac{z-1}{2}$ **b)** $\mathbf{r} = (4\lambda - 3)\mathbf{i} + (4 - 4\lambda)\mathbf{j} + (\lambda + 2)\mathbf{k}$, $\dfrac{x+3}{4} = \dfrac{4-y}{4} = z - 2$
4 c) $\mathbf{r} = (5\lambda + 4)\mathbf{i} + (2\lambda - 2)\mathbf{j} + (\lambda + 2)\mathbf{k}$, $\dfrac{x-4}{5} = \dfrac{y+2}{2} = z - 2$ **d)** $\mathbf{r} = 3\mathbf{i} + (\lambda + 3)\mathbf{j} + (2\lambda + 2)\mathbf{k}$, $x = 3$, $y - 3 = \dfrac{z-2}{2}$
4 e) $\mathbf{r} = -4\mathbf{i} + \lambda\mathbf{j} + 3\mathbf{k}$, $x = -4$, $z = 3$ **5 a)** $(-1, 0, -1)$ **b)** $(1, 0, 3)$ **c)** $(9, 0, 3)$ **d)** $(3, 0, -4)$ **e)** $(-4, 0, 3)$ **6 a)** 40.2°
6 b) 90° **7 a)** Parallel **b)** Intersect at $2.2\mathbf{i} - 0.6\mathbf{j} + 2.6\mathbf{k}$ **c)** Intersect at $18\mathbf{i} - 16\mathbf{j} - 3\mathbf{k}$ **d** Skew **e)** Intersect at $-3.5\mathbf{i} - 4\mathbf{j} - 4.5\mathbf{k}$

Exercise 2A

2 $\dfrac{-17\mathbf{i} + 18\mathbf{j} - 2\mathbf{k}}{\sqrt{617}}$ **5** $-\mathbf{i} + 4\mathbf{j} + 2\mathbf{k}$ **6 a)** 7.53 units² **b)** $\dfrac{-11\mathbf{i} + 5\mathbf{j} - 9\mathbf{k}}{\sqrt{227}}$ **7 a)** Points collinear **8** 7.37 units²

Exercise 2B

1 19 units³ **2 a)** $\frac{5}{6}$ units³ **b)** 7.58 units² **c)** 0.330 units **3** 3 units³ **5 a)** $-(\lambda + 2)\mathbf{i} + (\lambda - 1)\mathbf{j} + (2\lambda + 1)\mathbf{k}$, $\lambda = -\frac{1}{2}$, $1.5\sqrt{2}$ units
6 a) 0.535 units **b)** $1\frac{1}{3}$ **7** 6 units, 3 units, 9 units; P lies on shortest line joining two given lines **8** $1.5\sqrt{2}$ units
9 a) Do not intersect **b)** Intersect

Exercise 2C

1 $\mathbf{r} = 2(2\cos\theta + 1)\mathbf{i} + (4\sin\theta - 3)\mathbf{j}$ **2** $\mathbf{r} = 3\cos\theta\mathbf{i} + 3\sin\theta\mathbf{j}$
3 Parabola, vertex $(0\ 0\ -4)$, line of symmetry $\mathbf{r} = \lambda\mathbf{j} - 4\mathbf{k}$, lying in plane $z + 4 = 0$
4 Helix around z-axis, radius 3, pitch 4π **5** Helix around x-axis, radius 1, pitch 2

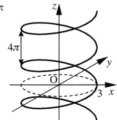
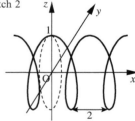

6 Ellipse at 45° to **ij** plane, major axis length $2\sqrt{2}$ in **ik** plane, minor axis length 2 along y-axis **7** Circle at 45° to **ij** plane, radius $\sqrt{2}$

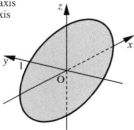
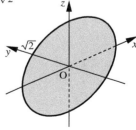

8 a) $\phi = 3\theta$ **b)** $3a\cos\theta\mathbf{i} + 3a\sin\theta\mathbf{j} + a\mathbf{k}$ **c)** $\mathbf{r} = a(3\cos\theta + \sin\theta\sin 3\theta)\mathbf{i} + a(3\sin\theta - \cos\theta\sin 3\theta)\mathbf{j} + a(1 - \cos 3\theta)\mathbf{k}$

ANSWERS

Exercise 2D

1 Force $(-3\mathbf{i} - 3\mathbf{j} - 3\mathbf{k})$ N acting along $3x = 3y - 11 = 3z - 6$ 2 b) Force $(\mathbf{i} - \mathbf{j})$ N, couple $(5\mathbf{i} + \mathbf{j} + \mathbf{k})$ N m
3 Force $(\mathbf{i} + \mathbf{j} + \mathbf{k})$ N acting along $3x = 3y + 2 = 3z + 4$, couple $\tfrac{1}{3}(\mathbf{i} + \mathbf{j} + \mathbf{k})$ N m 4 Force $(\mathbf{i} - \mathbf{k})$ N acting along $x = 1 - z$, $y = -1$, couple $(\mathbf{i} - \mathbf{k})$ N m 5 Force $(8\mathbf{i} + 10\mathbf{j} + 12\mathbf{k})$ N, couple $(12\mathbf{i} + 6\mathbf{j} - 6\mathbf{k})$ N m
6 Tension 70 N, reaction 190 N

Exercise 3A

1 $5\dfrac{dv}{dt} = 5g - kv$, $k = \dfrac{49}{60}$ 2 $\dfrac{dQ}{dt} = k(Q_{\text{MAX}} - Q)$ 3 $\dfrac{dV}{dt} = k(36\pi V^2)^{1/3}$ 4 $\dfrac{dh}{dt} = \dfrac{8}{\pi h^2}$ 5 $\dfrac{dN}{dt} = kn(N - n)$ 6 $\dfrac{dA}{dt} = 2k\sqrt{\pi A}$

Exercises 3B and 3C

1 a)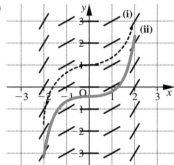

b)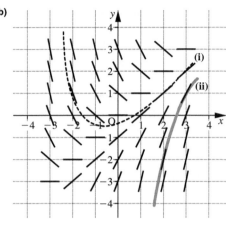

1 c)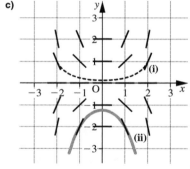

d)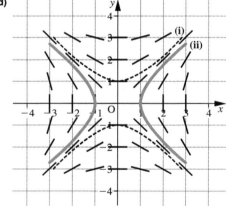

1 e)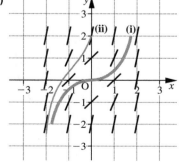

ANSWERS

2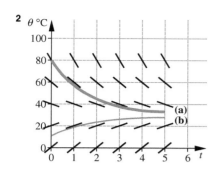

3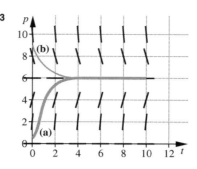

Exercise 3D

1 a) $y^2 = x^2 + c$ **b)** $y^3 - 3y^2 = 3x^2 + c$ **c)** $y = 3 + Ae^{\frac{1}{2}(x-2)^2}$ **d)** $y = \dfrac{2(1 + Ae^{2x^2})}{1 - Ae^{2x^2}}$ **2** $\sin y = 1 - \cos x$ **4 a)** $y = \dfrac{Ax^2 - 1}{Ax^2 + 1}$

4 b) $y = Ae^{\frac{1}{2}(x+2)^2}$ **c)** $\sec y = Ax$ **d)** $\sec y = A(1 - \cos x)$ **5 a)** $y = x + \dfrac{1 + Ae^{2x}}{1 - Ae^{2x}}$ **b)** $y = x + \dfrac{1 + e^{2x}}{1 - e^{2x}}$ **6 a)** $y = x\left(1 - \dfrac{2}{\ln Ax}\right)$

6 b) $y^2 - 2xy - x^2 = A$ **c)** $x^2 + 2y^2 \ln Ay = 0$ **d)** $y = \dfrac{x(2 \ln Ax - 1)}{\ln Ax}$ **7** $v = 10g(1 - e^{-0.1t})$ **8 a)** $\dfrac{400}{v} - v = 10\dfrac{dV}{dt}$

8 b) $v = 20\sqrt{1 - e^{0.2t}}$ **c)** 1.44 s

Exercise 3E

1 a) $y = Ae^{-2x} + 2$ **b)** $y = \dfrac{x^2}{3} + x + \dfrac{A}{x}$ **c)** $y = Ae^{-x} + x - 1$ **d)** $y = \tfrac{1}{2}e^{-x} + Ae^{-3x}$ **e)** $y = Ae^{-x^2} + 2$ **f)** $y = Ae^{-\cos x} - 1$

1 g) $y = Ae^{-x^3} + 6$ **h)** $y = \dfrac{x^2}{3} + \dfrac{A}{x}$ **i)** $y = Ax^2 - x(1 + \ln x)$ **j)** $y = (x + A)\cos x$ **2** $\dfrac{dV}{dt} + 0.2v = 10 - t$, $14.8\,\text{m s}^{-1}$

Exercise 3F

1 $y = Ae^{4x}$ **2 a)** $y = Ae^{-x}$ **b)** $x = Ae^{5t}$ **c)** $p = Ae^{-2t}$ **d)** $x = Ae^{0.5t}$ **e)** $y = Ae^{-3x}$ **f)** $z = Ae^{3y}$

Exercise 3G

1 $y = Ae^{-3x} + \tfrac{2}{9}(3x - 7)$ **2** $x = Ae^{2t} - \tfrac{2}{5}\cos t + \tfrac{1}{5}\sin t$ **3** $y = Ae^{-x} + \tfrac{1}{2}e^x$ **4** $p = Ae^{5t} - \tfrac{1}{5}t^3 - \tfrac{3}{25}t^2 - \tfrac{6}{125}t - \tfrac{6}{625}$
5 $z = Ae^{-2x} + \tfrac{1}{13}\sin 3x - \tfrac{8}{13}\cos 3x$ **6** $x = Ae^{4t} - \tfrac{1}{2}e^{-2t}$ **7** $y = Ae^{-4x} + x - \tfrac{1}{4} - \tfrac{1}{5}e^x$ **8** $x = Ae^t + e^{2t} + \tfrac{1}{5}\cos 2t - \tfrac{2}{5}\sin 2t$

Exercise 3H

1 $y = (A + 8x)e^{4x}$, $y = 4(1 + 2x)e^{4x}$ **2** $x = (A + 10t)e^{-5t} + \tfrac{2}{25}(5t - 1)$, $x = \left(\tfrac{52}{25} + 10t\right)e^{-5t} + \tfrac{2}{25}(5t - 1)$ **3** $y = (A - 6x)e^{2x}$,
3 $y = 2(1 - 3x)e^{2x}$ **4** $x = (A + 12t)e^{-3t} + t - \tfrac{1}{3}$, $x = \left(6\tfrac{1}{3} + 12t\right)e^{-3t} + t - \tfrac{1}{3}$ **5** $y = (A + 4x - 4x^2)e^{2x}$, $y = 4(2 + x - x^2)e^{2x}$
6 $x = \left(A + 9t - \tfrac{15}{2}t^2\right)e^{-3t} + 2t - \tfrac{2}{3}$, $x = \left(9\tfrac{2}{3} + 9t - \tfrac{15}{2}t^2\right)e^{-3t} + 2t - \tfrac{2}{3}$

Exercise 3I

1 $y = Ae^{-3x} + Be^x$ **2** $y = (Ax + B)e^{2x}$ **3** $x = e^{-2t}(A\cos 2t + B\sin 2t)$ **4** $y = (Ax + B)e^{-3x}$ **5** $x = e^t(A\cos 2t + B\sin 2t)$
6 $y = Ae^{-4x} + Be^{-2x}$ **7** $x = e^{-t}(A\cos t + B\sin t)$ **8** $y = Ae^{4x} + Be^{-x}$ **9** $y = (Ax + B)e^{4x}$ **10** $y = Ae^{-4x} + Be^{2x}$
11 $y = (Ax + B)e^{-x}$ **12** $x = e^{-2t}(A\cos 3t + B\sin 3t)$ **13** $y = Ae^{3x} + Be^{-3x}$ **14** $x = A\cos 2t + B\sin 2t$ **15** $y = A + Be^{3x}$
16 $y = A + Be^{-4x}$

Exercise 3J

1 $y = Ae^{-3x} + Be^x - \tfrac{1}{3}x^2 + \tfrac{5}{9}x + \tfrac{4}{27}$ **2** $y = (Ax + B)e^{2x} + 4e^{3x}$ **3** $x = e^{-2t}(A\cos 2t + B\sin 2t) + \tfrac{1}{2}t - \tfrac{5}{8}$
4 $y = (Ax + B)e^{-3x} + \tfrac{4}{169}(5\sin 2x - 12\cos 2x)$ **5** $x = e^t(A\cos 2t + B\sin 2t) + \tfrac{5}{8}e^{-t}$ **6** $y = Ae^{-4x} + Be^{-2x} + \tfrac{3}{325}(18\cos 3x - \sin 3x)$
7 $x = e^{-t}(A\cos t + B\sin t) + \tfrac{3}{2}e^{-2t} - t + 1$ **8** $y = Ae^{4x} + Be^{-x} - \tfrac{5}{4}x^2 + \tfrac{15}{8}x - \tfrac{97}{32}$ **9** $y = (Ax + B)e^{4x} + \tfrac{1}{2}e^{2x} + \tfrac{2}{289}(15x - 8\sin x)$
10 $y = Ae^{-4x} + Be^{2x} - \tfrac{1}{8}(e^{-2x} + 4x + 1)$ **11** $y = (Ax + B + \tfrac{3}{2}x^2)e^{-x}$
12 $x = e^{-2t}(A\cos 3t + B\sin 3t) + \tfrac{1}{145}(27\sin 2t - 24\cos 2t) - \tfrac{6}{169}(13t - 4)$ **13** $y = \left(A + \tfrac{2}{3}x\right)e^{3x} + Be^{-3x}$ **14** $x = \left(A - \tfrac{1}{2}t\right)\cos 2t + B\sin 2t$
15 $y = A + Be^{3x} + \tfrac{1}{2}e^{4x}$ **16** $y = A + Be^{-4x} + \tfrac{1}{8}x(2x + 5)$

ANSWERS

Exercise 3K

1 $y = \frac{1}{108}(11e^{-3x} - 27e^x - 36x^2 + 60x + 16)$ **2** $y = 2e^{2x}(2e^x - 1 - 3x)$ **3** $x = \frac{1}{8}[e^{-2t}(13\cos 2t + 27\sin 2t) + 4t - 5]$

4 $y = \frac{2}{169}[(221x + 24)e^{-3x} + 2(5\sin 2x - 12\cos 2x)]$ **5** $x = \frac{1}{8}e^t(13\sin 2t - 5\cos 2t + 5e^{-t})$

6 $y = \frac{80}{13}e^{-2x} - \frac{83}{25}e^{-4x} + \frac{3}{325}(18\cos 3x - \sin 3x)$ **7** $x = \frac{1}{2}e^{-t}(11\sin t - 5\cos t) + \frac{3}{2}e^{-2t} - t + 1$ **8** $y = \frac{69}{160}e^{4x} + \frac{23}{5}e^{-x} - \frac{5}{4}x^2 + \frac{15}{8}x - \frac{97}{32}$

9 $y = \frac{1}{578}[(316 - 349x)e^{4x} + 289e^{2x} + 60\cos x - 32\sin x]$ **10** $y = \frac{1}{24}(29e^{2x} + 25e^{-4x} - 3e^{-2x} - 12x - 3)$ **11** $y = \frac{3}{2}x^2 e^{-x}$

12 $x = \frac{e^{-2t}}{24\,505}(576\cos 3t + 52\,346\sin 3t) + \frac{1}{145}(27\sin 2t - 24\cos 2t) - \frac{6}{169}(13t - 4)$ **13** $y = \frac{1}{6}[(7 + 4x)e^{3x} + 11e^{-3x}]$

14 $x = 3\cos 2t + \left(\frac{\pi}{8} - 1 - \frac{1}{2}t\right)\sin 2t$ **15** $y = \frac{1}{6}(11 - 2e^{3x} + 3e^{4x})$ **16** $y = e^{-4x} + \frac{1}{8}x(2x + 5)$

Exercise 3L

1 a) $\mathbf{r} = A e^{-3t} + \frac{1}{3}((3t - 1)\mathbf{j} - \mathbf{k})$ **b)** $\mathbf{r} = A e^{\frac{1}{2}t^2} + \mathbf{i} + 3\mathbf{k}$ **c)** $\mathbf{r} = A t + \frac{1}{2}t^3\mathbf{i} + 2t^2\mathbf{j} + 3t\ln t\,\mathbf{k}$ **d)** $\mathbf{r} = A(t + 1)^{-2} + 3\mathbf{j}$

2 a) $\mathbf{r} = A e^{2t} - \frac{1}{4}(2t + 3)\mathbf{i} - \frac{3}{2}\mathbf{j} - \frac{1}{4}(2t + 1)\mathbf{k}$, $\mathbf{r} = \frac{1}{4}[(19e^{2t} - 2t - 3)\mathbf{i} + 2(e^{2t} - 3)\mathbf{j} + (e^{2t} - 2t + 1)\mathbf{k}]$

2 b) $\mathbf{r} = e^{-3t}(\mathbf{A} + t\mathbf{j})$, $\mathbf{r} = e^{-3t}[(t + 6)\mathbf{j} + \mathbf{k}]$

2 c) $\mathbf{r} = A e^t + \frac{1}{2}(\sin t - \cos t)\mathbf{i} + \frac{1}{5}(2\sin 2t - \cos 2t)\mathbf{j}$, $\mathbf{r} = \frac{1}{2}(e^t + \sin t - \cos t)\mathbf{i} + \frac{1}{5}(e^t + 2\sin 2t - \cos 2t)\mathbf{j}$

3 a) $\mathbf{r} = A e^t + \mathbf{B} e^{2t} + \frac{1}{2}(\mathbf{i} - \mathbf{j} + \mathbf{k})$, $\mathbf{r} = \frac{1}{2}[(e^{2t} - 2e^t + 1)\mathbf{i} + (2e^t - e^{2t} - 1)\mathbf{j} + (e^{2t} - 2e^t + 1)\mathbf{k}]$

3 b) $\mathbf{r} = (At + \mathbf{B})e^{-t} + (t - 1)\mathbf{i}$, $\mathbf{r} = [(t + 2)e^{-t} + t - 1]\mathbf{i} + (t + 1)e^{-t}\mathbf{j} + (t + 1)e^{-t}\mathbf{k}$

3 c) $\mathbf{r} = e^t(\mathbf{A}\cos 2t + \mathbf{B}\sin 2t) + (0.2\cos t - 0.1\sin t)\mathbf{j} + (0.1\cos t + 0.2\sin t)\mathbf{k}$

3 c) $\mathbf{r} = \frac{1}{20}\{[e^t(4\cos 2t - \sin 2t) + 4\cos t - 2\sin t]\mathbf{j} + [e^t(2\cos 2t - 3\sin 2t) + 2\cos t + 4\sin t]\mathbf{k}\}$

4 a) $\frac{d\mathbf{v}}{dt} + \mathbf{v} = 3\cos t\,\mathbf{i}$ **b)** $\mathbf{v} = -1.023\mathbf{i} + 0.404\mathbf{j} - 0.539\mathbf{k}$ **5 a)** $\frac{d^2\mathbf{r}}{dt^2} + 0.2\frac{d\mathbf{r}}{dt} + 0.1\mathbf{r} = 0$

5 b) $\mathbf{r} = 3.20\mathbf{i} - 0.711\mathbf{j} - 6.05\mathbf{k}$, $\mathbf{v} = 0.213\mathbf{i} - 0.676\mathbf{j} - 2.92\mathbf{k}$

Exercise 3M

1 a) $x = A e^{3t} + B e^{-t} + 2$, $y = -A e^{3t} + B e^{-t} + 1$ **b)** $x = -0.5(e^{3t} + 3e^{-t} - 4)$, $y = 0.5(e^{2t} - 3e^{-t} + 2)$

1 c)

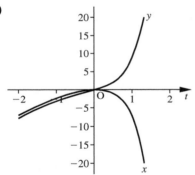

2 a) $x = e^t(A\cos 2t + B\sin 2t) + 2$, $y = -e^t(A\sin 2t - B\cos 2t) + 1$ **b)** $x = -e^t\sin 2t + 2$, $y = -e^t\cos 2t + 1$

2 c)

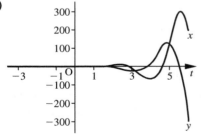

3 a) $x = e^{-2t}(A\cos 2t + B\sin 2t) - 1\frac{3}{4}$, $y = e^{-2t}[(A - 2B)\cos 2t + (2A + B)\sin 2t] - 3\frac{1}{4}$

3 b) $x = \frac{1}{4}[e^{-2t}(15\cos 2t + 11\sin 2t) - 7]$, $y = \frac{1}{4}[e^{-2t}(41\sin 2t - 7\cos 2t) - 13]$

3 c)

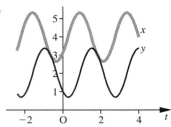

4 a) $x = A\cos 3t + B\sin 3t + 4$, $y = A\sin 3t - B\cos 3t + 2$ **b)** $x = 4 + \sin 3t - \cos 3t$, $y = 2 - \sin 3t - \cos 3t$

4 c)

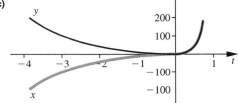

5 a) $x = Ae^{5t} + Be^{-t} + 7$, $y = Ae^{5t} - Be^{-t} - 6$ **b)** $x = 3e^{5t} - 5e^{-t} + 7$, $y = 3e^{5t} + 5e^{-t} - 6$

5 c)

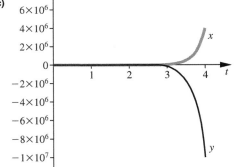

6 a) $x = e^{2t}\left(Ae^{t\sqrt{3}} + Be^{-t\sqrt{3}}\right)$, $y = -e^{2t}\left[(1 + \sqrt{3})Ae^{t\sqrt{3}} + (1 - \sqrt{3})Be^{-t\sqrt{3}}\right] + 1$

6 b) $x = e^{2t}\left[\left(1 - \frac{\sqrt{3}}{6}\right)e^{t\sqrt{3}} + \left(1 + \frac{\sqrt{3}}{6}\right)e^{-t\sqrt{3}}\right] = e^{2t}\left(2\cosh t\sqrt{3} - \frac{\sqrt{3}}{3}\sinh t\sqrt{3}\right)$

6 b) $y = -e^{2t}\left[\left(\frac{3 + 5\sqrt{3}}{6}\right)e^{t\sqrt{3}} + \left(\frac{3 - 5\sqrt{3}}{6}\right)e^{-t\sqrt{3}}\right] + 1 = -e^{2t}\left(\cosh t\sqrt{3} + \frac{5\sqrt{3}}{3}\sinh t\sqrt{3}\right) + 1$

6 c)

Exercise 3N

1 15.6   **2** 0.258   **3** 7.19   **4** −0.229   **5** 1.68   **6** 13.0

Exercise 3O

1 a) Using final two table values, $y(2) = 0.478\,267\,614$ **b)** $h = 0.000\,05$

2 b) Using final two table values, $y(0.2) = -2.4960$ **c)** $h = 0.000\,025$

ANSWERS

Exercise 4A

1 $48.1 \,\mathrm{m\,s^{-1}}$ **2** $3848 \,\mathrm{m\,s^{-1}}$ **4 a)** Spherical droplet, no air resistance **b)** $\frac{gt}{4} - \frac{250gr_0}{k}\left[1 - \left(\frac{1000r_0}{kt + 1000r_0}\right)^3\right]$, $\frac{g}{4} + \frac{3g}{4}\left(\frac{1000r_0}{kt + 1000r_0}\right)^4$

7 $201.7 \,\mathrm{m}$, $0.495 \,\mathrm{m\,s^{-1}}$ **8** $388 \,\mathrm{m\,s^{-1}}$ **9** $8.82 \,\mathrm{m\,s^{-1}}$ **10 b)** $\frac{gT^2(2\ln 2 - 1)}{2(\ln 2)^2}$

Exercise 4B

1 a) $v = 3t - 2t^2$ **b)** $v = t - \frac{1}{2}\ln(t+1)$ **c)** $v = -6\cos 3t$ **2 a)** $v = \sqrt{\frac{1}{2}t + 16}$ **b)** $v = 3(\mathrm{e}^{-0.5t} + 1)$ **c)** $v = \frac{3(3\mathrm{e}^{6t} - 1)}{2(3\mathrm{e}^{6t} + 1)}$

2 d) $v = \frac{10\mathrm{e}^{6t}}{5\mathrm{e}^{6t} - 4}$ **3 a)** $v^3 = 3s + 64$ **b)** $x = 6 - v - 3\ln\left|\frac{v-3}{3}\right|$ **c)** $v^2 = \frac{22\mathrm{e}^{-6s} + 5}{3}$ **d)** $v^2 = 8\ln\left(\frac{s}{12}\right)$ **e)** $v^2 = \frac{s - 24}{2s}$

4 a) $v = \frac{5(3\mathrm{e}^{-t} + 1)}{2}$, $v \to 2.5 \,\mathrm{m\,s^{-1}}$ as $t \to \infty$ **c)** $v = \frac{5 - 3\mathrm{e}^{-t}}{2}$, $v \to 2.5 \,\mathrm{m\,s^{-1}}$ from below **5 a)** $mv\frac{\mathrm{d}v}{\mathrm{d}x} = -\frac{k}{x^2}$

5 b) $x = \frac{2kd}{2k - mu^2d}$ **6 a)** $30 \,\mathrm{N\,s}$ **b)** $4 \,\mathrm{m\,s^{-1}}$ **7 a)** $-3 \,\mathrm{J}$ **b)** $9.165 \,\mathrm{m\,s^{-1}}$ **c)** $2.86 \,\mathrm{m}$ **8 a)** $mk\ln 4$ **9 a)** $v = \sqrt{\frac{P(1 - \mathrm{e}^{-Rt/m})}{R}}$

10 a) $V = \sqrt{\frac{P}{k}}$

Exercise 5A

1 $11.4 \,\mathrm{m\,s^{-1}}$ at $18.1°$ to cushion **2** $3.825 \,\mathrm{J}$ **3** $5.25 \,\mathrm{m\,s^{-1}}$ at $24.2°$ above horizontal **4 a)** 0.155

4 b) $2.48 \,\mathrm{m\,s^{-1}}$ at $53.8°$ to line of centres **5 a)** A: $-1.6\mathbf{i} + 3\mathbf{j}$, B: $2.4\mathbf{i} - 4\mathbf{j}$ **b)** $5.4 \,\mathrm{J}$ **6** $\frac{1}{3}$ **7** 0.5

8 $3.28 \,\mathrm{m\,s^{-1}}$, deflected through $112.4°$ **9 b)** ea **13** $u\sqrt{\cos^2\theta + e^2\sin^2\theta}$ at $\arctan(e\tan\theta)$ to cushion

14 A: $\frac{1}{2}u\sqrt{4\sin^2\theta + (1-e)^2\cos^2\theta}$ at $\arctan\left(\frac{2\tan\theta}{1-e}\right)$ to line of centres, B: $\frac{1}{2}u(1+e)\cos\theta$ along line of centres

Exercise 5B

1 a) $\frac{3}{8}\sqrt{2g}$ **b)** Inextensible string; smooth, light pulley **2** $0.782 \,\mathrm{s}$

3 A: $\frac{20\sqrt{2}}{21} \,\mathrm{m\,s^{-1}}$ in direction AB; B: $\frac{20\sqrt{10}}{21} \,\mathrm{m\,s^{-1}}$ at $18.4°$ to BC; C: $\frac{20}{7} \,\mathrm{m\,s^{-1}}$ along BC.

Impulsive tensions: $\frac{20\sqrt{2}}{7} \,\mathrm{N\,s}$ in AB, $\frac{80}{7} \,\mathrm{N\,s}$ in BC

5 B: $\frac{u\sqrt{3}}{4}$ in direction of string, A: $\frac{u\sqrt{7}}{4}$ at $49.1°$ to string

6 Impulsive tensions: $2\sqrt{3} \,\mathrm{N\,s}$ in AB, $8\sqrt{3} \,\mathrm{N\,s}$ in BC; speed of A $= \sqrt{3} \,\mathrm{m\,s^{-1}}$

7 Impulsive tensions: $5 \,\mathrm{N\,s}$ in AB, $10\sqrt{2} \,\mathrm{N\,s}$ in BC; speed of A $= 2.5 \,\mathrm{m\,s^{-1}}$ **9** $\frac{5J^2}{12m}$

10 A: $\frac{2u\sqrt{3}}{13}$ along string AB; B: $\frac{2u\sqrt{7}}{13}$ at $19.1°$ to string BC; C: $\frac{5u}{13}$ along string BC.

Exercise 6A

1 a) Oscillatory (weakly damped) with period $\frac{4\pi}{\sqrt{11}}$ **b)** Strongly damped **c)** Critically damped **d)** Strongly damped

1 e) Oscillatory (weakly damped) with period 2π **f)** Critically damped **2 a)** $\ddot{x} + 2\dot{x} + 5x = 0$ **b)** Period π **c)** $x = 3\mathrm{e}^{-t}\cos 2t$

2 d) Ratio $= \mathrm{e}^{-\pi}$ **3 a)** $x = \mathrm{e}^{-1.5t}\left[A\cos\left(\frac{t\sqrt{11}}{2}\right) + B\sin\left(\frac{t\sqrt{11}}{2}\right)\right]$ **b)** $x = 0.2\,\mathrm{e}^{-1.5t}\left[\cos\left(\frac{t\sqrt{11}}{2}\right) + \frac{3}{\sqrt{11}}\sin\left(\frac{t\sqrt{11}}{2}\right)\right]$ **c)** $1.54 \,\mathrm{s}$

4 a) $2.25 \,\mathrm{m}$ below A **b)** $x = 0.25\mathrm{e}^{-2t}(3\cos 6t + \sin 6t)$ **c)** $0.263 \,\mathrm{m}$ **5 a)** $2\ddot{x} + k\dot{x} + 40x = 20$ **b)** $x = \frac{1}{2}\left[1 - 2t\sqrt{5} + 1)\mathrm{e}^{-2t\sqrt{5}}\right]$

5 c) $1.499\,35 \,\mathrm{m}$ below A **6** $1.9957 \,\mathrm{m}$ **7 a)** $2 \,\mathrm{m}$ **b)** $\ddot{x} + \sqrt{3g}\dot{x} + gx = 0$ **c)** $\frac{4\pi}{\sqrt{g}}$ **d)** $\coth\left(\frac{\pi\sqrt{3}}{2}\right)$ **8 a)** $1.02 \,\mathrm{s}$ **b)** $0.809 \,\mathrm{m\,s^{-1}}$

8 c) $0.596 \,\mathrm{m}$

ANSWERS

Exercise 6B

1 a) $\ddot{x} + 25x = 2.5\sin 2t$ **b)** $x = \frac{1}{21}(2.5\sin 2t - \sin 5t)$ **c)**

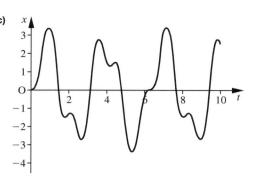

2 a) $\omega = 5$ **b)** $x = 0.1(\sin 5t - 5t\cos 5t)$ **3 a)** $\ddot{x} + 6\dot{x} + 25x = 2.5\sin 2t$ **b)** $x = \frac{1}{156}[e^{-3t}(8\cos 4t - \sin 4t) - 8\cos 2t + 14\sin 2t]$
4 a) $\ddot{x} + x = \sin t$ **5 a)** $\ddot{x} + 2k\dot{x} + x = \sin t$ **b)** $x = \frac{1}{2}[(t+1)e^{-t} - \cos t]$ **c)** Oscillates with $x = -\frac{1}{2}\cos t$ **7 a)** $0.75\,\text{m}$
7 b) $x = \frac{1}{3}gb(2\sin 2\sqrt{g}t - \sin 4\sqrt{g}t)$ **8 a)** $\frac{1}{2}\pi\,\text{s}$ **b)** $2 + \frac{1}{2}\pi V\,\text{m}$
9 a) $\ddot{x} + 16x = 2.4(\sin t + \sin 2t)$ **b)** $x = 0.16\sin t + 0.2\sin 2t - 0.14\sin 4t$
9 c) Coils of the section AC touch after approximately 2 s **10 a)** $3\frac{3}{4}a$ **b)** $a\ddot{x} + 4gx = 2ag\sin\left(t\sqrt{\frac{g}{a}}\right)$
10 c) $x = \frac{a}{3}\left[2\sin\left(t\sqrt{\frac{g}{a}}\right) - \sin\left(2t\sqrt{\frac{g}{a}}\right)\right]$ **d)** $\frac{a(45 - 2\sqrt{3})}{12}$

Exercise 7A

1 a) $4.37\,\text{s}$ **b)** $214\,\text{m}$ **c)** $22.4\,\text{m}$ **2 a)** $11.1\,\text{s}$ **b)** $541\,\text{m}$ **c)** $144\,\text{m}$ **3 a)** $55°$ **b)** $35°$ **4** $8°$ or $62°$ to plane
5 $7.2°$ or $102.8°$ to plane **7 b) i)** $0.0589U$ **ii)** $0.0506U^2$ **8 a)** $\dfrac{2U\sin\theta}{g\cos\alpha}$ **b)** $\dfrac{2U^2\sin\theta\cos(\theta - \alpha)}{g\cos^2\alpha}$
8 c) $\dfrac{U^2\sin^2\theta}{2g\cos^2\alpha}$ **d)** $\dfrac{U^2}{g(1 - \sin\alpha)}$ **9** $8.23\,\text{m}$ **10** $R = 0.0765U^2$, $\theta = 19.5°$ **11** $\sin\alpha\sqrt{2ga\tan\alpha}$
12 $16.7°$ below horizontal or $76.7°$ above horizontal **13 b)** $\dfrac{2V\sin\beta}{g\cos\alpha}$

Exercise 7B

1 a) $\dot{v} + 80v = 10$ **b)** $0.125\,\text{m s}^{-1}$ **c)** $v = \frac{1}{8}(1 - e^{-80t})$ **d)** $x = \frac{1}{8}t + \frac{1}{640}(e^{-80t} - 1)$ **2 a)** $m\dot{v} + kv^2 = 10m$ **b)** $\sqrt{\dfrac{10m}{k}}$
2 c) $v = \sqrt{\dfrac{10m}{k}}\tanh\left(t\sqrt{\dfrac{10k}{m}}\right)$ **d)** $kv^2 = 10m(1 - e^{-2kx/m})$ **3 a)** $m\dot{v} + kv^2 = 10m$ **b)** $\dfrac{10m}{V}$ **c)** $v = V(1 + 2e^{-10t/V})$
3 d) $vV + V^2\ln\left(\dfrac{v - V}{2V}\right) = 3V^2 - 10x$ **4 a)** $\dot{v} + 0.02v + 10 = 0$ **b)** $v = 530e^{-0.02t} - 500$ **c)** $2.91\,\text{s}$ **d)** $43.3\,\text{m}$ **e)** $34\,\text{J}$
5 a) $\dot{v} + 0.004v^2 + 10 = 0$ **b)** $v = \dfrac{30 - 50\tan 0.2t}{1 + 0.6\tan 0.2t}$ **c)** $2.70\,\text{s}$ **d)** $38.4\,\text{m}$ **e)** $\dot{v} - 0.004v^2 + 10 = 0$ **f)** Total flight time $5.54\,\text{s}$

Exercise 8A

1 $x^2 - 4xy + 4y^2 - 2x + y = 0$ **2** $y = 8\cos\sqrt{x}$ **3 a)** $x = t + \sin t$, $y = 1 - \cos t$ **b)** $3.37\,\text{m}$ **4** $9x^2 + 16y^2 = 144$
5 a) $x = 8e^{-t} - 4e^{-2t}$, $y = 4e^{-t} - 4e^{-2t}$ **6** $x = e^t\cos t$, $y = e^t\sin t$, $z = t$; helix with exponentially increasing radius **7** $y = \cos\sqrt{x}$
8 $x = \dfrac{2\sin t - 3\cos t - 2t + 3}{m}$, $y = \dfrac{3t + 2 - 2\cos t - 3\sin t}{m}$

Exercise 8B

1 a) $\dot{\mathbf{r}} = a\omega\hat{\mathbf{r}} + a\omega\theta\hat{\mathbf{s}}$, $\ddot{\mathbf{r}} = -a\omega^2\hat{\mathbf{r}} + 2a\omega^2\hat{\mathbf{s}}$ **b)** $\dot{\mathbf{r}} = -2a\omega\sin 2\theta\,\hat{\mathbf{r}} + a\omega\cos 2\theta\,\hat{\mathbf{s}}$, $\ddot{\mathbf{r}} = -5a\omega^2\cos 2\theta\,\hat{\mathbf{r}} - 4a\omega^2\sin 2\theta\,\hat{\mathbf{s}}$
1 c) $\dot{\mathbf{r}} = (2\cos\theta - \sin\theta)\omega\hat{\mathbf{r}} + (2\sin\theta + \cos\theta)\omega\hat{\mathbf{s}}$, $\ddot{\mathbf{r}} = -2(\sin\theta + \cos\theta)\omega^2\hat{\mathbf{r}} + 2(2\cos\theta - \sin\theta)\omega^2\hat{\mathbf{s}}$
1 d) $\dot{\mathbf{r}} = \omega(1 + \theta)^{-1}\hat{\mathbf{r}} + \omega\ln(1 + \theta)\hat{\mathbf{s}}$, $\ddot{\mathbf{r}} = -\omega^2((1 + \theta)^{-2} + \ln(1 + \theta))\hat{\mathbf{r}} + 2\omega^2(1 + \theta)^{-1}\hat{\mathbf{s}}$ **3** $2ma\omega^2 e^\theta$
4 a) $\dot{\mathbf{r}} = (a\cos\theta - b\sin\theta)\omega\hat{\mathbf{r}} + (a\sin\theta + b\cos\theta)\omega\hat{\mathbf{s}}$, $\ddot{\mathbf{r}} = -2(a\sin\theta + b\cos\theta)\omega^2\hat{\mathbf{r}} + 2(a\cos\theta - b\sin\theta)\omega^2\hat{\mathbf{s}}$ **b)** Speed $\omega\sqrt{a^2 + b^2}$
5 $\ddot{\mathbf{r}} = \dfrac{8r^4 - 9\theta^4}{r}\hat{\mathbf{r}} + \dfrac{6\theta^2(3\theta + r^2)}{r}\hat{\mathbf{s}}$ **7** $a\omega^2(4\cos 2\theta\cos\theta - 5\sin 2\theta\sin\theta)$ **9** $r = Ae^\theta + Be^{-\theta}$, $F = 2m\omega^2\sqrt{r^2 - 4AB}$

ANSWERS

Exercise 8C
1 a) $2\,\text{m s}^{-1}, 4\sqrt{3}\,\text{m s}^{-1}$ **3 c)** $7\sqrt{7}:8$ **4** $r=\sqrt{3t^2+2t+1}$ **6 b)** 0.594

Exercise 8D
1 $0.4\sqrt{2}\,\text{N}$ **2 a)** $1.35\,\text{m s}^{-1}$ **b)** $-1.05\,\text{m s}^{-2}, 0.675\,\text{m s}^{-2}$ **3** $0.48\,\text{m s}^{-2}, 0.128\,\text{m s}^{-2}$ **4 b)** $2mg\sqrt{2}$

Exercise 9A
1 $\dfrac{a(3m+4\lambda a)}{3(m+\lambda a)}$ **2** $\dfrac{a\sin\alpha}{\alpha}$ from centre **3** $\dfrac{a\sin^2\alpha}{2(1-\cos a)}$ from centre **4** $\left(\dfrac{6\sqrt{3}-7}{20}, \dfrac{2+\sqrt{3}}{20}\right)$ **5** $\left(\dfrac{3a}{4}, 0\right)$ **7** $\left(\dfrac{2a\sin\alpha}{3\alpha}, 0\right)$
8 $\left(\dfrac{4a\sin^3\alpha}{2\alpha-\sin 2\alpha}, 0\right)$ **9** 2.16 m **10** $(1\tfrac{23}{35}, 0)$ **11** (0.218, 0) **12** (0.574, 0.410)

Exercise 9B
1 a) $mgl\sin(\theta+a)$, taking zero GPE level with B **2 a)** $0.000\,025g(30-7\cos\theta)$, taking zero GPE at ground level
3 b) $\theta=36.9°$ (stable) or $90°$ (unstable) **4 a)** $\tfrac{1}{2}ka^2(2\cos\theta-1)^2 - 4Wa\cos\theta$, taking zero GPE level with A
4 b) Only one, stable, position $\theta=0$ if $k<4W/a$.
4 b) If $k>4W/a$, $\theta=0$ is unstable and there is a second, stable, position $\theta = \arccos\left(\dfrac{2W+ka}{2ka}\right)$
5 a) $W(r\cos\theta+r\theta\sin\theta+b\cos\theta)$, taking zero GPE at centre of cylinder
6 a) $mgl(2\operatorname{cosec}\theta-\cot\theta-2)$, taking zero GPE with middle mass at pulley level **b)** $\theta=60°$
8 a) $2Wr(\sec\theta-\sin\theta)$, taking zero GPE at centre of cylinder
9 If θ is angle between spring and upward vertical at A, there is stable equilibrium for $\theta=16.3°$ and $143.7°$, and unstable equilibrium for $\theta=66.1°$

Exercise 10A
1 a) $2.5\,\text{rad s}^{-2}$ **b)** $96\,\text{N m}$ **c)** $1470\,\text{J}$ **2 a)** $160\,\text{kg m}^2$ **b)** $5760\,\text{J}$ **3** $600\pi\,\text{N m}$ **4** $12.5\pi\,\text{N m}$
5 a) Particles at $0.128\,\text{m s}^{-2}$, pulley at $0.428\,\text{rad s}^{-2}$ **b)** $2.39\,\text{rad s}^{-1}$ **c)** $58.8\,\text{J}$ **6** $0.064\,\text{m s}^{-2}$ **7** $1715\,\text{J}$ **8** $\dfrac{4mg}{4m+M}$
9 $\dfrac{(M-m)ga}{(M+m)a^2+I}$

Exercise 10B
3 $\dfrac{Ma^2}{2}$ **4** $\dfrac{Ma^2}{2}$ **5 a)** $\dfrac{Ma^2}{6}$ **b)** $\dfrac{Ma^2}{2}$ **6** $\dfrac{5Ma^2}{2}$ **8** $\dfrac{5Ma^2}{2}$ **9** $7:5$ **10** $\dfrac{Ma}{3}$ **11** $\dfrac{Ma^2}{4}$

Exercise 10C
1 a) $35\,\text{kg m}^2$ **b)** $11\tfrac{2}{3}\,\text{kg m}^2$ **2 a)** $\dfrac{Ma^2}{2}$ **b) i)** $3Ma^2$ **ii)** $\dfrac{Ma^2}{2}$ **iii)** $\dfrac{5Ma^2}{2}$ **3 a)** $0.393\,\text{kg m}^2$ **b)** $0.783\,\text{kg m}^2$ **c)** $3\,\text{kg m}^2$
4 a) $\dfrac{Ma^2}{6}$ **b)** $\dfrac{Ma^2}{6}$ **c)** $\dfrac{2Ma^2}{3}$ **5** $a\sqrt{\dfrac{2}{3}}$ **6 a)** $\dfrac{Ma^2}{4}$ **b)** $\dfrac{Ma^2}{4}$ **7** $\dfrac{M(a^2+b^2)}{2}$ **8 a)** $a\sqrt{2.5}$ **b)** $a\sqrt{2.375}$ **c)** $a\sqrt{4.875}$

Exercise 11A
1 $\sqrt{\dfrac{3g}{2}}\,\text{rad s}^{-1}$ **2 a)** $5.55\,\text{N m}$ **b)** $1.54\,\text{rad s}^{-1}$ **3** $36.5\,\text{rad s}^{-1}$ **4** $\sqrt{\dfrac{2g}{a}}$ **5** $0.419\,\text{N}$ **6** $11.1\,\text{m}$ **7** $10.4\,\text{N}$ **8** $2.15\,\text{rad s}^{-1}$
9 $6.14\,\text{rad s}^{-1}$ **10 a)** $2\sqrt{g(1-\cos\theta)}$ **b)** Slips when $\theta=38.1°$

Exercise 11B
1 $\dfrac{8\pi}{\sqrt[3]{g}}\,\text{s}, 1\tfrac{7}{9}\,\text{m}$ **2** $\sqrt{5}:\sqrt{6}$ **3** $\dfrac{8\pi}{\sqrt{15g}}\,\text{s}$ **4** $a\sqrt{\dfrac{2}{3}}$ **5 a)** $\dfrac{\pi}{2}\sqrt{\dfrac{17a}{g}}$ **b)** Reduced by 2.23% **6 a)** $\dfrac{k^2+h^2}{h}$
9 a) $2\pi\sqrt{\dfrac{2k^2+h_1^2+h_2^2}{(h_1+h_2)g}}$

ANSWERS

Exercise 11C

1 a) $12g$ N upwards **b)** 44.7 N at $80.5°$ to vertical **c)** 81.1 N at $55.5°$ to vertical **2 a)** $\sqrt{\frac{8g}{3}}$ rad s^{-1} **b)** 67.3 N at $76.6°$ to vertical
3 $29.0M$ N at $17°$ to vertical **5 a)** $\frac{Mg(1+6\sin^2\theta)}{3}$, $Mg\sin 2\theta$ **b)** Min $\frac{Mg}{3}$ N, max $\frac{7Mg}{3}$ N **6** 68.1 N at $73.7°$ to vertical
7 $8.75°$ **8** 0.7

Exercise 11D

1 404 rad s^{-1} **2** $4(\sqrt{2}-1)$ N m **3** 7.5π N **4** 5.44 s **5** $\frac{5g\ln 10}{2a}$ **6 a)** $M\sqrt{\frac{2ga}{3}}$ **b)** $-M\sqrt{\frac{ga}{6}}$
7 Reaction $\frac{J(19a-12x)}{19a}$, centre of percussion $x=\frac{19a}{12}$ **8** $\frac{1}{\sqrt{2}}$ **9** $\frac{2\omega}{3}$ **10** $5\sqrt{g}$ rad s^{-1} **12** $\frac{4a\omega}{5}$
13 $m=2M/27$, speed $6a\omega$ in opposite direction

Exercise 12

1 a) 2.39 m s^{-2} **2** Disc wins by 5 m **3 a)** 13.1 m s^{-1} **b)** 5.11 m s^{-1} **4** $\frac{2g}{3}, \frac{mg}{3}$ **9 b)** $F=\frac{2mg\sin\theta}{7}$, $R=\frac{mg(17\cos\theta-10)}{7}$
9 c) $43.6°$ **10** 20 m s^{-1}, -10 m s^{-1} **11** $2a/3$ **12 a)** Disc rotates at $\omega/3$ about fixed point
13 A has speed 5 m s^{-1} perpendicular to AB, B has speed 0 m s^{-1} **14** $\frac{11v}{15a}$ **15** $5:1$
16 A: speed $\frac{2v}{15}$, direction AB; B: speed $\frac{2v\sqrt{13}}{15}$ at $13.9°$ to BC; C: speed $\frac{7v}{15}$, direction BC
17 C: speed $\frac{13v}{21}$, direction BC; rod: speed $\frac{4v}{21}$, direction BC, and rotates at $\frac{v\sqrt{3}}{7a}$ **18** A: $\frac{10v}{11}$, B: $\frac{-2v}{11}$ **20** $2\pi\sqrt{\frac{a}{g}}$ **21 b)** $2\pi\sqrt{\frac{35a}{2g}}$

Examination questions

Chapters 1 to 4

1 a) $6\mathbf{i}+12\mathbf{j}+6\mathbf{k}$ **b) i)** 7.35 units^2 **ii)** $x+2y+z=9$ **c)** $\mathbf{r}=(8+\lambda)\mathbf{i}+2(1+\lambda)\mathbf{j}+(3+\lambda)\mathbf{k}$ **d)** $(7, 0, 2)$, $\sqrt{6}$ units **2 a)** $5\sqrt{2}$ N
2 b) $\mathbf{r}=(\lambda-4.8)\mathbf{i}-\lambda\mathbf{j}$ or similar **3 a)** $\mathbf{r}=(2+\lambda)\mathbf{i}+(2\lambda-1)\mathbf{j}+(2-\lambda)\mathbf{k}$ or similar **b)** $4\mathbf{i}+3\mathbf{j}$ **c)** $\frac{2}{3}\sqrt{6}(-3\mathbf{i}+4\mathbf{j}+5\mathbf{k})$
4 b) $\mathbf{r}=(0.5\mathbf{i}-0.5\mathbf{j})+\lambda(2\mathbf{i}+\mathbf{j}-\mathbf{k})$ or similar **5 a)** $p=4, q=-2, r=0$ **b)** $\sqrt{221}$ N m **6** $\sqrt{29}$ N m
7 i) $P=0, Q=-2, R=-5$; forces not in equilibrium because total moment not zero **ii)** $\mathbf{i}+\mathbf{j}-\mathbf{k}, 2\mathbf{i}-7\mathbf{j}+\mathbf{k}$ **iii)** $\beta=4$
7 iv) $-7\mathbf{i}+9\mathbf{j}-6\mathbf{k}$ **v)** For example, $\mathbf{F}_4=2\mathbf{j}+3\mathbf{k}$. Forces not in equilibrium because resultant not zero
8 i) Point A must move perpendicularly to OA as body is rigid **ii)** B's velocity relative to A must be perpendicular to AB as body is rigid
8 iii) $\omega_1=9, \omega_2=-2, \omega_3=7$ **iv)** $\lambda=1, \mu=-2, \nu=9, \boldsymbol{\omega}=-\mathbf{i}-3\mathbf{j}-\mathbf{k}$ **9 i)** Circle, centre $(0, 0)$, radius $1/m$
9 ii), iii)   **iv)** 1.75 **v)** 1.6966

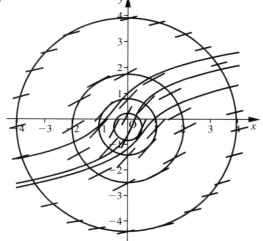

10 i) $I=A\mathrm{e}^{-20t}+B\mathrm{e}^{-5t}+0.25\cos 20t+0.15\sin 20t$ **ii)** $I=\frac{1}{60}(17\mathrm{e}^{-20t}-32\mathrm{e}^{-5t}+15\cos 20t+9\sin 20t)$
10 iii) $I=0.25\cos 20t+0.15\sin 20t$; unchanged because independent of A and B **iv)** $0.292\sin(20t+1.03)$ **11 i)** $p=5.6, q=7$

ANSWERS

11 ii) $p = Ae^{-10t} + Be^{-3t} + 5.6$, $q = -5Ae^{-10t} + 2Be^{-3t} + 7$, $p = -0.6e^{-10t} - 5e^{-3t} + 5.6$, $q = 3e^{-10t} - 10e^{-3t} + 7$

11 iv)

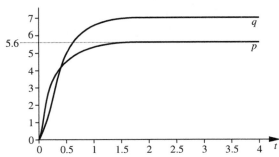

12 $r = e^{-2t}i + (2e^{-2t} - 1)j$ **13** $9x^2 + 4y^2 = 36$ **15 ii)** Because when $v = \sqrt{\dfrac{g}{k}}$, $\dfrac{dv}{dt} = 0$

15 iii) $v\dfrac{dv}{dx} + kv^2 = g$, $x = \dfrac{1}{2k}\ln\left(\dfrac{g}{g - kv^2}\right)$ **iv)** 36.8 m **16 c)** $v^2 = \dfrac{2g[(1 + \lambda x)^3 - 1]}{3\lambda(1 + \lambda x)^2}$ **17** $U - c\ln k$ **18** $\lambda = \dfrac{u}{cT^2}$

19 0.001 42 m s^{-1}

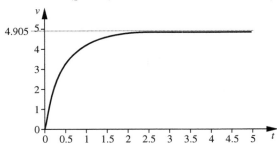

20 b) Integrating factor e^{2t}, $v = e^{3t} + Ae^{-2t}$ **c)** 19.95 m s^{-1}

Chapters 5 to 8

1 b) $u\sqrt{3}/2$ **2** $17\frac{1}{3}$ m s^{-1} **3** $8/\pi$ m s^{-1} **4 a)** P moves in direction of impulse, along line of centres **b)** $0.669u$
5 a) $2i - 0.8j$, $-i + 0.2j$ **b)** 1.8 J **6 ii)** $7/6\sqrt{3}$ **iii)** 3.12 N s **iv)** 0.142 s **7 a)** $\frac{5}{9}u\cos\alpha$ **8 b)** $x = e^{-6t}(A\cos 8t + B\sin 8t)$
8 c) $x = e^{-6t}(0.1\cos 8t + 0.075\sin 8t)$ **9 ii)** 3.13 m **10 b)** $x = ae^{-\frac{1}{2}nt}\left[\cos\left(\dfrac{n\sqrt{3}}{2}t\right) + \dfrac{1}{\sqrt{3}}\sin\left(\dfrac{n\sqrt{3}}{2}t\right)\right]$
11 i) $0.2M$ **ii)** $M\ddot{x} + 2\sqrt{g}\dot{x} + 5gx = 0$ **iii)** $M < 0.2$ **iv)** Resistive force proportional to velocity **12** $x = \frac{1}{9}e^{-t}(3 + \sin 3t)$
18 a) $6a\sqrt{\dfrac{3ag}{5}}$ **c)** $2g$ towards A **19 a)** Radial $mr\omega^2(k^2 - 1)$, transverse $2mkr\omega^2$

Chapters 9 to 12

1 b) 26.6° **c)** Stable **2 a)** If string not vertical, horizontal force component acts on ring, causing it to move
2 c), d) Unstable when $\theta = 0$ and π, stable when $\theta = 2\pi/3$ **3 b), c)** Stable when $\theta = -1.003$ rad, unstable when $\theta = 0.635$ rad
6 $6ma^2$ **7** $\dfrac{6g}{7}$ **8** $0.003\,34$ rad s^{-1} **9 b)** 12 N m **10 i)** $\dfrac{148ma^2}{3}$ **ii)** $k = \dfrac{2mga}{\pi}$ **11** 53.4 rad s^{-1} **13 a)** 72π J
13 b) $\sqrt{6\pi}$ rad s^{-1} **14 ii)** $\dfrac{3g\sin\theta}{4a}$, $S = \dfrac{3mg\sin\theta}{4}$ **15 c)** $\dfrac{5mg}{4}$ **16 b)** $\dfrac{12g}{17a}$ **c)** $\dfrac{10mg}{17}$ **17** $5\sqrt{ga}$ **18 a) ii)** $4v$
20 ii) Unstable when $\theta = 0$, stable when $\theta = 0.841$ rad **v)** $\pi\sqrt{\dfrac{23a}{10g}}$ **21** 18 m

Index

acceleration
  angular, relation to torque 239
  unifrom 204
  normal component of 176
  polar resolutes of 159
  radial component of 159
  tangential component of 176
  transverse component of 159
air resistance 145
angle
  between two lines 4
  between vectors 4
angular momentum 203, 249
  conservation of 242, 248
  relation to impulsive torque 239
arc
  centre of mass of 191
  length of 157
area
  swept by position vector 165
  using vector product 21
auxiliary equation 51

boundary conditions 45, 65

cartesian component of vector 1
cartesian equation of line 9
catenary 179
central force 165
centre of percussion 241
complementary function 52, 57
centre of mass 188
  of arc 191
  motion of 247
  of shell 193
centre of curvature 175
compound pendulum 231
  period of 231
conservation
  of angular momentum 242, 248
  of mechanical energy 226
  of moment of momentum 242
couple
  relation to angular acceleration 203
  work done by 225
critical damping 120
cross product *see* vector product
curvature
  centre of 175
  circle of 175
  radius of 175

curves
  motion of particle on 159
  vector equations of 25
damping constant 117
damping of oscillations
  critical 120
  harmonic 115
  linear 116
  strong 119
  weak 121
dashpot 16
differential equations 35
  auxiliary equation 51
  complementary function 52, 57
  degree of 36
  first-order linear 48
  homogeneous 45
    linear 50
  integrating factor 48
  numerical solution of 76
    Euler's method 76
  order of 35
  particular integral 52, 60
  second-order linear 56
  simultaneous 70
  variables separable 43
  vector equations 67
direction cosines 2
direction ratios 2
distance
  of a point from a line 22
  between points 6

equation of rotational motion 203
equilibrium
  neutral 196
  stable 196
  unstable 196
equivalent force systems 29
equivalent simple pendulum 232
Euler's method 76

force
  at axis of rotation 235
  central 165
  non-coplaner systems 29
  variable 88
  vector moment of 27
  work done by 92

helix
  pitch 26
  vector equation 25

inertia 201
  see also moment of inertia
impact, oblique 103
impulse 103
  of a torque 239
  of a variable force 91

impulsive tension 109
inclined plane
  maximum distance above 139
  maximum distance from 138
  maximum range on 140
  projectiles on 137
  range on 138
initial conditions 45, 64
integrating factor 48
intersection of lines 10
intrinsic coordinates 175
isocline 38

kinetic energy
  of rotation 201
  total 250

lines
  angle between 11
  cartesian equations of 9
  distance of a point from 22
  pairs, intersecting 10
    parallel 11
    skew 11
    shortest distance between 23
  vector equations of 7

mass, variable 81
moment of a force 27
moment of inertia 201
  calculation of 209
  of composite bodies 218
  of non-uniform bodies 216
  summary of results 222
moment of momentum 203
  conservation of 242
motion
  damped harmonic 115
  of particle in a plane 155
  of particle on a curve 159
  planetary 170
  under central force 165
  under variable forces 88
  with variable mass 81

normal component
  of acceleration 176
  of velocity 176
numerical solutions, of differential equations 76

oscillations
  forced 126
  harmonic motion 115
  small 256
  see also damping of oscillations, equivalent simple pendulum, resonance

parallel axes theorem 219
parallelepiped 21
particular integral 52, 60
pendulum *see* compound pendulum, equivalent simple pendulum
perpendicular axes theorem 220
planetary motion 170
  period of orbit 173
polar resolutes of acceleration 159
power 93
projectile on inclined plane 137
  maximum distance above plane 139
  maximum distance from plane 138
  maximum range 140
  range 138

radius of curvature 175
radius of gyration 221
range, on inclined plane 138
reaction, at axis of rotation 235
resonance 133
rocket, motion of 85
rotation
  about fixed axis 224
  kinetic energy of 201
  with translation 248

scalar product 3
  properties of 4
scalar triple product 20
separation of variables 43
shell, centre of mass of 193
skew lines 11
steady state 61, 130

tangent field 38
torque
  impulse of 239
  relation to angular acceleration 203
  work done by 225
transient 61, 130

vector equation of line 7
  conversion to cartesian form 9
vector equation of curve 25
  circle 26
  helix 26
vector moment of a force 27
vector product
  application to area 18
  to volume 21
  associativity 15

INDEX

calculation of 14, 17
commutativity 14
distributivity 15
of parallel vectors 14
of perpendicular
  vectors 14
see also scalar product,
  scalar triple product

vectors
  angle between 4
  cartesian components of 1
  differential equations 67
  direction cosines of 2
  direction ratios of 2
  position vectors 6
  resolution of 1

resultant of 1
unit vectors 3
velocity
  radial component of
    160
  tangential component of
    176
  terminal 146

transverse component
  of 160
volumes, using vector
  product 21
work done
  by couple (torque) 225
  by variable force 92
work–energy equation 225